Modeling and Simulation of Polymerization Processes

Modeling and Simulation of Polymerization Processes

Editors

Eduardo Vivaldo-Lima
Yousef Mohammadi

MDPI • Basel • Beijing • Wuhan • Barcelona • Belgrade • Manchester • Tokyo • Cluj • Tianjin

Editors
Eduardo Vivaldo-Lima
Facultad de Química
Universidad Nacional
Autónoma de México
Ciudad de México
Mexico

Yousef Mohammadi
Centre for Advanced
Macromolecular Design
(CAMD), School of Chemical
Engineering
The University of New South
Wales
Sydney
Australia

Editorial Office
MDPI
St. Alban-Anlage 66
4052 Basel, Switzerland

This is a reprint of articles from the Special Issue published online in the open access journal *Processes* (ISSN 2227-9717) (available at: www.mdpi.com/journal/processes/special_issues/polymerization_processes).

For citation purposes, cite each article independently as indicated on the article page online and as indicated below:

LastName, A.A.; LastName, B.B.; LastName, C.C. Article Title. *Journal Name* **Year**, *Volume Number*, Page Range.

ISBN 978-3-0365-4812-8 (Hbk)
ISBN 978-3-0365-4811-1 (PDF)

Cover image courtesy of Eduardo Vivaldo-Lima
Photo of the reactors room of Prof. Eduardo Vivaldo-Lima's lab at FQ-UNAM (Lab D-324B), in Mexico City.

© 2022 by the authors. Articles in this book are Open Access and distributed under the Creative Commons Attribution (CC BY) license, which allows users to download, copy and build upon published articles, as long as the author and publisher are properly credited, which ensures maximum dissemination and a wider impact of our publications.

The book as a whole is distributed by MDPI under the terms and conditions of the Creative Commons license CC BY-NC-ND.

Contents

About the Editors vii

Preface to "Modeling and Simulation of Polymerization Processes" ix

Eduardo Vivaldo-Lima, Yousef Mohammadi and Alexander Penlidis
Special Issue: Modeling and Simulation of Polymerization Processes
Reprinted from: *Processes* **2021**, *9*, 821, doi:10.3390/pr9050821 1

Zhengxiang Zhang, Haibin Zhou, Wentao Li and Chao Tang
Molecular Simulation of Improved Mechanical Properties and Thermal Stability of Insulation Paper Cellulose by Modification with Silane-Coupling-Agent-Grafted Nano-SiO$_2$
Reprinted from: *Processes* **2021**, *9*, 766, doi:10.3390/pr9050766 5

Miguel Ángel Vega-Hernández, Gema Susana Cano-Díaz, Eduardo Vivaldo-Lima, Alberto Rosas-Aburto, Martín G. Hernández-Luna and Alfredo Martinez et al.
A Review on the Synthesis, Characterization, and Modeling of Polymer Grafting
Reprinted from: *Processes* **2021**, *9*, 375, doi:10.3390/pr9020375 15

Luis Valencia, Francisco Enríquez-Medrano, Ricardo López-González, Priscila Quiñonez-Ángulo, Enrique Saldívar-Guerra and José Díaz-Elizondo et al.
Ethylene Polymerization via Zirconocene Catalysts and Organoboron Activators: An Experimental and Kinetic Modeling Study
Reprinted from: *Processes* **2021**, *9*, 162, doi:10.3390/pr9010162 109

Enrique Saldívar-Guerra, Ramiro Infante-Martínez and José María Islas-Manzur
Mathematical Modeling of the Production of Elastomers by Emulsion Polymerization in Trains of Continuous Reactors
Reprinted from: *Processes* **2020**, *8*, 1508, doi:10.3390/pr8111508 131

Ikenna H. Ezenwajiaku, Emmanuel Samuel and Robin A. Hutchinson
Kinetics and Modeling of Aqueous Phase Radical Homopolymerization of 3-(Methacryloylaminopropyl)trimethylammonium Chloride and its Copolymerization with Acrylic Acid
Reprinted from: *Processes* **2020**, *8*, 1352, doi:10.3390/pr8111352 165

Ali Seyedi, Mohammad Najafi, Gregory T. Russell, Yousef Mohammadi, Eduardo Vivaldo-Lima and Alexander Penlidis
Initiator Feeding Policies in Semi-Batch Free Radical Polymerization: A Monte Carlo Study
Reprinted from: *Processes* **2020**, *8*, 1291, doi:10.3390/pr8101291 183

Federico Florit, Paola Rodrigues Bassam, Alberto Cesana and Giuseppe Storti
Solution Polymerization of Acrylic Acid Initiated by Redox Couple Na-PS/Na-MBS: Kinetic Model and Transition to Continuous Process
Reprinted from: *Processes* **2020**, *8*, 850, doi:10.3390/pr8070850 203

Almendra Ordaz-Quintero, Antonio Monroy-Alonso and Enrique Saldívar-Guerra
Thermal Pyrolysis of Polystyrene Aided by a Nitroxide End-Functionality. Experiments and Modeling
Reprinted from: *Processes* **2020**, *8*, 432, doi:10.3390/pr8040432 223

Anete Joceline Benitez-Carreón, Jesús Guillermo Soriano-Moro, Eduardo Vivaldo-Lima, Ramiro Guerrero-Santos and Alexander Penlidis
New Aspects on the Modeling of Dithiolactone-Mediated Radical Polymerization of Vinyl Monomers
Reprinted from: *Processes* **2019**, *7*, 842, doi:10.3390/pr7110842 . . . . **249**

About the Editors

Eduardo Vivaldo-Lima

Professor Vivaldo-Lima is Full Professor at the Department of Chemical Engineering of the Faculty of Chemistry of the National Autonomous University of Mexico (UNAM). He is member of the Mexican National Researchers System with the highest level (3). He is also external academic member of Institute for Polymer Research of the University of Waterloo.

His research work in polymer science and engineering, with emphasis on polymer reaction engineering, focuses on controlled radical polymerization, polymerizations in compressed fluids, the production of polymer networks, and the development of biorefining processes from lignocellulosic biomasses. He is a member of the editorial board of {Journal of Macromolecular Science}, {Part A: Pure and Applied Chemistry}, and {Processes}.

He has received several awards, including the "2010 Heberto Castillo Award", UNAM's 2013 "National University Award", and the 2014 "Ing. Estanislao Ramírez Ruiz"IMIQ's Award, to name a few.

Yousef Mohammadi

Dr. Yousef Mohammadi obtained a PhD from Amirkabir University of Technology-Tehran Polytechnic in Biomedical Engineering and Biomaterials in 2007. He worked for the Petrochemical Research and Technology Company (NPC-rt) in Tehran, Iran, for 14 years. He presently works for the Centre for Advanced Macromolecular Design (CAMD), School of Chemical Engineering of the University of New South Wales, Sydney, Australia.

He has been a senior researcher and a technical project manager in areas such as chemical and polymer engineering, as well as the production of nanofibrous materials.

As a data scientist, he has been able to combine knowledge of high-performance programming, deterministic/stochastic mathematical algorithms/tools and computational intelligence (AI, Machine Learning and Data Mining) to design/develop first-principle and/or data-driven modelers and optimizers capable of precisely pre-processing, processing and post-processing big and complex datasets.

Preface to "Modeling and Simulation of Polymerization Processes"

Polymer reaction engineering (PRE) is the branch of engineering that deals with the technology of large-scale polymer production and the manufacture of polymer products through polymerization processes. PRE is a broad and multidisciplinary area, relatively young and developing fast, combining polymer science, chemistry, and technology with the principles of process engineering. The practical history of PRE started and evolved during the first half of the twentieth century. The 1930s were rich with theoretical findings in polymer science and engineering and with the commercial production of several new polymers. These investigations would transform our understanding of polymer manufacture and culminate in the development of several continuous polymerization processes and the establishment of PRE as a new area of research in the 1940s. The period from 1950 to 1990 saw the continued growth and evolution of process technologies, largely stimulated by the combination of PRE principles with the fundamental understanding of polymerization kinetics developed in earlier years. These principles include the development of mathematical models for polymerization processes and their solution using mathematical packages or specialized chemical engineering or polymerization software. The modeling and simulation of polymerization processes (MSPP) has been fundamental in the development of polymerization technologies since the early stages of PRE to date.

The importance of MSPP has already been recognized by MDPI Processes. A few related issues have been published in the last few years: "Computational Methods for Polymers", Masoud Soroush, August 2019; "Modeling, Simulation and Control of Chemical Processes", José Carlos Pinto, 2019; "Renewable Polymers: Processing and Chemical Modifications", Marc A. Dubé and Tizazu Mekonnen, March 2019; "Process Modelling and Simulation: Cesar de Prada, Costas Pantelides and Jose Luis Pitarch, February 2019; and "Polymer Modeling, Control and Monitoring", Masoud Soroush, February 2016.

The previous issues of *Processes* on related PRE topics have focused on recent specialized topics. This Special Issue on "Modeling and Simulation of Polymerization Processes" aimed to address both new findings on basic topics as well as the modeling of emerging aspects of product design and polymerization processes. It provides a nice view of the state of the art with regard to the modeling and simulation of polymerization processes. The use of well-established methods (e.g., the method of moments) and relatively more recent modeling approaches (e.g., Monte Carlo stochastic modeling) to describe the polymerization processes of long-standing interest in industry (e.g., rubber emulsion polymerization) to polymerization systems of more modern interest (e.g., RDRP and plastic pyrolysis processes) are comprehensively covered in this Special Issue.

We are indebted to Prof. Alexander Penlidis, from the University of Waterloo, in Canada, who suggested our names for this Special Issue, and for his continuous help and guidance in key aspects of our academic paths.

Eduardo Vivaldo-Lima and Yousef Mohammadi

Editors

Editorial

Special Issue: Modeling and Simulation of Polymerization Processes

Eduardo Vivaldo-Lima [1,*], Yousef Mohammadi [2,*] and Alexander Penlidis [3]

1. Departamento de Ingeniería Química, Facultad de Química, Universidad Nacional Autónoma de Mexico, CU, Mexico City 04510, Mexico
2. Centre for Advanced Macromolecular Design (CAMD), School of Chemical Engineering, The University of New South Wales, Sydney, NSW 2052, Australia
3. Department of Chemical Engineering, Institute for Polymer Research (IPR), University of Waterloo, Waterloo, ON N2L 3G1, Canada; penlidis@uwaterloo.ca
* Correspondence: vivaldo@unam.mx (E.V.-L.); ymhitech@gmail.com (Y.M.)

Citation: Vivaldo-Lima, E.; Mohammadi, Y.; Penlidis, A. Special Issue: Modeling and Simulation of Polymerization Processes. *Processes* **2021**, *9*, 821. https://doi.org/10.3390/pr9050821

Received: 26 March 2021
Accepted: 29 April 2021
Published: 8 May 2021

Publisher's Note: MDPI stays neutral with regard to jurisdictional claims in published maps and institutional affiliations.

Copyright: © 2021 by the authors. Licensee MDPI, Basel, Switzerland. This article is an open access article distributed under the terms and conditions of the Creative Commons Attribution (CC BY) license (https://creativecommons.org/licenses/by/4.0/).

This Special Issue (SI) of Processes on Modeling and Simulation of Polymerization Processes (MSPP), and the associated Special Issue reprint, contain papers that deal with this very important area of scientific investigation in polymer science and engineering, both in academic and particularly industrial environments.

Polymer reaction engineering (PRE), also known as macromolecular reaction engineering (MRE), is the branch of engineering that deals with the technology of large-scale polymer production and the manufacture of polymer products through polymerization processes. PRE is a broad and multidisciplinary area, which combines polymer science, chemistry, and technology with the principles of process engineering. The practical history of PRE started and evolved during the first half of the twentieth century. The 1930s were rich with theoretical findings in polymer science and engineering, and with the commercial production of several new polymers. These investigations would transform our understanding of polymer manufacture and culminate in the development of several continuous polymerization processes and the establishment of PRE as a new area of research in the 1940s. The period from 1950 to 1990 saw the continued growth and evolution of process technologies, largely stimulated by the combination of PRE principles with the fundamental understanding of polymerization kinetics developed in earlier years. These principles include the development of mathematical models for polymerization processes, and their solution using mathematical packages or specialized chemical engineering or polymerization software. MSPP has been fundamental in the development of polymerization technologies since the early stages of PRE to date.

Previous issues of Processes on PRE have focused on recent specialized topics [1–6]. This Special Issue on MSPP aimed to address both new findings on basic topics as well as modeling of emerging aspects of product design and polymerization processes. In spite of rather adverse times for academic paper publishing (the COVID-19 pandemic slowed down academic and all types of activities globally), and the call for special issues on PRE topics in other journals that appeared soon after our call had been released (e.g., [7], where we also contributed), and the fact that potential contributions from emerging countries were indirectly discouraged by Processes (by not providing fee discounts to contributions from these countries), nine interesting papers were published in this SI nonetheless.

A brief description of content and contributions of each of the nine contributed papers from this SI is offered below.

The first contributed paper addresses the reversible deactivation radical polymerization (RDRP) of methyl methacrylate (MMA) using dithiolactone controllers, which in principle should follow a reversible addition-fragmentation chain transfer (RAFT) radical polymerization mechanism [8]. The authors claim that the system can be well represented by reversible addition, thus neglecting fragmentation, a mechanism simplification which has been questioned. Although the agreement between calculated and experimental data was only moderate, considering a single set of parameters for a wide window of reaction conditions, there are several interesting ideas and arguments in the discussion: (i) model discrimination in RAFT polymerization is not a straightforward issue and typical experimental data are not enough to discriminate among rival models; (ii) although it should be standard practice, the authors remind us that robust mathematical models should provide adequate description of broad windows of operating conditions with a single set of parameters, and not tuning the parameters for each experimental condition; (iii) full molecular weight distributions (MWD) should be used for model validation purposes; and (iv) as much as possible, independent experimental measurements should be sought (for instance, radical concentration measurements by EPR provided insightful information for the analysis of model completeness).

In the second contribution, Ordaz-Quintero et al. [9] compared the pyrolysis of polystyrene (PS) and a modified PS obtained by adding a nitroxide-end moiety (e.g., TEMPO) (PS-T), hypothesizing that by using PS-T, the initiation process to produce monomer (unzipping) during the pyrolysis process at temperatures above the ceiling temperature of PS (310 °C) would be enhanced due to the tendency of PS-T to activate at the nitroxide end at high temperatures. To test this hypothesis, PS-T was prepared in their lab by nitroxide-mediated polymerization (NMP), and regular PS was synthesized by conventional free-radical polymerization (FRP) and the two types of materials were subjected to their thermal pyrolysis process. They found that the yield of styrene increases from $\sim$33% in the case of dead polymer, to $\sim$38.5% for PS-T (a yield enhancement of 15–16%). Not only did they provide an innovative use of NMP for adding value to plastic waste by pyrolysis of the waste material, but they also presented mathematical models for the pyrolysis of PS and PS-T. They also carried out parameter sensitivity analyses and data fitting, observing adequate agreement with the experimental trends observed.

The third contribution deals with the detailed modeling of the polymerization of non-ionized acrylic acid in aqueous solution [10]. The polymerization process considered is initiated by a persulfate/metabisulfate redox couple and the kinetic scheme considers the possible formation of mid-chain radicals and transfer reactions. The proposed model is validated using experimental data collected in a laboratory-scale discontinuous reactor. The developed kinetic model is then used to intensify the discontinuous process by shifting it to a continuous one based on a tubular reactor with intermediate feeds. One of the experimental runs is selected to show how the proposed model can be used to assess the transition from batch to continuous processes and allows a faster scale-up to industrial scale. This is a nice example of innovative changes in operation of a reasonably well-established polymerization processes. Particularly, the initiating system is non-standard, and the approach adopted to transfer the discontinuous production to a continuous one is relatively new and thus represents an interesting application of the proposed methodology.

In the fourth contribution, Seyedi et al. present a comprehensive study on the effect of initiator feeding policies on polymerization rate and molecular weight development in the semibatch homopolymerizations of styrene and MMA using a highly effective and flexible Monte Carlo algorithm developed in their work [11]. This is also a nice example of gaining deeper understanding and new knowledge about apparently fully understood processes, using novel mathematical approaches. The approach is general and can easily be applied to a variety of polymerizations.

The radical homopolymerization kinetics of 3-(methacryloylaminopropyl) trimethylammonium chloride (MAPTAC) and its batch copolymerization with nonionized acrylic acid (AA) in aqueous solution are extensively investigated and modeled in the fifth contribution of this issue [12]. This contribution is part of a series of papers on the copolymerization kinetics of cationic monomers published by the authors in different journals to broaden and generalize the knowledge of these systems. The novelty in this case is the use of a different cationic monomer (MAPTAC), compared to the earlier studies from the same group. One of the unique aspects of this contribution is the way in which an otherwise conventional free radical copolymerization kinetics modeling framework is soundly adapted to the presence of cationic monomers and their ion-related features.

The sixth contribution of this issue deals with the mathematical modeling of the production of elastomers by emulsion polymerization in trains of continuous reactors [13]. A single mechanistic model was built and used to describe the production of both SBR and NBR elastomers in trains of emulsion polymerization CSTRs. Some novel aspects of this model compared to previous models are the following: (i) radical compartmentalization was considered in more detail, both in the particle kinetics description and in the calculation of the moments of the MWD, and for the first time, a 0-1-2 model was used to describe these two important aspects in these processes; (ii) the monomer partitioning model was as simple as possible, and its parameters could be estimated a priori based on published physicochemical data; and (iii) a single modeling framework was presented in a unified form for the SBR and NBR systems. This is another nice example of how detailed knowledge and experience about an important industrial emulsion polymerization processes significantly benefit from deeper analysis and model improvements of systems considered "well-understood".

In the seventh contribution, Valencia et al. addressed the analysis of the polymerization of ethylene via zirconocene catalysts and organoboron activators, using experimental and mathematical modeling tools [14]. The novel aspects of this contribution include the proposal of a mathematical model for this type of catalytic system. Highlights include: (i) a systematic investigation of the effect of the different types of catalyst, activators, solvents, and operating conditions on ethylene polymerization rate and catalyst activity; (ii) estimations of ethylene concentrations in the liquid solvent based on Duhem's theorem in a VLE system; (iii) estimation of kinetic rate constants by optimization protocols and the method of moments; and (iv) description of the kinetic behavior of different species during the polymerizations. This is an example on the use of PRE tools for new systems, or for systems not previously described with these tools.

In the eighth contribution, Vega-Hernández et al. presented a critical review on the synthesis, characterization, and modeling of polymer grafting [15]. Although the motivation behind this review stemmed from grafting synthetic polymers onto lignocellulosic biopolymers, a comprehensive overview is also provided on the chemical grafting, characterization, and processing of grafted materials of different types, including synthetic backbones. Although polymer grafting has been studied for many decades, so has the modeling of polymer branching and crosslinking for that matter, thereby reaching a good level of understanding to describe existing branching/crosslinking systems, polymer grafting has remained behind in modeling efforts. Areas of opportunity for further study are suggested within this review.

Zhang et al. presented a study on the improvement of mechanical properties and thermal stability of a composite material made of cellulose reinforced with clusters of silane coupling agents grafted onto nano-SiO_2 [16]. Mechanical property-related indicators of the composite material, such as tensile modulus (E), shear modulus (G), bulk modulus (K) and Poisson's ratio (V), are calculated using molecular dynamics simulations. This contribution complements the types of mathematical approaches used in PRE for product or process modeling. This special issue manages to provide a nice view of the state-of-the-art on modeling and simulation of polymerization processes. The use of well-established methods (e.g., the method of moments) and relatively more recent modeling approaches (e.g., Monte Carlo stochastic modeling) to describe polymerization processes of long-standing interest in industry (e.g., rubber emulsion polymerization) to polymerization systems of more modern interest (e.g., RDRP and plastic pyrolysis processes) are comprehensively covered in this issue.

One can locate and read these papers via the following link: https://www.mdpi.com/journal/processes/special_issues/polymerization_processes (accessed on 1 May 2021).

Funding: This research received no external funding.

Conflicts of Interest: The authors declare no conflict of interest.

References

1. Soroush, M. Special Issue on "Computational Methods for Polymers" (Editorial). *Processes* **2020**, *8*, 386. [CrossRef]
2. Pinto, J.C.; Feital, T.; Castor, C. Special Issue "Modeling, Simulation and Control of Chemical Processes". *Processes* **2019**. Available online: https://www.mdpi.com/journal/processes/special_issues/chemical_processes (accessed on 26 March 2021).
3. Mekonnen, T.; Dubé, M.A. Special Issue "Renewable Polymers: Processing and Chemical Modifications" (Editorial). *Processes* **2019**, *7*, 398. [CrossRef]
4. De Prada, C.; Pantelides, C.; Pitarch, J.L. Special Issue on "Process Modelling and Simulation" (Editorial). *Processes* **2019**, *7*, 511. [CrossRef]
5. Soroush, M. Special Issue "Polymer Modeling, Control and Monitoring" of Processes (Editorial). *Processes* **2016**, *4*, 24. [CrossRef]
6. Penlidis, A. Special Issue: "Water Soluble Polymers" (Editorial). *Processes* **2017**, *5*, 31. [CrossRef]
7. McKenna, T.; DesLauriers, P.J. Dedicated to the 60th Birthday of Joao Soares (Editorial). *Macromol. React. Eng.* **2020**, *14*, 2000050. [CrossRef]
8. Benitez-Carreón, A.; Soriano-Moro, J.; Vivaldo-Lima, E.; Guerrero-Santos, R.; Penlidis, A. New Aspects on the Modeling of Dithiolactone-Mediated Radical Polymerization of Vinyl Monomers. *Processes* **2019**, *7*, 842. [CrossRef]
9. Ordaz-Quintero, A.; Monroy-Alonso, A.; Saldívar-Guerra, E. Thermal Pyrolysis of Polystyrene Aided by a Nitroxide End-Functionality. Experiments and Modeling. *Processes* **2020**, *8*, 432. [CrossRef]
10. Florit, F.; Rodrigues Bassam, P.; Cesana, A.; Storti, G. Solution Polymerization of Acrylic Acid Initiated by Redox Couple Na-PS/Na-MBS: Kinetic Model and Transition to Continuous Process. *Processes* **2020**, *8*, 850. [CrossRef]
11. Seyedi, A.; Najafi, M.; Russell, G.; Mohammadi, Y.; Vivaldo-Lima, E.; Penlidis, A. Initiator Feeding Policies in Semi-Batch Free Radical Polymerization: A Monte Carlo Study. *Processes* **2020**, *8*, 1291. [CrossRef]
12. Ezenwajiaku, I.; Samuel, E.; Hutchinson, R. Kinetics and Modeling of Aqueous Phase Radical Homopolymerization of 3-(Methacryloylaminopropyl)trimethylammonium Chloride and its Copolymerization with Acrylic Acid. *Processes* **2020**, *8*, 1352. [CrossRef]
13. Saldívar-Guerra, E.; Infante-Martínez, R.; Islas-Manzur, J. Mathematical Modeling of the Production of Elastomers by Emulsion Polymerization in Trains of Continuous Reactors. *Processes* **2020**, *8*, 1508. [CrossRef]
14. Valencia, L.; Enríquez-Medrano, F.; López-González, R.; Quiñonez-Ángulo, P.; Saldívar-Guerra, E.; Díaz-Elizondo, J.; Zapata-González, I.; Díaz de León, R. Ethylene Polymerization via Zirconocene Catalysts and Organoboron Activators: An Experimental and Kinetic Modeling Study. *Processes* **2021**, *9*, 162. [CrossRef]
15. Vega-Hernández, M.; Cano-Díaz, G.; Vivaldo-Lima, E.; Rosas-Aburto, A.; Hernández-Luna, M.; Martinez, A.; Palacios-Alquisira, J.; Mohammadi, Y.; Penlidis, A. A Review on the Synthesis, Characterization, and Modeling of Polymer Grafting. *Processes* **2021**, *9*, 375. [CrossRef]
16. Zhang, Z.; Zhou, H.; Li, W.; Tang, C. Improvement on the mechanical properties and thermal stabil-2ity of insulation paper cellulose modified by silane coupling 3 agents grafted Nano-SiO$_2$. *Processes* **2021**, *9*, 766. [CrossRef]

Article

Molecular Simulation of Improved Mechanical Properties and Thermal Stability of Insulation Paper Cellulose by Modification with Silane-Coupling-Agent-Grafted Nano-SiO$_2$

Zhengxiang Zhang [1], Haibin Zhou [1], Wentao Li [1] and Chao Tang [2,*]

1 Extra-High Voltage Transmission Company, China Southern Power Grid, Guangzhou 510663, China; Zhangzhengxiang@mail.ehv.csg.cn (Z.Z.); zhouhaibin@mail.ehv.csg.cn (H.Z.); liwenta@mail.ehv.csg.cn (W.L.)
2 College of Engineering and Technology, Southwest University, Chongqing 400715, China
* Correspondence: swutc@swu.edu.cn; Tel./Fax: +86-023-68251265

Abstract: Cellulose is an important part of transformer insulation paper. Thermal aging of cellulose occurs in long-term operation of transformers, which deteriorates the mechanical properties and thermal stability of cellulose, resulting in a decrease in the transformer life. Therefore, improvement of the mechanical properties and thermal stability of cellulose has become a research hotspot. In this study, the effects of different silane coupling agents on the mechanical properties and thermal stability of modified cellulose were studied. The simulation results showed that the mechanical parameters of cellulose are only slightly improved by KH560 (γ-glycidyl ether oxypropyl trimethoxysilane) and KH570 (γ-methylacrylloxy propyl trimethoxy silane) modified nano-SiO$_2$, while the mechanical parameters of cellulose are greatly improved by KH550 (γ-aminopropyl triethoxy silane) and KH792 (N-(2-aminoethyl)-3-amino propyl trimethoxy silane) modified nano-SiO$_2$. The glass-transition temperature of the composite model is 24 K higher than that of the unmodified model. The mechanism of the change of the glass-transition temperature was analyzed from the point of view of free-volume theory. The main reason for the change of the glass-transition temperature is that the free volume abruptly changes, which increases the space for movement of the cellulose chain and accelerates the whole movement of the molecular chain. Therefore, modifying cellulose with KH792-modified nano-SiO$_2$ can significantly enhance the thermal stability of cellulose.

Keywords: nano-SiO$_2$; silane coupling agent; thermal stability; mechanical parameter; molecular simulation

1. Introduction

Oil–paper insulation systems guarantee safe operation of oil-immersed transformers, but they are affected by various factors in long-term transformer operation [1–5]. Improvement of the mechanical properties and thermal stability of cellulose has become a research hotspot. In recent years, the rise of nanotechnology has prompted research on nanoparticle-modified cellulose insulating paper. Many researchers have made progress in this field [6–10]. Liu used nano-Al$_2$O$_3$ to modify insulating paper and studied the depth and degree of the traps [11]. The results showed that with increasing trap depth and density, the breakdown field strength of the modified insulating paper increases, and the conductivity decreases. The trap characteristics are the main reasons for the change of the dielectric properties of the modified insulating paper. Liao used nanoparticles to modify cellulose insulating paper, added SiO$_2$ hollow microspheres and soaked the modified insulating paper, and tested and analyzed its dielectric constant and breakdown voltage [12]. When the content of SiO$_2$ hollow microspheres was 5%, the dielectric constant of the modified insulating paper was 34% lower than that of the unmodified insulating paper, and the breakdown voltage of the oil–paper composite insulation system increased by 15.5%. However, nanoparticles agglomerate in the process of modification. At present, the main method to avoid agglomeration of nanoparticles is to modify the nanoparticles with silane coupling agents. Wang investigated the influences of different types of silane coupling agents and grafting rate on the interface between nano-SiO₂ and cellulose [13]. They found that the interface effect was different for silane coupling agents with different chain lengths and groups. Therefore, it is necessary to study the mechanical properties and thermal stability of cellulose modified by different types of silane coupling agents.

In this study, the nano-SiO₂ content in all of the models was 5% [9,12,14]. The mechanical properties (tensile modulus, shear modulus, volume model, and Poisson's ratio) and thermal stability (glass-transition temperature and free-volume fraction) of cellulose modified by nano-SiO₂ modified by different silane coupling agents were investigated and compared through a molecular dynamics simulation based on the same grafting rate of the silane coupling agent. The mechanism of the change of the properties and thermal stability was also investigated.

2. Molecular Dynamics Simulation

Because of the high surface activity of nano-SiO₂, it is easily oxidized and produces hydroxyl groups on the nano-SiO₂ surface. First, the surface of the nano-SiO₂ model needs to be hydroxylated (the unsaturated O atoms on the surface must be treated with H and the unsaturated Si atoms with -OH). Second, the hydroxyl groups at one end of the silane coupling agent combine with the hydroxyl groups on the surface of nano-SiO₂ to form Si–O–Si so that the silane coupling agent can randomly connect to the O atoms on the surface of nano-SiO₂. There are three hydroxyl groups in a general silane coupling agent. One of the hydroxyl groups of the silane coupling agent bonds to a hydroxyl group on the surface of nano-SiO₂, and the remaining two hydroxyl groups form condensation with other grafted silane coupling agent molecules or exist in a free state. To simplify the model, we considered that the other two hydroxyl groups in the model do not participate in the condensation reaction.

Wang et al. [13] found that the grafting ratio is relatively good when the grafting quantity is 4, and the influence of the size of the model on the simulation time has also been investigated [15–17]. Therefore, different types of silane coupling agents with grafting quantity of 4 were grafted on the surface of nano-SiO₂, and the particle size of nano-SiO₂ was 5 Å. First, nano-SiO₂ was placed in a cell with dimensions of 40 × 40 × 40 Å, and its centroid was set to coincide with the centroid of the cell. The cellulose chain (degree of polymerization = 10) was filled in [18] using the packing function in the AC module, and the initial density was set to 0.6 g/cm³. By applying pressure to different models, the density can reach 1.45 g/cm³. Subsequently, the relevant parameters were calculated and analyzed. The composite models constructed in this study are shown in Figure 1.

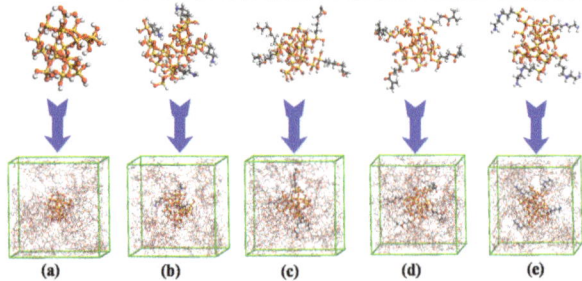

Figure 1. Composite models: (**a**) unmodified model, (**b**) KH550 model, (**c**) KH560 model, (**d**) KH570 model, and (**e**) KH792 model.

In this study, we used Molecular Simulation (MS) software. First, 5000 steps of geometric optimization were performed for the established amorphous model of nano-SiO₂ and cellulose. Second, annealing of the model was performed with annealing cycle of 5, initial temperature of 300 K, and heating ramps per cycle of 13. Finally, 100 steps of the annealed convergence model were performed in the NVT ensemble. The dynamic simulation of 200 ps was performed in the NPT ensemble, and then the rationality criterion of the model after the dynamic simulation was obtained. After the system reached a stable configuration, the relevant parameters were analyzed. The pressure was set to 0.0001 GPa (standard atmospheric pressure), the Ewald method was used for the electrostatic interactions, the atom-based method was used for the van der Waals interactions, and the COMPASS force field was used. A Nose–Hoover thermostat was used to collect the dynamic information every 0.5 ps [19,20].

3. Results and Discussion

3.1. Mechanical Properties

The mechanical properties of materials are important indicators for characterizing the mechanical strength. In this study, the relevant mechanical parameters of cellulose were obtained by static elastic constant analysis (Table 1). The tensile modulus (E) is the ratio of the stress to the strain. The material has higher rigidity and greater ability to resist deformation due to the larger E. The shear modulus (G) is the ratio of the shear stress to the strain. The bulk modulus (K) is the incompressibility of the material. Poisson's ratio (V) is the plasticity of the material, with a high value indicating high plasticity [21,22]. The mechanical parameters of the models are shown in Figure 2. With increasing temperature, the modulus values (E, G, and K) of each model gradually decrease, while Poisson's ratio (V) is relatively stable. This indicates that the temperature has a relatively large effect on the deformation resistance, shear deformation resistance, and incompressibility of cellulose, while the effect of temperature on the plasticity of cellulose is relatively small. Comparing the mechanical properties of the models with four commonly used silane coupling agents, the mechanical parameters (E, K, and V) of cellulose are only slightly improved by KH560- and KH570-modified nano-SiO_2, but the mechanical parameters (E and K) of cellulose are greatly improved by KH550- and KH792-modified nano-SiO_2, although Poisson's ratio (V) is not significantly improved. Therefore, compared with KH560 and KH570, nano-SiO_2 modified with silane coupling agents containing amino groups (KH550 and KH792) can significantly improve the deformation resistance, shear deformation resistance, and incompressibility of cellulose, with little effect on the plasticity.

Table 1. Free volume of the unmodified nano-SiO_2 and cellulose composite model.

Temperature	Free Volume (A3)	Occupied Volume (A3)	FFV
303 K	7631.24	54,868.76	0.139082
323 K	7701.92	54,398.08	0.141584
343 K	7744.04	54,615.96	0.141791
363 K	7739.88	54,460.12	0.14212
383 K	7784.60	54,315.4	0.143322
403 K	7833.52	54,466.48	0.143823
423 K	8251.68	54,760.64	0.150686
443 K	8332.32	54,267.68	0.153541
463 K	8371.68	54,428.32	0.153811
483 K	8436.20	54,863.80	0.153766
503 K	8631.24	54,868.76	0.157307

3.2. Cross Energy

The energy is the most important property for the stability of the system. The energy of the system can be expressed as

$$E_{total} = (E_{internal} + E_{cross}) + E_{nonbond} \tag{1}$$

where $E_{\text{internal}} + E_{\text{cross}}$ is the bond interaction term and E_{nonbond} is the non-bonding interaction term. The changes of the overall potential energy and non-bonding interaction energy of the models with temperature are shown in Figure 3.

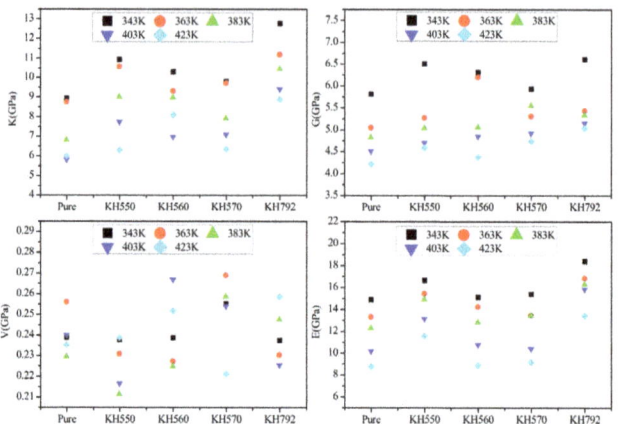

Figure 2. Mechanical parameters of the models.

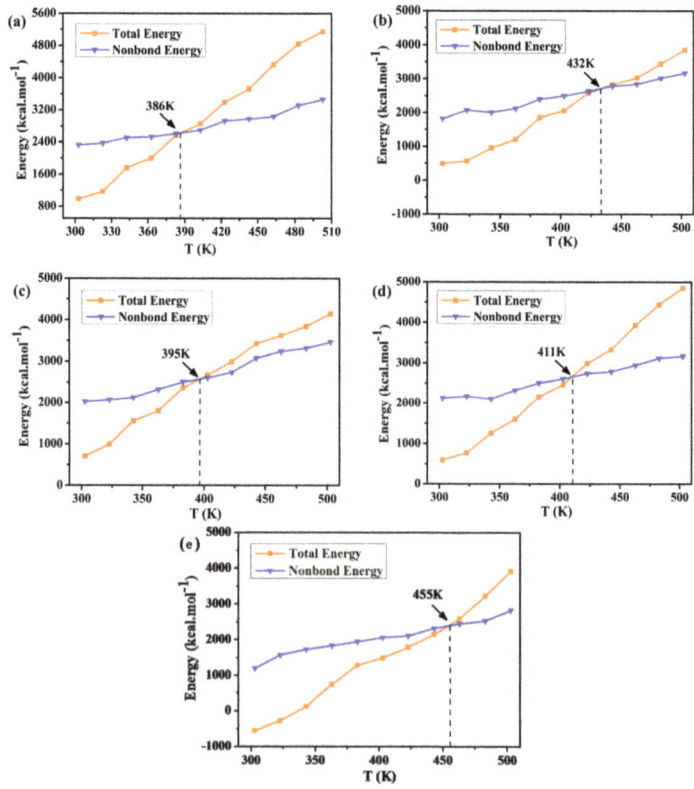

Figure 3. Energies of the models: (**a**) ungrafted model, (**b**) KH550 model, (**c**) KH560 model, (**d**) KH570 model, and (**e**) KH792 model.

From Figure 3, the total potential energy and non-bonding interaction energy of all of the models essentially linearly increase, but the growth rate of the non-bonding interaction energy is significantly lower than that of the total potential energy. In the ungrafted model and KH550, KH560, KH570, and KH792 composite models, the starting temperatures of the total potential energies are 386, 395, 411, 432, and 455 K, respectively. From Equation (1), when the total potential energy of the system becomes greater than the non-bonding interaction energy, the bond energy interaction term in the system changes from the previously negative value to a positive value, while the repulsion force between the atoms is greater than the attraction force between the atoms in the molecule, and the chemical bonds become unstable. When the temperature gradually increases, some active bonds and glycoside bonds are easy to break, which is the phenomenon of cellulose thermal degradation. Once cellulose thermal degradation occurs, the degree of polymerization of cellulose decreases, which leads to deterioration of the mechanical properties of cellulose. Therefore, the main reason for the decrease of the mechanical properties of cellulose is that the internal atomic force changes from attractive to repulsive, which makes the chemical bonds unstable. The order of the temperatures corresponding to the cross-energy points of the different models is KH792 > KH550 > KH560 > KH570 > pure.

3.3. Glass-Transition Temperature

The glass-transition temperature is the temperature of the transition from the glass state to the high-elastic state. As a glassy substance, cellulose has a glassy state, a high-elastic state, and a viscous-flow state, which are the three states of the polymer related to temperature. The transition points between these states are called the glass-transition temperature and melting point. The thermodynamic properties of the material obviously change before and after the glass-transition temperature. Therefore, it is of great importance to study the glass-transition characteristics of cellulose at high temperature to improve its thermal stability. When studying the glass-transition temperatures of polymers, the most commonly used and reliable method is the specific volume–temperature curve method [23–26]. The specific volume is the reciprocal of the density, and it is the most commonly used physical quantity in determination of the glass-transition temperature. During the glass transition, the specific volume and other properties of the material significantly change along with molecular segment movement, and the curve of the specific volume against the temperature abruptly changes. In this method, the specific volume obtained by the NPT molecular dynamics simulation is plotted against the temperature. The inflection point of the specific volume change with temperature is determined, and linear fitting is performed before and after the inflection point. The intersection of the two fitted curves is the turning point of vitrification, and the corresponding temperature is the vitrification temperature. The obtained fitted curves of the vitrification temperature for the unmodified and KH792-modified models are shown in Figure 4.

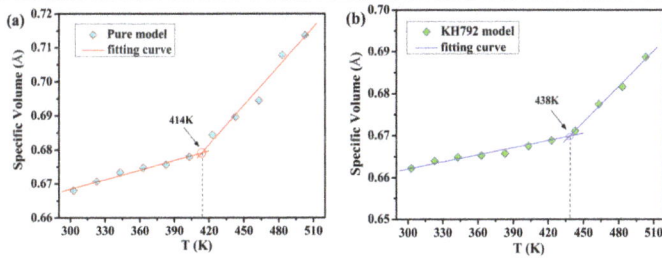

Figure 4. Specific volume–temperature curves of the models: (a) unmodified model and (b) KH792 model.

From Figure 4, the glass-transition temperature of the cellulose and unmodified nano-SiO$_2$ composite model is 414 K, while that of the KH792-modified nano-SiO$_2$ and cellulose composite model is 438 K. Thus, the glass-transition temperature of the KH792-modified nano-SiO$_2$ and cellulose composite model increases by 24 K. The glass-transition temperature range of cellulose in the polymer manual is 250 to 580 K, so the value of the glass-transition temperature obtained from simulation analysis in this study is reasonable. However, it is only for reference, and it does not affect comparison between the glass-transition temperatures of the modified model and unmodified model. The glass-transition temperature of the KH792-modified nano-SiO$_2$ and cellulose composite model is higher than that of the unmodified nano-SiO$_2$ and cellulose composite model, which shows that the glass state to high-elastic state transition of the cellulose chain modified by KH792 is slower than that of the unmodified model, which effectively delays the cellulose chain transition from the glass state to the high-elastic state. Therefore, compared with the unmodified composite model, for the cellulose chain modified by KH792, the modified nano-SiO$_2$ and cellulose composite model can maintain better performance, the thermodynamic performance is less affected by temperature, and the thermal stability is significantly enhanced.

3.4. Free-Volume Fraction

In the molecular dynamics simulation, the free volume can be determined by the interstices between the chains in the model system. The free volume and occupied volume of the system can be measured by the Atom Volumes and Surface Tool in MS, and the free-volume fraction (FFV) of the system can be obtained by the ratio of the free volume to the total volume of the system (the sum of the free volume and occupied volume). Because of the different expansion coefficient, the occupied volume of the polymer linearly increases with temperature, and the free volume abruptly increases near the glass-transition temperature. Therefore, the mechanism of the glass-transition temperature was analyzed using free-volume theory. When calculating the free volume, the hard ball probe method was used, and the probe radius was set to 1 Å [26–28]. The free-volume distributions of the unmodified composite model and KH792-modified composite model are shown in Figure 5. The blue parts of the figures are the free-volume distribution areas. The free volumes of the unmodified composite model and KH792-modified composite model are given in Tables 1 and 2, respectively.

Table 2. Free volume of the KH792-modified nano-SiO$_2$ and cellulose composite model.

Temperature	Free Volume (A3)	Occupied Volume (A3)	FFV
303 K	5424.80	56,672.80	0.087359
323 K	5422.48	57,077.52	0.086760
343 K	5492.64	56,907.36	0.088023
363 K	5543.68	56,756.32	0.088984
383 K	5624.60	56,532.08	0.090491
403 K	5667.92	56,775.40	0.090769
423 K	5687.43	56,301.52	0.091749
443 K	6198.48	56,689.52	0.098564
463 K	6147.60	56,352.40	0.098362
483 K	6309.80	56,190.20	0.100957
503 K	6511.24	56,908.76	0.102669

From Table 1, for the unmodified nano-SiO$_2$ composite model, when the temperature is 303–403 K, the change of the free volume is relatively small, However, above 423 K, the free volume starts to increase, and a clear increasing trend appears. From Table 2, for the KH792-modified nano-SiO$_2$ composite model, when the temperature is 303–423 K, the change of the free volume is small. However, above 443 K, the free volume starts to increase, and the free volume also shows a significant increasing trend. The sudden increase of the free volume increases the space for movement of the cellulose chain, leading to increased movement. The cellulose changes from the glassy state to the high-elastic state, and the thermal stability of the cellulose chain sharply decreases after it changes to the high-elastic state. The free volume increase of the unmodified nano-SiO_2 composite model occurs at 403–423 K, which is essentially the same as the glass-transition temperature of the unmodified nano-SiO_2 composite model obtained by the specific-volume method in Section 3.3 (413 K). The free volume increase of the KH792-modified nano-SiO_2 composite model occurs at 423–443 K. This is consistent with the glass-transition temperature of the modified composite model obtained by the specific-volume method in Section 3.3 (438 K). Therefore, this section explains the microscale mechanism of improvement of the thermal stability of cellulose modified by KH792-modified nano-SiO_2 from the perspective of the free volume.

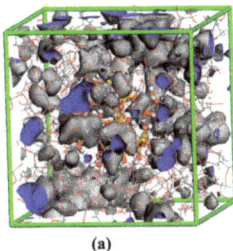 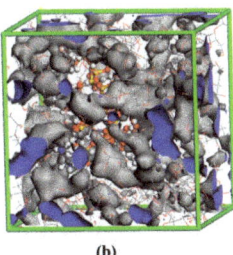

(a) (b)

Figure 5. Free volume of the models: (a) unmodified model and (b) KH792 model.

Above all, hydroxyl groups are produced after hydrolysis of silane coupling agent. Therefore, the silane coupling agent can react with the silicon hydroxyl groups on the surface of nano-SiO_2, so that one end of the silane coupling agent can be connected with the surface of nano-SiO_2. The organic functional group of KH550 is amino, the organic functional group of KH560 is epoxy, the organic functional group of KH570 is acyloxy, and the organic functional group of kh792 is aminopropyl. Organic functional groups determine the binding ability between nano-SiO_2 modified by silane coupling agent and cellulose. The addition of silane coupling agent enhances the compatibility between organic phase and inorganic phase to a certain extent. At the same time, the addition of silane coupling agent creates more hydrogen bonds between nano-SiO_2 and cellulose, which makes the structure of cellulose more compact.

4. Conclusions

In this study, the effects of silane coupling agents (KH550, KH560, KH570, and KH792) on the mechanical properties and thermal stability of cellulose were investigated with molecular dynamics simulation. It was found that the mechanical properties and thermal stability of cellulose can be improved by silane coupling agents.

The temperature has a great effect on the deformation resistance, shear deformation resistance, and incompressibility of cellulose, but it has little effect on its plasticity. The mechanical parameters (E, K, and V) of cellulose are only slightly improved by KH560- and KH570-modified nano-SiO_2. However, the mechanical parameters (E and K) of cellulose are significantly improved by KH550- and KH792-modified nano-SiO_2, while Poisson's ratio (V) is only slightly improved. Therefore, this shows that silane-coupling-agent-modified nano-SiO_2 mainly affects the deformation resistance, shear deformation resistance, and incompressibility of cellulose, but it has little effect on its plasticity. Silane coupling agents containing amino groups significantly improve the deformation resistance, shear deformation resistance, and incompressibility of cellulose. The mechanism of the change of the mechanical properties was analyzed from the point of view of the energy. The reason for the decrease of the mechanical properties of cellulose is that the internal atomic force changes from attractive to repulsive, which makes the chemical bonds unstable. The overall potential energy obtained from energy analysis is higher than the starting temperature of the non-bonding energy, and the analysis results of the mechanical properties are essentially the same. Therefore, nano-SiO_2 modified by silane coupling agents can enhance the mechanical properties of cellulose, and nano-SiO$_2$ modified by KH792 most significantly improves the mechanical properties of cellulose.

The glass-transition temperature of the nano-SiO$_2$-modified cellulose composite model without surface modification is 414 K, and that of the KH792-modified nano-SiO$_2$ composite model is 438 K. Thus, the glass-transition temperature of the composite model increases by 24% after surface modification of nano-SiO$_2$ by KH792. The mechanism of the change of the glass-transition temperature was analyzed from the point of view of free-volume theory. The main reason for the change of the glass-transition temperature is that the free volume abruptly changes, which increases the space for cellulose chain movement and intensifies the overall movement of the molecular chain. The range of the glass-transition temperature obtained by free-volume theory is essentially the same as that obtained by the specific-volume method. Therefore, the thermal stability of cellulose is significantly enhanced by modifying cellulose by KH792-modified nano-SiO$_2$.

Author Contributions: Conceptualization, Z.Z. and C.T.; formal analysis, W.L. and C.T.; software, H.Z.; validation, Z.Z., H.Z. and C.T.; data curation, Z.Z., H.Z. and C.T. All authors have read and agreed to the published version of the manuscript.

Funding: This research was funded by the science and technology project of EHV Power Transmission Company, China Southern Power Grid.

Institutional Review Board Statement: Not applicable.

Informed Consent Statement: Not applicable.

Data Availability Statement: The data that support the findings of this study are available from the corresponding author, C.T., upon reasonable request.

Conflicts of Interest: The funders had no role in the design of the study; in the collection, analyses, or interpretation of data; in the writing of the manuscript, or in the decision to publish the results.

References

1. Liao, R.; Guo, P.; Zhou, N. The thermal aging characteristics of the new anti-aging mixed oil-pressboard insulation. *Trans. China Electrotech. Soc.* **2015**, *22*, 222–230.
2. Xia, G.; Wu, G.; GAO, B. A new method for evaluating moisture content and aging degree of transformer oil-paper insulation based on frequency domain spectroscopy. *Energies* **2017**, *8*, 1195. [CrossRef]
3. Leibfried, T.; Jaya, M.; Majer, N. Postmortem investigation of power transformers-profile of degree of polymerization and correla-tion with furan concentration in the oil. *IEEE Trans. Power Deliv.* **2013**, *2*, 886–893. [CrossRef]
4. Tang, C.; Zhang, S.; Zhang, F. Simulation and Experimental about the Thermal Aging Performance Improvement of Cellulose Insulation Paper. *Trans. China Electrotech. Soc.* **2016**, *10*, 68–76.
5. Rafiq, M.; Lv, Y.; Li, C. Effect of Al$_2$O$_3$ nanorods on the performance of oil-impregnated pressboard insulation. *Electr. Eng.* **2019**, *102*, 715–724. [CrossRef]
6. Wei, G.; Liu, J.; Peng, J. Trangsformer oil breakdown and dielectric frequency response characteristic of oil-paper insulation modified by nanoparticles. *High Volt. Eng.* **2013**, *39*, 3082–3087.
7. Liao, R.; Lv, C.; Wu, W. Insulating properties of insulation paper modified by nanao-Al$_2$O$_3$ for power transformer. *J. Eiectric Power Sci. Technol.* **2014**, *1*, 3–7.
8. Yan, S.; Liao, R.; Lv, Y. Influence of nano-Al$_2$O$_3$ on electrical properties of insulation paper under thermal aging. *Trans. China Electrotech. Soc.* **2017**, *11*, 225–232.
9. Tang, C.; Zhang, S.; Li, X.C. Molecular Dynamics Simulations of the Effect of Shape and Size of SiO$_2$ Nanoparticle Dopants on Insulation Paper Cellulose. *AIP Adv.* **2016**, *6*, 125106. [CrossRef]
10. Tang, C.; Zhang, S.; Xie, J. Molecular Simulation and Experimental Analysis of Al$_2$O$_3$ -nanoparticle-modified Insulation Paper Cellulose. *IEEE Trans. Dielectr. Electr. Insul.* **2017**, *2*, 1018–1026. [CrossRef]
11. Liu, Q.; Chi, M.; Chen, Q. Analsis of dielectric characteristics of nano-Al$_2$O$_3$ modified insulation pressboard. *Proc. CSEE* **2017**, *14*, 4246–4253.
12. Zhang, F.; Liao, R.; Yuan, Y. Preparation for low-permittivity insulation paper and its breakdown performance. *High Volt. Eng.* **2012**, *3*, 691–696.
13. Zheng, W.; Tang, C.; Xie, J. Micro-scale effects of nano-SiO$_2$ modification with silane coupling agents on the cellulose/nano-SiO$_2$ interface. *Nanotechnology* **2019**, *30*, 445701. [CrossRef]
14. Tang, Y.; Tang, C.; Hu, D. Effect of aminosilane coupling agents with different chain lengths on thermo-mechanical properties of cross-linked epoxy resin. *Nanomaterials* **2018**, *8*, 951. [CrossRef]

15. Yang, L.; Ming, D.; Jianzhuo, D. Molecular Dynamics Simulation on Impact of Temperature on Viscosity of Nano-modified Transformer Oil. *Insul. Mater.* **2017**, *50*, 465106.
16. Zhang, S.; Tang, C.; Hao, J. Thermal stability and dielectric properties of nano-SiO_2-doped cellulose. *Appl. Phys. Lett.* **2017**, *111*, 012902. [CrossRef]
17. Zhang, J.; Tang, C.; Wang, Q. Analysis of nano-SiO_2 affecting the acids diffusion in the interface betweenoil and cellulose paper. *Chem. Phys.* **2019**, *529*, 110557. [CrossRef]
18. Wang, X.; Tang, C.; Wang, Q. Selection of optimum polymerization degree and force field in the molecular dynamics simulation of insulating paper cellulose. *Energies* **2017**, *10*, 1377. [CrossRef]
19. Xie, Q.; Fu, K.; Liang, S. Micro-structure and thermome-chanical properties of crosslinked epoxy composite modified by nano-SiO_2: A molecular dynamics simulation. *Polymers* **2018**, *7*, 801. [CrossRef]
20. Du, D.; Tang, C.; Zhang, J. Effects of hydrogen sulfide on the mechanical and thermal properties of cellulose insulation paper: A molecular dynamics simulation. *Mater. Chem. Phys.* **2019**, *240*, 122153. [CrossRef]
21. Mazeau, K.; Heux, L. Molecular dynamics simulations of bulk native crystalline and amorphous structures of cellulose. *J. Phys. Chem. B* **2003**, *10*, 2394–2403. [CrossRef]
22. Fu, Y.; Liu, Y.; Lan, Y. Molecular simulation on the glass transition of polypropylene. *Polym. Mater.* **2009**, *10*, 53–56.
23. Tang, C.; Zheng, W.; Wang, L. Thermal stability of polyphenylsilsesquioxane-modified metaaramid insulation paper. *High Volt.* **2020**, *3*, 264–269. [CrossRef]
24. Chang, K.; Chung, Y.; Yang, T. Free volume and alcohol transport properties of PDMs membranes: Insights of nano-structure and interfacial affinity from molecular modeling. *J. Membr.* **2012**, *11*, 119–130. [CrossRef]
25. Li, X.; Tang, C.; Wang, J. Analysis and mechanism of adsorption of naphthenic mineral oil, water, formic acid, carbon dioxide, and methane on meta-aramid insulation paper. *J. Mater. Sci.* **2019**, *11*, 8556–8570. [CrossRef]
26. Li, X.; Tang, C.; Qian, W. Molecular simulation research on the micro effect mechanism of interfacial properties of nano SiO_2/meta-aramid fiber. *Int. J. Heat Technol.* **2017**, *1*, 123–129. [CrossRef]
27. Tian, W.; Tang, C.; Wang, Q. The effect and associate mechanism of nano SiO_2 particles on the diffusion behavior of water in insulating oil. *Material* **2018**, *11*, 2373. [CrossRef]
28. Du, D.; Tang, Y. Effects of Different Grafting Density of Amino Silane Coupling Agents on Thermomechanical Properties of Cross-Linked Epoxy Resin. *Polymers* **2020**, *8*, 1662. [CrossRef]

Review

A Review on the Synthesis, Characterization, and Modeling of Polymer Grafting

Miguel Ángel Vega-Hernández [1], Gema Susana Cano-Díaz [1], Eduardo Vivaldo-Lima [1,2,*], Alberto Rosas-Aburto [1], Martín G. Hernández-Luna [1], Alfredo Martinez [3], Joaquín Palacios-Alquisira [4], Yousef Mohammadi [5] and Alexander Penlidis [2,*]

[1] Departamento de Ingeniería Química, Facultad de Química, Universidad Nacional Autónoma de México, Ciudad de México 04510, Mexico; angeluz_alchemist@hotmail.com (M.Á.V.-H.); suscad@unam.mx (G.S.C.-D.); alberto_rosas_aburto@comunidad.unam.mx (A.R.-A.); martinhl@unam.mx (M.G.H.-L.)
[2] Department of Chemical Engineering, Institute for Polymer Research, University of Waterloo, Waterloo, ON N2L 3G1, Canada
[3] Instituto de Biotecnología, Universidad Nacional Autónoma de México, Cuernavaca, Morelos 62210, Mexico; alfredo.martinez@mail.ibt.unam.mx
[4] Departamento de Fisicoquímica, Facultad de Química, Universidad Nacional Autónoma de México, Ciudad de México 04510, Mexico; polylab1@unam.mx
[5] Petrochemical Research and Technology Company (NPC-rt), National Petrochemical Company (NPC), Tehran P.O. Box 14358-84711, Iran; mohammadi@npc-rt.ir
* Correspondence: vivaldo@unam.mx (E.V.-L.); penlidis@uwaterloo.ca (A.P.); Tel.: +519-888-4567 (ext. 36634) (A.P.)

Abstract: A critical review on the synthesis, characterization, and modeling of polymer grafting is presented. Although the motivation stemmed from grafting synthetic polymers onto lignocellulosic biopolymers, a comprehensive overview is also provided on the chemical grafting, characterization, and processing of grafted materials of different types, including synthetic backbones. Although polymer grafting has been studied for many decades—and so has the modeling of polymer branching and crosslinking for that matter, thereby reaching a good level of understanding in order to describe existing branching/crosslinking systems—polymer grafting has remained behind in modeling efforts. Areas of opportunity for further study are suggested within this review.

Keywords: polymer grafting; polymer synthesis; polymer characterization; mathematical modeling; polymer reaction engineering; reversible deactivation radical polymerization

1. Introduction

Graft copolymers consist of branches of polymer segments covalently bonded to primary polymer chains. Graft copolymers containing a single branch are known as *miktoarm* star copolymers. The backbone and branches can be homo- or copolymers with different chemical structures or compositions [1]. However, if the polymer molecule is a homopolymer, the reaction route to produce the branches is known as polymer branching; polymer grafting is usually considered as a chemical route to produce materials whose branches are chemically different from the backbone or primary polymer chain. The branches typically have the same chain size and are randomly distributed throughout the backbone's length as a consequence of the synthetic route used synthesize them. However, more efficient methods that allow the synthesis of graft copolymers with equidistant and same-length branches, with which the microstructure and composition can be controlled to a remarkable level, have been developed [1]. From a surface-chemistry perspective, this definition of polymer grafting is extended to composites in which the main chain constitutes a diverse array of materials, ranging from brick and fiberglass to paper and wood [2]. Materials with improved or simply different polymer properties from mechanical, thermal, melt flow or dilute solution perspectives can be synthesized by polymer grafting [1,3–7]. The structure–properties relationship has been an important issue in the analysis of polymer grafting [1].

Some of the first reports on polymer grafting available in the open literature (e.g., the oldest records available through Web of Science) include the grafting of polystyrene (PSty) [8] and poly(methyl methacrylate) (PMMA) [9] onto "government rubber styrene" (GRS) [8], which is a synthetic copolymer of butadiene and styrene, or onto natural rubber [9]; grafting of PSty, poly(butyl methacrylate) (PBMA), poly(lauryl methacrylate) (PLMA), poly(methyl acrylate) (PMA), and poly(ethyl acrylate) (PEA) onto PMMA with pendant mercaptan groups [10]; grafting of polyacrylamide (PAM) onto polyacrylonitrile (PAN), or the other way around (PAN onto PAM) [11]; grafting of PMMA onto PAN [12]; grafting of PSty onto polyethylene (PE) [13]; and grafting of several polymers, such as PAN, PMMA, PSty, poly(acrylic acid) (PAA), and poly(vinylidene chloride) (PVDC), onto cellulose [14–16], to name a few. A more complete literature review on the chemistry of polymer grafting is summarized in Table 1.

Table 1. Overview of the synthesis and characterization of grafted copolymers with an emphasis on the period 1950 to 1970, plus some additional, more recent ones.

Backbone	Functionalization Method	Graft Chains	Grafting Technique	Grafting Conditions	Measured Properties and Characterization Methods	Ref.
Rubber (GRS or natural)	Generation of internal free radicals by CTP.	PSty or PMMA	Grafting from	95–180 °C; Mass FRP (rubber dissolved in monomer, in presence of initiator); solvent-non-solvent fractionation.	Determination of vinyl unsaturation (peracid and infrared methods); molecular weight by intrinsic viscosity; DMA; mechanical properties: tensile strength, elongation-at-break, hardness, modulus at 100 and 300% elongation, as well as tear at 20, 80, and 120 °C.	[8,9]
PMMA (a copolymer of MMA and small content of GMA)	Incorporation of mercaptan groups by reaction of GMA and hydrogen sulfide in presence of sodium ethoxide catalyst.	PSty, PBMA, PLMA, or PMA	Grafting from	Mass or solution polymerization of monomer in presence of PMMA with pendant mercaptan groups (grafting occurs by chain transfer to mercaptan groups).	Solvent extraction of ungrafted polymer, measured by UV analysis; grafting efficiency calculated with the aid of a kinetic model.	[10]
Either PAN or PAM	Two methods used: (a) CTP; (b) photolysis of a copolymer containing a few per cent of ACN.	PAM, PAN; PMMA (onto PAN) [12]	Grafting from	(a) SP in sodium perchlorate at 55 °C, using persulphate-bisulphite; (b) SP in sodium perchlorate at 25 to 35 °C in a quartz tube under a G.E. Sun Lamp.	Composition by IR; molecular weight by intrinsic viscosity calibrated from light scattering data; phase contrast microscopy; measurement of softening points; analysis of X-ray scattering curves [12].	[11,12]
PE	UV irradiation of surface of sensitized PE.	PSty	Grafting from	(a) Sensitized PE irradiated one minute, stand a week, and then proceed to mass polymerization in Sty at 70 °C; (b) irradiation of plastic in presence of Sty.	Mass difference or ability of the surface to adhere to pressure-sensitive tape under load.	[13]
Cellulose	γ-ray pre-irradiation technique.	PSty	Grafting from	Pre-irradiation by Co^{60} γ-rays in water or in a H$_2$O$_2$ solution; grafting in a 20 vol. Sty solution (methanol/water) at 50 °C.	Degree of grafting by weight gain; estimation of active sites by the ferrous ion method; solubilization of material by acetylation and acetolysis, followed by IR spectroscopy.	[14]
Cellulose	Binding initiators or components of initiation systems by means of ion exchange with such materials.	Various polymers (PAN, PMMA, PSty, PAA, PVDC)	Grafting from	Starting material contacted with a dilute solution of catalyst cation salt; the exchanged cellulose was then placed in the monomer or monomer solution; the mixture was heated or irradiated for the required time.	Swelling by centrifuging; grafting efficiency by solvent extraction; mechanical properties (elongation, moduli; toughness); chemical properties (basic/acid dying and hydrolysis, bromination, oxidation, complex formation, etc.); wetting; rotproofing.	[15]

Table 1. Cont.

Backbone	Functionalization Method	Graft Chains	Grafting Technique	Grafting Conditions	Measured Properties and Characterization Methods	Ref.
Cellulose	Formation of free radicals in cellulose by exposure to high energy electrons or to γ-rays from Co[60].	PSty	Grafting from	Sty brought into intimate contact with cellulose by the inclusion technique; then, (a) perform irradiation with high energy electrons using a 2-M.e.v. Van de Graaff accelerator; or (b) induce grafting by γ-ray irradiation.	Degree of grafting and grafting efficiency from extraction curves (mass determination); drastic hydrolysis of cellulose backbone and molecular weight determination of isolated PSty chains by intrinsic viscosity; calculation of grated chains per cellulose chain.	[16]
Terpolymer of ethylene–ethyl acrylate-maleic anhydride (EEAMA)	Step-growth polymerization (SGP) between maleic anhydride (MAnh) from EEAMA and amine groups from PDMS.	Polydi-methyl-siloxane (PDMS)	Grafting to	Melt reactive mixing proceeded in a Haake Rheocord 3000 batch mixer; $T = 140\ °C$.	Composition and evidence of grafting by ^1H and ^{13}C nuclear magnetic resonance. Molecular weight distributions (MWDs) were determined using a GPC or SEC chromatograph (Alliance GPCV 2000, Waters) with refractive index viscometer detectors. Linear viscoelastic properties were determined using a rheometer (AR 2000, TA Instrument) with a cone-and-plate configuration.	[17]
Cores of different polymers: polystyrene substrates; arborescent poly(γ-benzyl l-glutamate) (PBG)	Typically, successive anionic grafting reactions of pre-formed side chains onto substrates randomly functionalized with coupling sites (e.g., alkyne-azide click chemistry coupling).	poly(2-vinyl pyridine), polyisoprene, poly(tert-butyl methacrylate), and poly(ethylene oxide); poly(γ-benzyl L-glutamate) (PBG); polyglycidol, polyethylene oxide) (PEO), or poly(l-glutamic acid) (PGA)		"Dendrigraft polymers" synthesized by different chemical routes.	MWD determined by SEC and static light scattering; morphology of arborescent polystyrene molecules determined by small-angle neutron scattering (SANS); film formation of isoprene copolymers on mica surfaces investigated using atomic force microscopy (AFM) after spin-casting from different solvents; morphologies of core-shell-corona copolymers studied by transmission electron microscopy (TEM).	[18–26]

Abbreviations. ACN: α-chloroacrylonitrile; CTP: chain transfer to polymer; DMA: dynamic mechanical analysis; FRP: conventional free-radical polymerization; GMA: glycidyl methacrylate; IR: infrared spectroscopy; PAA: poly(acrylic acid); PALA: poly(allyl acrylate); PAM: polyacrylamide; PAN: polyacrylonitrile; PBMA: poly(butyl methacrylate); PE: polyethylene; PLMA: poly(lauryl methacrylate); PMA: poly(methyl acrylate); PMMA: poly(methyl methacrylate); PSty: polystyrene; PVDC: poly(vinylidene chloride); REX: reactive extrusion; SP: solution polymerization.

The renewed emphasis on the use of biobased monomers and biopolymers as a viable route to decrease (synthetic) polymer waste and disposal issues has invigorated the research efforts on the development of improved materials with important contents of biopolymers (frequently as backbones); grating is part of the synthetic procedure of such materials. These trends are in the scope of some recent review papers focused on polymer grafting, which include the grafting of polymers onto cellulose [27,28], chitin/chitosan [29,30], or polysaccharides in general [2,31].

The use of lignocellulosic waste as raw material for biorefining processes aimed at producing value-added chemicals or materials (e.g., bioethanol, cellulose, xylose, or hybrid materials, to name a few) has increased significantly since the start of the present century. Biorefineries from lignocellulosic waste require multistep processes, starting with pretreatment of the biomass. In this way, the constituent biopolymers are available for subsequent reactive processes [32–35]. The synthesis of value-added materials from lignocellulosic waste biomasses by using polymer grafting onto lignocellulose itself [36] or onto its individual components (cellulose [27], hemicellulose [37], or lignin [38]) represents an important route in the concept of biorefineries.

Although a few early studies focused on the mathematical descriptions of polymer grafting under specific circumstances—such as the calculation of grafting efficiency and molecular weight development of the grafted branches onto a pre-formed polymer containing pendant mercaptan groups capable of acting as effective chain transfer agents, based on a comprehensive kinetic model including chain transfer to polymer and bimolecular polymer radical termination [39], or the theoretical calculation of molecular weight distributions of vinyl polymers grafted onto solid polymeric substrates by irradiation, also based on a kinetic description of the growing of the grafted branches [40], and a few comprehensive recent models for other specific situations (e.g., the detailed description of free-radical polymerization (FRP)-induced branching in reactive extrusion of PE [41])—are indeed available, the fact is that the cases addressed by mathematical models are by far less common than the available experimental systems. The purpose of the present review is to first offer a rather detailed summary of what is known from a polymer chemistry angle about polymer grafting, with an emphasis on what backbones and grafts are used, how active sites on backbones are generated, and how polymer branches are grown or grafted, among other process details. The second objective is to review what polymer grafting situations have been modeled, which tools have been used, and what limitations persist. By doing that, we can show areas of opportunity. Do keep in mind that the system that motivated this study was the grafting of synthetic polymers onto lignocellulosic biopolymers.

2. Chemistry of Polymer Grafting

The main chemical routes for polymer grafting are the following: "grafting onto" (also referred as "grafting to"), "grafting from," and the macromonomer or macromer (or "grafting through") method [1,2,27,28]. There are general reviews focused on the synthesis of grafted copolymers [1,2]. The ranges of backbones and grafts, backbone activating methods, graft growing (polymerization) routes, characterization techniques, quantification methods of grafting and branching molar mass distributions, and applications are so vast that reviews on specific aspects or subtopics related to these issues have been written. For instance, there are reviews focused on the grafting of polymer branches onto natural polymers [42] and biofibers [43]; grafting onto cellulose [27] or cellulose nanocrystals [28]; grafting onto chitin/chitosan [29,30]; microwave-activated grafting [31,44]; laccase-mediated grafting onto biopolymers and synthetic polymers [2]; radiation-induced RAFT-mediated graft copolymerization [45]; and polymer grafting onto inorganic nanoparticles [46] to name a few.

Herein, brief descriptions of such chemical routes are provided. In Table 1 we provide an overview of the grafted copolymer materials synthesized in the 1950–1970 period, plus some additional more recent cases, considering backbone structure, functionalization or active site generation techniques or procedures, grafted arm structure, polymer grafting technique, polymer grafting conditions, measured properties and characterization methods, and related references. The literature on polymer grafting onto cellulose, chitin/chitosan, lignocellulosic biopolymers, other polysaccharides and natural biopolymers, inorganic materials, and metallic surfaces is addressed in the subsequent sections of this review.

2.1. Types of Polymer Grafting

As stated earlier, polymer grafting can proceed by the "grafting to" technique, where a polymer molecule with a reactive end group reacts with the functional groups present in the backbone; by "grafting from," where polymer chains are formed from initiating sites within the backbone; and by "grafting through," where a macromolecule with a reactive end group copolymerizes with a second monomer of low molecular weight. Simplified representations of these grafting techniques are shown in Figure 1.

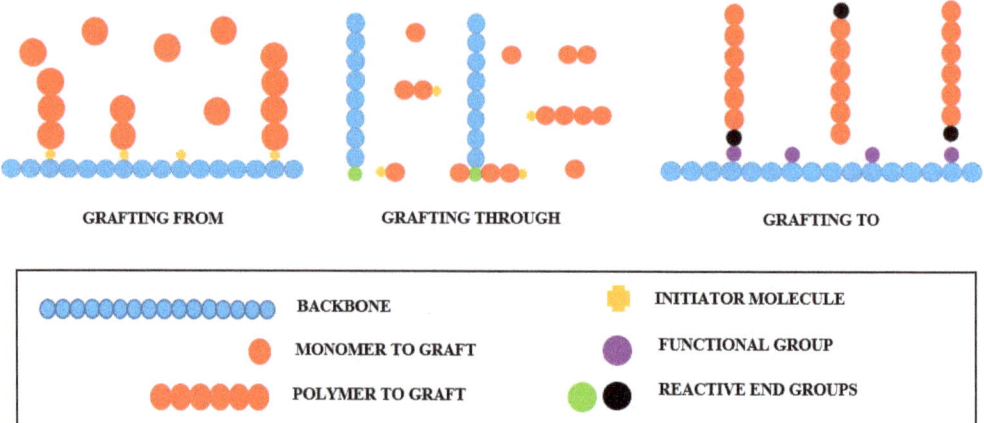

Figure 1. Polymer grafting chemical routes.

"Grafting to" and "grafting from" are the most common polymer grafting chemical routes. Better defined graft segments are obtained by the "grafting to" technique since the polymerization is independent of the union between the backbone and grafts. In contrast, materials of higher grafting densities can be produced by the "grafting from" route due to the lack of steric hindrance restrictions [47].

However, each polymer grafting route has its own advantages and disadvantages in terms of chemical nature, density, dispersity, and length of the grafts obtained, and the ease and efficiency of the chemical reactions involved. Interestingly, different polymer grafting routes can be combined to produce specific grafted materials [48].

2.2. Main Backbones Used in Polymer Grafting

A polymer backbone is a polymer molecule that supports polymeric side chains, called branches or grafts. Side chains can be inserted onto the backbone during the synthesis of the backbone (copolymerization situation) or as a post-production process of the backbone [49]. In the first case, polymers with homogeneous bulk properties are obtained. The second case is very attractive since it allows the modification of many polymeric materials, including natural and synthetic fibers, or inorganic and metal particles. Backbones processed by polymer modification do not usually show significant changes in bulk properties. Surface modification is often carried out following a "grafting from" technique; that is why this method is also known as surface initiated polymerization (SIP) [50]. The backbones used for polymer grafting can be synthetic polymers, biopolymers, or inorganic and metal surfaces.

Synthetic polymers are human-made polymers and include a wide variety of materials, such as polyolefins, vinyl and fluorinated polymers, nylons, etc. The applications of synthetic graft copolymers include the synthesis of antifouling membranes [51], stimuli-response materials [52], and biomedical applications [53].

Biopolymers are produced by the cells of living organisms. Polysaccharides have become important lately because of their characteristics of availability, biocompatibility, low cost, and non-toxicity, making them candidates for substitution of petroleum-based materials [47]. Polysaccharide-based graft copolymers are used as drug delivery carrier, food packaging and wastewater treatment [54]. Some of the most studied polysaccharides are cellulose [27,28], lignin [55], chitin/chitosan [29,56–58], starch [59], and various gums [42].

Surface functionalization of inorganic and metallic particles that allow the incorporation of polymer shells by polymer grafting has also become important, since polymer coatings alter the interfacial properties of the modified particles. Zhou et al. reviewed different applications for inorganic and metallic particles grafted with biopolymers [50]. One important inorganic surface modified by polymer grafting is silica [60].

2.3. Backbone Functionalization Methods

Several chemical modification procedures have been developed due to the wide variety of backbones of interest. Chemical modification reactions depend on the functional groups (or absence thereof) along the backbone. Two main chemical routes used to attach, grow, or graft polymer molecules onto lignin have been proposed [55]: (a) creation of new chemically active sites, and (b) functionalization of hydroxyl groups.

The introduction of functional groups into a polymer backbone increases its reactivity, making it accessible for forward polymerization or coupling reactions. Functionalization reactions are therefore required to generate the end functional pre-formed polymer or the reactive end of the macromolecule species involved in the "grafting to" and "grafting through" polymer grafting techniques, respectively. Functionalization is also required in the formation of the macromolecular species, such as macro-initiators and macro-controllers, involved in the "grafting from" polymer grafting technique [27,61]. The most important functionalization reactions involved in polymer grafting include sulfonation, esterification, etherification, amination, phosphorylation, and thiocarbonation, among others.

2.4. Backbone Activation Methods

Another way to generate grafting sites within the polymer backbone is to use polymer grafting activators, such as free-radical initiators. As shown in Figure 2, polymer grafting activators can be classified into physical, chemical, and biological. The main characteristics of these activators are highlighted in Sections 2.4.1–2.4.4.

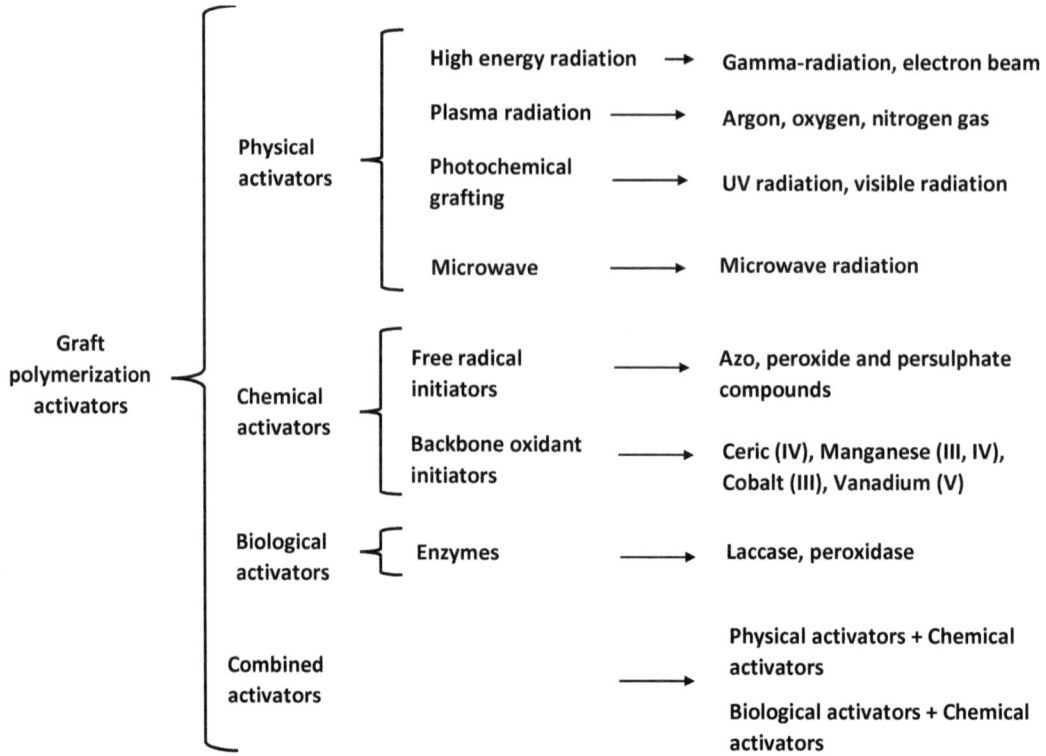

Figure 2. Backbone activators used for polymer grafting.

2.4.1. Physical Activators

High energy radiation, also referred to as ionizing radiation, includes γ-beam and electron-beam radiations. Radiation-promoted grafting may follow one of three possible routes: (a) pre-irradiation of the backbone in the presence of an inert gas to generate free radicals before placing the backbone in contact with monomers; (b) pre-irradiation of the backbone in an environment containing air or oxygen to produce hydroperoxides or diperoxides in its surface, followed by high temperature reaction with monomer; and (c) the mutual irradiation technique, where backbone and monomer are irradiated simultaneously to generate free radicals [62].

Plasma is a partially ionized gas where free electrons, ions, and radicals are mixed. Different functional groups can be introduced, or free radicals can be generated on backbones by this process, depending on the gas used. Polymer grafting reactions carried out in plasma are sometimes classified as high energy radiation reactions [63].

The absorption of UV light on the surface of the material generates free radicals that serve as nucleation sites. The surface is then placed in contact with monomer for subsequent polymerization [64].

Microwave irradiation consists of direct interaction of electromagnetic irradiation with polar molecules and ionic particles, promoting very fast non-contact internal heating, which enhances reaction rates and leads to higher yields. Singh et al. carried out a successful polymerization of acrylamide on guar gum under microwave irradiation [65]. They proposed a mechanism in which free radicals are produced within the polysaccharide backbone by the effect of microwave irradiation on the hydroxyl groups of the biopolymer [65].

2.4.2. Chemical Activators

As shown in Figure 1, chemical activators include free radical and backbone oxidant initiators. Free radical initiators are compounds that present either direct or indirect homolytic fission. The first case involves the initiator itself and the second one requires participation of another molecule from the environment [66].

Oxidant initiators react directly with functional groups from the backbone, generating activation sites. Polymer grafting of polysaccharides using oxidant initiators has been reported in the literature [66].

2.4.3. Biological Activators

Enzymes catalyze polymer modification reactions through functional groups located at chain ends, along the main chain, or at side branches, promoting highly specific non-destructive transformations on backbones, under mild reaction conditions. Successful grafting of lignin by oxidation of its phenolic structures using laccases has been reported recently [2].

2.4.4. Combined Activators

Combinations of physical and chemical activators for polymer grafting have been successfully carried out. For instance, microwave assisted polymerization (MAP) has been combined with the use of chemical activators for the production of hydrogels synthesized by crosslinking graft copolymerization, taking advantage of the short reaction times required to obtain high yields [67,68]. Enzymes are also used in combination with radical initiators for more effective grafting copolymerization processes [69,70].

2.5. Polymer Grafting by Free-Radical Polymerization

As stated earlier, the "grafting through" and "grafting from" techniques require a polymerization reaction to bond the polymer grafts to the backbone. Different polymerization methods have been used for polymer grafting, but the most effective ones use free radical methods (e.g., FRP, RDRP, and REX), due to their versatility to work with different chemical groups, and their tolerance to impurities. A short overview on free-radical polymerization reactions is presented in Table 2. Polymer grafting by FRP, and other reactions, is affected by several factors, including the chemical nature of the components contained in the system—backbone, monomer, initiator, and solvent—and the interactions among them. Other aspects related to polymer grafting, including temperature and the use of additives, need to be considered [67]. The synthetic routes and activators used in graft polymerization provide a variety of interesting and versatile routes for this type of polymer modification.

Table 2. Free-radical polymerization methods.

Polymerization Method	General Description	Type of Reaction
Conventional free-radical polymerization (FRP)	Three steps involved: (1) initiation, with formation of free radicals; (2) propagation, where free radicals react with monomer; and (3) termination of polymer radicals by either combination, disproportionation, or chain transfer to small molecules. The simultaneous participation of these reactions leads to broad molar mass distributions.	Chain transfer reaction: free radicals generated in the system tend to react with backbones by CTP, thereby activating them. Direct generation of free radicals along the backbone: activators generate free radicals from reaction with functional groups placed along the backbone, which correspond to the initiating step of a SIP [27].

Table 2. Cont.

Polymerization Method	General Description	Type of Reaction
Reversible deactivation radical polymerization (RDRP)	A group of polymerization techniques based on free radical technology that controls the growth of polymer molecules during the polymerization. Polymer radicals are reversibly deactivated by effect of controllers that act under some relatively new chemical routes. These techniques allow the development of advanced materials, with various architectures, and well-defined microstructures. Each technique has its own mechanism and conditions that favor them. SIP can proceed by any of the known RDRP techniques.	Atom transfer radical polymerization (ATRP): It is a catalytic process where an alkyl halide macromolecule reacts with the catalyst, allowing the formation of a radical that propagates until it reacts again with the catalyst, in a reversible way [71]. Nitroxide mediated polymerization (NMP): A stable nitroxide free radical acts as controller, reversibly deactivating the propagating and polymer radicals forming dormant polymer molecules with alkoxyamine end functionalities [50]. Reversible addition-fragmentation chain transfer (RAFT) polymerization: Thiocarbonilthio compounds are used as chain transfer agents which control molecular weight development by reversible activation-deactivation reactions [50].
Reactive extrusion (REX)	REX is a set of techniques designed to produce and modify polymers, typically carried out in single or twin extruders. Five main types of reactive polymerizations caried out in extruders have been reported: bulk polymerization, polymer grafting, polymer functionalization, controlled degradation, and reactive blending [72]. Reactions proceed in melt phase. Examples of polymer grafting by REX include polyolefin [73] and starch modifications [74].	Polymer modification by free-radical polymerization: Free radical initiators such as peroxides are used to generate activate sites within the backbone [73,75]. Polymer modification by insertion of active pendant groups: It consists of the copolymerization of monomers who have no functional groups with co-monomers possessing pendant which make polymer grafting easier to accomplish [49].

3. Backbones and Supports Used in Polymer Grafting

As explained earlier, grafted materials consist of side chains or arms attached to primary polymers referred to as backbones. The purpose of polymer grafting is to combine chemical, mechanical, interfacial, electrical, or other polymer properties between the constituent materials. The diversity of backbones and the ways in which side chains are attached to them through polymer grafting will be briefly overviewed in this section.

3.1. Cellulose, Lignin, and Lignocellulosic Biomasses as Backbones

Lignocellulosic biopolymers are abundant in nature. They are made of cellulose, hemicellulose, and lignin. They also contain moisture, extractive organic compounds, and ashes from inorganic compounds in lesser amounts. Each of these components has distinct characteristics. The extractive organic compounds present in lignocellulosic biopolymers are oligomers and oligosaccharides of low molecular weight, sugars, fatty acids, resins, etc. [76].

Cellulose, hemicellulose, and lignin can be modified by polymer grafting leading to new promising materials with interesting properties. However, the extractables are not useful for this purpose since they are not part of a skeleton or stiff structure that may provide support or mechanical stability. Extractables also consume reactants required for the grafting process. They are usually removed prior to the polymer grafting process, although in some studies, they remain in the system during the formation of grafted arms [77].

Polymer grafting of xylan onto lignin has been studied since the early 1960s. Early reports on the topic reported the grafting of organic polymers, such as 4-methyl-2-oxy-3-oxopent-4-ene and methyl methacrylate polymers [78,79], xylan [80], ethylbenzene, and styrene [81–84], onto lignin or lignin derivatives. The topic of polymer grafting of synthetic polymers onto lignocellulosic biopolymers has gained renewed relevance in the last two decades due to environmental and sustainability issues [27,85–94].

Table 3. Overview of grafting of synthetic polymers onto cellulose and natural fibers.

Backbone	Functionalization Method	Grafted Chains	Grafting Technique	Grafting Conditions	Measured Properties and Methods	Refs.
Cotton fabric	FRP by a cellulose thiocarbonate–AIBN redox system.	PMMA, PAA, PAN, PAM	Grafting from	T = 60–80 °C.	Degree of grafting (GP), gravimetric method.	[95]
Cellulose	FRP by a KPS-FAS REDOX system.	PSty, PAN	Grafting from	T = 60 °C; t = 3 h; [STY] = 0.65 M [KPS] = 0.14 M; [FAS] = 0.01 M.	GP, gravimetric method; FTIR and TGA to corroborate GP.	[96]
Cellulose fabrics	FRP by KPS initiation.	PIA	Grafting from	T = 55–80 °C; t = 2.5–5 h; [KPS] = 0.05–0.5 M; [IA] = 0.5–4 M.	GP, gravimetric method; FTIR, XRD, TGA and SEM to corroborate grafting.	[97]
Cellulose	FRP by CAAC initiation.	PMBA, P(N-VP)	Grafting from	T = 30–70 °C; [CAAC] = 2 − 30 × 10^{-5} M.	Grafting yield (GY) and other grafting parameters; gravimetric method.	[98,99]
Cellulose	FRP by CAN-NAC REDOX system.	PEA, PNIPAAM, PAAM-PEA, (AAM-EMA), P(AAM-MA), P(AN-EMA)	Grafting from	T = 10–60 °C; t = 24 h; [CAN] = 1.5 − 32 × 10^{-3} M; [NAC] = 2.5–8 × 10^{-2} M.	GY and other grafting parameters, gravimetric method; FTIR and TGA to corroborate grafting. FTIR and EA for composition. MMass by viscometric method and GPC.	[100–105]
Cellulose microfibers	Redox initiation.	PAA	Grafting through	KPS = 0.1–0.4% respect to fiber weight. t = 3 h, T = 75 °C	GP, gravimetric method; FTIR and TGA to corroborate GP.	[106]
Cellulose powder	Co(acac)$_3$	N'N'-MBA [98] or N-VP [99]	Grafting from	Cellulose washed with CH$_3$OH, C$_3$H$_6$O, and water; then dried. Reaction under nitrogen atmosphere. Different temperatures, 30–60 °C [98], or 40–50 °C [99]; reaction carried out in water. Kinetic data from 0–150 min [86], or 0–120 min [99].	Percent grafting (% G), true grafting (% GT), grafting efficiency (% GE), homopolymer conversion (% CH), cellulose conversion (% CC), and total conversion (% CT) by gravimetric methods.	[98,99]
Cotton linter Cellulose powder	Ceric ammonium sulfate, 1% sulfuric acid.	AcN EA MMA	Grafting from	Cellulose treated with sodium hydroxide 5–30 % wt./vol at 25 °C. Grafting temperatures: 30, 40 and 60 °C, sodium bisulfite clay as initiator. Polymerization in diluted HCl for 2 h.	Percent grafting (% G), true grafting (% GT), grafting efficiency (% GE), by gravimetric methods.	[107]

Table 3. *Cont.*

Backbone	Functionalization Method	Grafted Chains	Grafting Technique	Grafting Conditions	Measured Properties and Methods	Refs.
Hydroxypropyl cellulose (HPC)	Steglich esterification of PABTC onto HPC using DCC and DMAP.	EA NIPAAM	Grafting from	Steglich esterification using DCC and DMAP in chloroform at 40 °C, in presence of HPC and PABTC. 6 days for 50% conversion. Polymerization of EA and NIPAAM at 60 °C using AIBN; 94% conversion with free polymer in DMAc.	Tg by DSC at 45, 55 and 135 °C for PNIPAAM, and Tm at 157 °C. TGA from ambient to 600 °C at 10 °C/min, nitrogen atmosphere for HPC, PNIPAAM and the grafted HPC-g-PNIPAAM. ^{1}HNMR for degree of substitution. SEC for HPC Macro CTA and HPC-g-PNIPAAM.	[108]
Cellulose chloroacetate (CellClAc)	Macro initiator, Cu(I)Cl/2'2'BIPI catalytic system via ATRP controller.	4NPA MMA	Grafting from	ATRP of 4NPA and MMA carried out in DMF at 130 °C for 24 h, in the presence of CellClAc as macro initiator, Cu(I)Cl/2,2'BIPI catalytic system. Grafting conversion under 15%.	NMR, FTIR, TGA, and elemental analysis.	[109]
Cellulosic *Grewia optiva* fibers	Redox initiation using FAS-H$_2$O$_2$ for grafting of MA, and KPS for polymerization of MA.	MA	Grafting from	1 g of mercerized *Grewia optiva* fibers was set in distilled aqueous solution with NaOH for 24 h, followed by addition of redox initiator (FAS-H$_2$O$_2$). Solution was stirred for 10 min; MA was then added. Solution was polymerized using microwave irradiation at different times.	FTIR, SEM, TGA, swell index.	[110]
Cellulose acetate	Solvents: DMSO, PDX, DMAc, C$_3$H$_6$O. Initiators for grafting and polymerization: CAN, Sn(Oct)$_2$ and BPO.	MMA	Grafting from	Focus on solvent effect. 1.25 g of cellulose acetate were dissolved in 125 mL of solvent. CAN or Sn(Oct)$_2$ or BPO (0.3–0.5 g) were added with MMA (1.25–2 mL). Nitrogen atmosphere, 2–6 h, 30–80 °C, except acetone, which proceeded at 55 °C.	Grafting yield (GY), total monomer conversion (TC), grafting efficiency (GE) and number of grafting chains per cellulose acetate molecule were obtained by gravimetric methods. TGA, GPC, FTIR, ^{1}HNMR.	[111]
Cellulose chloroacetate (CellClAc)	Macro initiator, Cu(I)Cl/2'2'BIPI catalytic system via ATRP controller.	NCA MMA	Grafting from	ATRP of NCA and MMA in DMF at 130 °C, in the presence of CellClAc.	FTIR, TGA, and elemental analysis.	[112]

Table 3. Cont.

Backbone	Functionalization Method	Grafted Chains	Grafting Technique	Grafting Conditions	Measured Properties and Methods	Refs.
Cellulose cotton fibers	Na$_2$CO$_3$ and thermal activation.	MTC-b-CD	Grafting to	Grafting of MCT-β-CD onto cotton fabric carried out in alkaline medium; 1 mg of cellulose fiber was impregnated with solutions of 50–150 g/L of MCTβ-CD and 20–80 g/L Na$_2$CO$_3$. Solvent was eliminated at room temperature before heating at 100–160 °C, for 10, 15 and 20 min. Sample were washed to obtain neutral pH. Silver nitrate solutions were added in situ for some samples.	FTIR and microbiological tests [113]. Gravimetric methods and analytical modeling [114]. MODDE software was used to study the relationship between 3 significant independent variables and degree of grafting.	[113,114]
Cotton linter cellulose	CTP using APS as initiator.	MMA	Grafting from in situ polymer formation (embedded)	Cellulose pre-swelling: Cellulose was pre-swollen in DMAc at 160 °C for 0.5 h. Pre-swollen cellulose was filtered. A solution of LiCl in DMAc (8%, w/w) was prepared. The pre-swollen cellulose was added to the DMAc/LiCl solution. The mixture was stirred at 100 °C for 2 h and purged with gaseous N$_2$. MMA polymerization was carried out at 70–90 °C using APS and DMSO.	TGA-DTA, FTIR, SEM, XRD.	[115]
Microcrystalline cellulose	Ring-opening polymerization (ROP) of L-LA with DMAP in an ionic liquid AmimCl.	PLLA	Grafting from	A 4% (w/w) microcrystalline cellulose/AmimCl solution was prepared and stirred at 60 °C, in N$_2$ environment for 1 h. L-LA and DMAP were then added. The sample was degassed 3 times in vacuum/N$_2$ during 1 h cycles. ROP proceeds at 90 °C in presence of N$_2$ atmosphere during 11 h.	Controlled release of vitamin C. Characterization by ^{1}HNRM, UV analysis, XRD, TEM, and HPLC.	[116]

Table 3. Cont.

Backbone	Functionalization Method	Grafted Chains	Grafting Technique	Grafting Conditions	Measured Properties and Methods	Refs.
Cellulose (DP=1130)	CTP, APS and MMA embedded.	MMA	Grafting from	Cellulose is pre-swollen in DMAc during 30 min, at 160 °C. 5 g of pre-swollen cellulose are mixed with 95 g of water and proper amounts of APS and MMA. Grafting proceeds at 80 °C. Excess of PMMA was washed with acetone.	Characterization: FTIR, WARD-XRD, SEM, and TGA-DTA.	[117]
Cellulose nanocrystal (CNC)	Macromolecular initiator obtained from the reaction between Br-iBuBr and CNC using TEA as catalyst, via SI-ATRP.	STY. Cast in MMA	Grafting from	Macromolecular initiator: Br-iBuBr, CNC, and TEA were dissolved in DMF, and stirred under N_2 at 70 °C for 24 h. CNC, 2-bromoisobutyrate, styrene and copper(I)bromide were dissolved in anisol; PMDETA was then added. The reaction proceeded at 100 °C for 12 h and N_2 atmosphere via SI-ATRP. The remanent PS homopolymer was eliminated using methanol, and centrifuging. The modified crystals were dosed into PMMA nanocomposites by solution casting.	Characterization: TEM, SEM, FTIR, UV-Vis., XRD, ^{13}C-NMR, TGA, DSC, and tensile test	[118]
Cellulose nanocrystal (CNC)	L-LA in situ polymerization using MgH$_2$ as redox agent. PLLA-g-CNC particles were casted in PLLA.	PLLA	Grafting from	1.50 g of L-LA and 0.05 g of CNCs were pre-mixed. 0.03 g of MgH$_2$ were then added and stirred during 30 min. The mixture was heated to 110 °C for 3, 6, 10, 12, 18 and 24 h, under nitrogen atmosphere. Casting method: 1.0 g of PLLA and 0.1 g of CNC or 1.3 g of PLLA-g-CNC were dispersed in 10 mL of chloroform or toluene solution and mixed for 10 h. The mixtures/solutions were poured into Teflon® plates and let stand for 18 h at 80 °C. The films were dried under vacuum at 60 °C for 72 h at room temperature.	Characterization: FTIR, ^{1}HNRM, ^{13}C-NMR, XRD, XPS, TGA, DSC and DMA.	[119]

Table 3. Cont.

Backbone	Functionalization Method	Grafted Chains	Grafting Technique	Grafting Conditions	Measured Properties and Methods	Refs.
Cellulose cotton fiber pulp	(a) Esterification of maleic anhydride grafted onto PHA (through double bond); (b) PHA-g-MA is grafted onto cellulose cotton fiber pulp (through the anhydride group).	PHA-g-MA	Grafting to	Cotton fibers were defibered into pulp to 25°SR using a PL4-2 speed governing beater; 30 g/m² paper film was then formed in a RK3-KWTjul Rapid-Koethen sheet former. PHA-g-MA was dissolved in refluxing dichloromethane at 60 °C, and the paper film was dipped into the solution of PHA. The PHA/CF composite film was then washed with dichloromethane and dried.	Characterization: FTIR; XRD; SEM; surface roughness; surface hydrophobicity, with contact angle; tensile test, and TGA.	[120]
Cellulosic filter papers	Radiation-induced graft copolymerization (RIGCP) of AcN.	Acrylonitrile	Grafting from	The reaction took place in a glass tube which contained a Whatman filter paper (W1) and a 45% AcN solution in DMF. The solution was irradiated with cobalt-60 g-rays at a dose rate of 4 kGy/h, in air atmosphere. The grafted material was washed with DMF and water, followed by drying (T = 80 °C, 2 h).	Degree of grafting (G%) was determined. Characterization: FTIR, XRD, XRF, TGA and SEM.	[121]
CellClAc	ATRP grafting of the studied monomers using CuCl, 2'2'BIPI as catalyst.	NCHA, 4VP, DA, DAAM.	Grafting from	Cellulose chloroacetate (ATRP macroinitiator) and any of the monomers (NCHA, 4VP, DAAM or DA) were added to 10 mL of DMF; an inert environment was created using argon. The grafting reaction proceeded at 130 °C for 24 h. The proportion of CellClAc, CuCl, 2'2'BIPI and each monomer was 1:1:3:100. The reacting mass proceeded to filtering and washing using acetonitrile, DMF, chloroform, a mixture of water-ethanol-HCl, pure water, ethanol, acetone, and diethyl ether. The product was dried under vacuum.	Characterization: FTIR, UV-Visible, TGA, elemental analysis, and electrical conductivity.	[122]

Table 3. *Cont.*

Backbone	Functionalization Method	Grafted Chains	Grafting Technique	Grafting Conditions	Measured Properties and Methods	Refs.
Regenerated cellulose fibers (rayon)	Photo-chemical grafting of PETA without photoinitiator.	PETA	Grafting to	Fibers were washed with a solution of PETA in isopropanol. Monomer content of 1% and 5% was used. The cellulosic material was irradiated with a broadband Hg lamp (emission band of 200 and 300 nm), at 50 W/cm. Layer deposition from homo-polymerization and subsequent grafting-to process onto the fiber took place.	Fiber volume content; tensile and fatigue tests; SEM.	[123]
Cellulose nanofibrils	Nitroxide TEMPO insertion and NMP of HEMA	HEMA	Grafting from	Preparation of TEMPO-oxidized cellulose nanofibrils. A suspension was prepared using HEMA (0.15 g), H_2O_2 (0.0075 g), $CaCl_2$ (0.0075 g) and TCNF (0.3 wt.%, 500 g), followed by stirring and heating at 50 °C for 4 h. Either freezer freezing, or gradient freezing in liquid nitrogen were used. The nanofibril suspension was set in a freezer (−25 °C). The frozen material pieces were 3 cm height. Drying in a lyophilizer to form aerogels then took place.	Characterization: XRD; FTIR; XPS; SEM; stress-strain under compression analyses; BET analysis and electrical resistivity.	[124]

Cellulose can be extracted from lignocellulose and used as such or modified for other applications. Table 3 provides an overview of grafting of synthetic polymers onto cellulose and natural fibers. (See the tables of Section 6 for explanation of abbreviations and symbols.)

Lignin follows cellulose in abundance on earth, providing a primary natural source of aromatic compounds [125]. Several industrial applications have been attempted for lignin [55, 125–127] but not all of them have succeeded due to different reasons [125,128–130].

Marton [8] described fifty-four different constituents that can be found in lignin based on interpretation of experimental data from biochemical degradation, oxidation, and other ways of decomposition of different types of lignin materials. The combinations and proportions among these structures lead to different properties of lignin materials. Three decades later, Lewis and Sarkanen [130] organized these fifty-four structures into a map that they called phenylpropanoid pathway. As observed in Figure 3, lignins and lignans are monolignol derived compounds. Sharma and Kumar described lignin as a complex material consisting mostly of three single unit lignol precursors, coniferyl alcohol, p-coumaryl alcohol, and sinapyl alcohol, along with other atypical monolignol constitutive units in trace amounts [55].

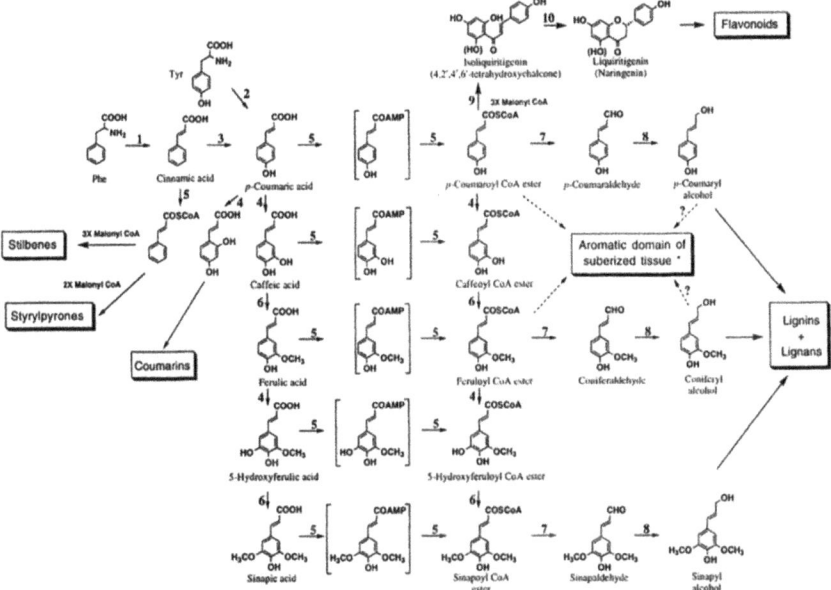

Figure 3. Main steps of the phenylpropanoid pathway: lignins and lignans are monolignol derived. 1, phenylalanine ammonia-lyase; 2, tyrosine ammonia-lyase (mostly in grasses); 3, cinnamate-4-hydroxylase; 4, hydroxylases; 5, CoA ligases involving AMP and CoA ligation, respectively; 6, O-methyltransferases; 7, cinnamoyl-CoA:NADP oxidoreductases; 8, cinnamyl alcohol dehydrogenases; 9, chalcone synthase; 10, chalcone isomerase. (Note: conversions from 7-coumaric acid to sinapic acid and corresponding CoA esters are marked in boxes since dual pathways seem to take place; *: may also involve 7-coumaryl and feruloyl tyramines, and small amounts of single unit lignols). Source: Adapted with permission from Lewis N. G. and Sarkanen S. (1998). Lignin and Lignan Biosynthesis, Washington, D.C.: Oxford University Press pp. 6–7 [130] Copyright © 2021 by American Chemical Society.

The process used for lignin extraction and the final properties of the material depend on the type of biomass employed [55]. Lignin is obtained from woods, which can be hard, soft, bushes, rinds, husks, corncobs, either products or residues. A pulp is obtained from these materials. The yield of lignin extraction depends on temperature, time, dispersion media, extraction method, and the amount of lignin present in the raw material. Lignin extraction methods can be biological or enzymatic, physical, or chemical. Integrated solutions are employed at the end to remove impurities from lignin so it can be bleached [55]. The complex structure of lignin contains specific surface moieties that provide reactive sites where polymers and other species can be synthesized, bonded, or modified [55]. These moieties were recognized as hydroxyl, carboxyl, carbonyl, and methoxyl groups.

There are two main routes for grafting of polymer chains onto lignin-based biopolymers [55]: (a) synthesis of new reactive sites within lignin's structure; and (b) modification or functionalization of lignin's hydroxyl groups. Route (a) allows lignin to become more reactive, both at the surface, and within the bulk. Polymer modification by route (a) improves both, the properties of lignin and those of the modified materials.

In route (b), a good number of functional groups can be placed in the end groups of lignin (what is sometimes referred to as the surface of lignin). Katahira et al. [131] identified seven side chain structures in the end groups of lignin: p-coumarate, ferulate, hydroxycinnamyl alcohol, hydroxycinnamaldehyde, arylglycerol, dihydrocinnamyl alcohol, and guaiacylpropane-1,3-diol end-units, as shown in Figures 4 and 5 [131]. However, it has been proposed that the phenolic hydroxyl groups shown in Figure 4, and the aliphatic hydroxyl functional groups corresponding to C-α and C-γ positions of the side molecule fragment shown in Figure 5, are the most reactive [55]. Both routes allow one to produce grafted materials, mainly polymers, and most of them come from route (b) above [85,132,133]. Further reports on lignin treatments and grafting can be found elsewhere [55,91,125,128].

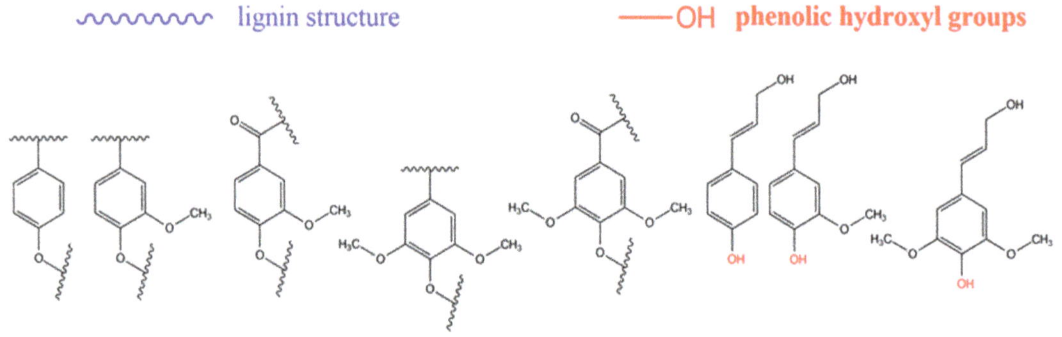

Figure 4. Repeating units in lignin. From left to right: p-hydroxyphenyl, guaiacyl, metoxy guaiacyl, syringyl, metoxi syringyl, p-coumaryl alcohol, coniferyl alcohol, synapyl alcohol. Source: Adapted with permission from Katahira et al. (2018). Lignin Valorization. Emerging Approaches: Croydon UK pp. 3 [131]. Copyright © 2021 The Royal Society of Chemistry.

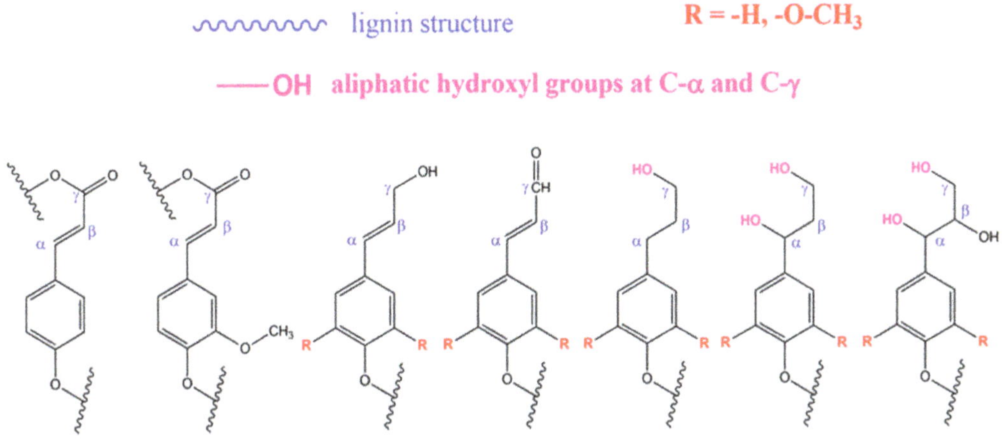

Figure 5. Side chain structure in end-groups in lignin. From left to right: p-coumarate, ferulate, hydroxycinnamyl alcohol, hydroxycinnamaldehyde, dihydroycinnamyl alcohol, arylpropane-1,3-diol, arylglycerol end units. Source: Adapted with permission from Katahira et al. (2018). Lignin Valorization. Emerging Approaches: Croydon UK pp. 5 [131] Copyright © 2021 The Royal Society of Chemistry.

The overview on grafting of synthetic polymers onto cellulose and other lignocellulosic biopolymers presented in Table 3 is further expanded in Table 4 to include other examples of lignocellulosic biomasses, and other natural biopolymers, such as polysaccharides, chitin, and chitosan. Examples of recent research reports (2020–2021) on synthesis of grafted polymers are provided in Table 5. Tables 6–13 contain extensive information related to characterization of polymer grafting. Table 14 summarizes the literature on modeling of polymer grafting. Table 15 shows the polymerization scheme of FRP including CTP and crosslinking. Finally, Tables 16–19 provide information on the many symbols and abbreviations used throughout the review.

Table 4. Overview of grafting of synthetic polymers onto natural polymers: lignin, hemicellulose, polysaccharides, chitosan, and chitin backbones.

Functionalization Method	Grafted Chains	Grafting Technique	Grafting Conditions	Refs.
Steglich esterification of RAFT controller onto organosolv lignin to produce a microcontroller. Subsequent polymerization of soybean oil derived methacrylate monomers.	Poly(soybean oil methacrylate) derivatives. PSBMA, PSBMAH, PSBMAEO	Grafting from	4-cyano-4-(phenylcarbonothioylthio) pentanoic acid was coupled with lignin via Steglich esterification using DCC and DMAP at ambient conditions. [Monomer]/[Lignin-RAFT]/[AIBN] molar ratio = 100:1:0.3; Temperature: 70–80 °C; Time: 24–48 h; Conversion: 74–94%.	[134]
Phosphorylation. S_N2 reaction mechanism.	Imidazole and $POCl_3$ react with (1H-imidazol-1-yl) phosphonic group, which reacts with lignin.	Grafting to	Reacted imidazole-$POCl_3$. DMF, 80 °C, 6h. Lignin-g-(1H-imidazol-1-yl)phosphonic group. 95 °C, 12 h, pH = 3–4 in DMF, yield = 92%. Composites were blended with PP-block-PE, MFI of 1.39 g/10 min (T = 230 °C; standard weight of 2.16 kg), ethylene content 17.8 wt %, from China Petrochemical Co. in a Thermo Haake torque rheometer (180 °C: 10 min; 60 rpm).	[125,135]
	Phosphorus(V) oxide	Grafting to	P_4O_{10}. Temperature = 20–25 °C, time = 7–8 h. Solvent: THF.	[125,136,137]
Hemicellulose grafting using TBD and ε-caprolactone monomer.	PCL - Hemicellulose (HC-g-PCL)	Grafting from	5 g of dried hemicellulose (5.00 g) were introduced into a reactor containing 10 mL of DMSO at 80 °C. Agitation at 60 rpm took place until a homogeneous viscous solution was obtained. Grafting temperature was 110 °C. ε-caprolactone monomer was added to the hemicellulose solution and agitated during 30 min. TBD, 1% of the total mass, was added. The reaction proceeded during 4 hrs. in a system with flow of nitrogen. Drying at 60 °C under vacuum proceeded for 2 days. FTIR, ^{1}HRM, ^{13}CNMR, HSQC, GPC, DSC, TGA-FTIR, Tensile Test, Contact Angle, Biodegradation test.	[37]
Several techniques presented in the review.	Monomers or polymers used in: ATRP: NIPAAM, DMAEMA, STY, MMA, BA, EG; RAFT: AM, AA, others; ROP: PCL, L-LA, EG; Click chemistry: polymer having alkyne or azide groups for the azide-alkyne cycloaddition, EG, PCL, PLLA.	Grafting, from; grafting to; network copolymer grafting	Grafting from: ATRP, RAFT, ROP, and FRP. Grafting onto: Thiol-ene reaction with photo-redox catalysis, click chemistry, epoxide group-mediated reactions, condensation. Network copolymer grafting: Polycondensation, free radical crosslinking.	[9]

Table 4. Cont.

Functionalization Method	Grafted Chains	Grafting Technique	Grafting Conditions	Refs.
Insertion of ACX onto lignin to synthesize a RAFT macrocontroller which polymerized AM and AA.	RAFT polymerization of AM and AA.	Grafting from	0.1 g of RAFT macroinitiator and 0.005 g of AIBN were added to a solution of monomer (0.3 g) in DMF (4 mL). The reactor was degassed with nitrogen for 30 min at ambient conditios. Reaction temperature was 70 °C, target conversion was 98% in 24 h for both monomers. The product was obtained by precipitation, filtration, washing and drying. Grafting was measured using ^{1}H NMR. 0.2 g Lignin-g-Pam were dissolved in KOH solution (0.1 g in 1 mL of water), followed by heating (T = 70 °C, t = 12 hrs.), followed by neutralization with HCl. The solution was then precipited into diethyl ether and dried in air. Ungrafted PAm was withdrawn by extraction with CH_2Cl_2. The product was vacuum dried, and analyzed by ^{1}H NMR and GPC, using adequate solvents.	[138,139]
Several techniques described in the review. Monomers copolymerized with natural extracts.	Several procedures described.	Grafting from Grafting through	Monomers and polymers grafted by: **FRP:** Guaiacol-AM, Vainillin-LMA. **RAFT:** Syringyl methacrylate, STY, MMA, 4-propylsyringol, 4-propylguaiacol. **ADMET (Acyclic Diene Metathesis Polymerization) and ROP:** Ferulic acid, isorbide, butanediol.	[140]
ATRP	NIPAAM → PNIPAAM	Grafting from	Backbone: Kraft lignin (alkali); Catalyst: CuBr/PMDETA; Solvent: water/DMF; T = 50 °C; Thermoresponsive material.	[140,141]
ATRP	DAEA → PDAEA	Grafting from	Backbone: Organosolv lignin; Catalyst: CuBr/Me$_6$TREN; Solvent: THF; T = 65 °C; Hydrophobic polymer composites.	[140,142]
ATRP	PEG-A; NIPAAM	Grafting from	Backbone: Kraft lignin (alkali); Catalyst: CuBr/HMTETA. Solvent: PDX; T = 60–70 °C; Thermogelling material.	[140,143]
ATRP	MMA → PMMA BMA → PBMA	Grafting from	Backbone: Kraft lignin; Catalyst: CuBr/PMDETA; Solvent: water/DMF; T = 80 °C; Thermoplastic elastomers.	[140,144]
SI-ATRP	NIPAAM → PNIPAAM	Grafting from	Backbone: Softwood Kraft lignin; Catalyst: CuCl/HMTETA; Solvent: water; Room temperature; Ionic responsive nanofibrous material.	[140,145]
ATRP	MMA → PMMA STY → PS	Grafting from	PMMA Backbone: Kraft lignin (alkali); Lignin + STY → lignin-g-PS; Catalyst: CuCl/2'2'BIPI; Solvent: DMF; T = 100 °C, 14 days. PS (PSty) Backbone: Kraft lignin (alkali); Lignin + MMA → Lignin-g-PMMA; Catalyst: CuBr/PMDETA; Solvent: DMF; T = 70 °C, 24 h; Thermoplastic lignin composites.	[140,146]

Table 4. Cont.

Functionalization Method	Grafted Chains	Grafting Technique	Grafting Conditions	Refs.
ATRP	MMA → PMMA	Grafting from	Backbone: Kraft lignin (alkali); Lignin + MMA → Lignin-g-PMMA; Catalyst: CuBr/PMDETA; Solvent: DMF; T = 70 °C, 24 h; Thermoplastic lignin composites.	[140,147]
ATRP	PEGMA	Grafting from	Backbone: Kraft lignin (alkali); Lignin-Br + PEGMA → Lignin-g-PEGMA; Catalyst: CuBr/HMTETA; Solvent: Acetone; Room temperature, overnight; Supramolecular hydrogels, self-healing materials.	[140,148]
ATRP	STY → PS	Grafting from	Backbone: Kraft lignin (alkali); Lignin-Br + STY → lignin-g-PS; Catalyst: FeCl$_3$ 6H$_2$O/PPh$_3$/ascorbic acid; Solvent: DMF; T = 110 °C, 24 h; Novel polymerization method.	[140,149]
ATRP	MMA → PMMA	Grafting from	Backbone: Kraft lignin (alkali); Lignin + MMA → lignin-g-PMMA; Catalyst: FeCl$_3$ 6H$_2$O/PPh$_3$/ascorbic acid; Solvent: DMF; T = 90 °C, 24 h; Novel polymerization approach.	[140,149]
ATRP	DMAEMA → PDMAEMA	Grafting from	Backbone: Kraft lignin (alkali); Lignin-Br + DMAEMA → lignin-g-PDMAEMA; Catalyst: CuBr/HMTETA; Solvent: PDX; T = 65 °C, 48 h; Gene delivery.	[150]
RAFT	AM → PAM	Grafting from	Backbone: Kraft lignin from softwood sources, prepared using KEX, 2-Bromopropionic acid, and thionyl chloride in dry THF in N$_2$ atmosphere at 70 °C, overnight. Lignin-KEX + AM → Lignin-KEX-g-PAM; RAFT controller: Lignin Macrocontroller; Initiator: AIBN; Solvent: DMF; T = 70 °C, overnight; Application: Pickering emulsions.	[151]
RAFT	AM → PAM	Grafting from	Backbone: Kraft lignin. Prepared with KEX, 2-Bromopropionic acid, and thionyl chloride in dry THF in N$_2$ atmosphere at 70 °C, overnight. Lignin-KEX + AM → Lignin-KEX-g-PAM; RAFT controller: Lignin Macrocontroller; Initiator: AIBN. Solvent: DMF; T = 70 °C; Application: Plasticizer for Portland cement paste.	[152]
Esterification via epoxy group	Lignin + GMA → Lignin-GMA + AM → Lignin-GMA-g-PAM	Grafting from	Backbone: Kraft lignin was first functionalized by reacting with GM through the epoxide ring; Lignin + GMA → Lignin-GMA; Lignin-GMA + AM → Lignin-GMA-g-PAM; RAFT controller: Lignin Macrocontroller; Initiator: AIBN; Solvent: DMF; T = 70 °C; Application: Plasticizer for Portland cement paste.	[153]

Table 4. Cont.

Functionalization Method	Grafted Chains	Grafting Technique	Grafting Conditions	Refs.
RAFT	Lignin + ACX → Lignin-ACX + AM → Lignin-ACX-g-PAM	Grafting from	Backbone: Kraft lignin; Lignin + ACX → Lignin-ACX; Lignin-ACX + AM → Lignin-ACX-g-PAM; RAFT controller: Lignin Macrocontroller; Initiator: AIBN; Solvent: DMF; T = 70 °C; Application: Cationic flocculant.	[153]
RAFT	Lignin + XCA → Lignin-XCA + DMC → Lignin-XCA-g-PDMC	Grafting from	Backbone: Kraft lignin; Lignin + XCA → Lignin-XCA; Lignin-XCA + DMC → Lignin-XCA-g-PDMC; RAFT controller: Lignin; Macrocontroller prepared by Steglich esterification using DCC and DMAP for 48 h; Initiator: AMBN. Solvent: DMF; T = 70 °C, under nitrogen atmosphere; Application: Cationic flocculant.	[154]
ROP	Lignophenol from Japanese cedar + EOX → Lignin-g-PEOX	Grafting from	Backbone: Lignophenol from Japanese cedar; Solvent containing benzyl bromide groups; Monomer: EOX; T = 100 °C, 12 h; Application: Composite materials.	[155]
ROP	Sulfonated Lignin + MOX → Lignin-g-PMOX	Grafting from	Backbone: Lignophenol from Japanese cedar; Initiator: Tosylated lignin; Monomer: MOX; Solvent: DMSO; T = 100 °C, 10 h; Application: Anti-ineffective ointment.	[156]
ROP	Indulin AT lignin + L-LA → Indulin AT lignin-g-PLLA	Grafting from	Backbone: Indulin AT lignin; Initiator: Triazabicyclodecene; Monomer: L-LA; Reaction carried out in bulk; T = 135 °C; Application: composites.	[157]
ROP	Biobutanol lignin + CL → Biobutanol lignin -g-PCL	Grafting from	Backbone: Biobutanol lignin; Initiator: Triazabicyclodecene; Monomer: CL; Reaction carried out in bulk; T = 135 °C; Application: composites.	[158]
ROP	Lignin (organosolv) + CL + L-LA → Lignin-g-PCL/PLLA	Grafting from	Backbone: Lignin orgaosolv; Initiator: Sn(Oct)$_2$ (tin(II)2-ethylhexanoate) and alkyl alcohol functional groups on lignin; Monomer: CL and L-LA (ratio 2:1); Solvent: Toluene; T = 120 °C, under Ar atmosphere, during 45 h; Application: composites.	[159]
ROP	Alkali Lignin + B-BL → Lignin-g-PHB	Grafting from	Backbone: Lignin alkali; Initiator: Sn(Oct)$_2$ (tin(II)2-ethylhexanoate) and alkyl alcohol functional groups on lignin; Monomer: B-BL; Reaction carried out in bulk; T = 130 °C, under nitrogen atmosphere, during 24 h; Application: Nano-fibers for medical applications.	[160]
ROP	Lignin alkali + CL → Lignin-g-PCL	Grafting from	Backbone: Lignin Alkali; Initiator: Sn(Oct)$_2$ (tin(II)2-ethylhexanoate) and alkyl alcohol functional groups on lignin; Monomer: CL; Reaction carried out in bulk; T = 50 °C, under nitrogen atmosphere, during 14–62 h; Application: Nano-fibers for medical applications.	[161]

Table 4. Cont.

Functionalization Method	Grafted Chains	Grafting Technique	Grafting Conditions	Refs.
ROP	Lignin alkali + CL → Lignin-g-PCL	Grafting from	Backbone: Lignin alkali; Initiator: Sn(Oct)$_2$ (tin(II)2-ethylhexanoate) and alkyl alcohol functional groups on lignin; Monomer: CL; Reaction carried out in bulk; T = 50 °C, under nitrogen atmosphere, during 14-62 h; Application: Nano-fibers for medical applications.	[162]
ROP	Softwood Kraft lignin from UPM BioPiva + EOX → Lignin-g-PEOX	Grafting from	Backbone: Softwood Kraft lignin from UPM BioPiva; Initiator: P4-t-Bu; Monomer: CL; Solvent: Toluene; T = 180 °C, under Ar atmosphere, during 16 h; Application: Non-ionic surfactants.	[163]
Embedded polymerization and further crosslinking.	Chitosan; N-VP; 4VP; Crosslinking agent: N'N'-MBA	Grafting from	1.5 g of chitosan were dried at 40-50 °C during 24 h and suspended in 150 mL of 2% CH$_3$COOH for 1 to 2 h. 2 mL of monomers in a 1:1 ratio were incorporated and stirred for 30 min. 0.15 mol/L AIBN were then added and stirred for 2 more hours. T = 65 °C, under air environment. Product washed using several solvents, and dried at 50 °C. The yield was 78-86%. A solution of grafted chitosan in acetic acid (0.1 % v/v) was heated at 70 °C; N'N'-MBA was then added and allowed to react during 30 min.	[164]
RAFT; Steglich esterification of phthalic anhydride and 4,4-Azobis(4-cyanovaleric acid).	Chitosan + Phthalic Anhydride→ (DCC/DMAP) → Chitosan-Phthaloylated Chitosan-Phthaloylated + 4,4-Azobis(4-cyanovaleric acid)→ (DCC/DMAP) → Chitosan-Phthaloylate-RAFT. Chitosan-Phthaloylate-RAFT + NIPAAM → Chitosan-Phthaloaylte-RAFT-g-NIPAAM.	Grafting from	Phthaloylation of chitosan: A solution of phthalic anhydride (5.5 mmol) in 6 mL of DMF/H$_2$O (95/5 v/v %) was prepared, followed by the addition of chitosan powder in 1.86 mmol. The system was purged by N$_2$ for 0.5 h. Heating at 120 °C was maintained during 8 h. The phthaloylated chitosan (1 mmol) was mixed with 1.2 mmol and 4,4-azobis(4-cyanovaleric acid) in 30 mL of dried DMF in the presence of carbonyl activating reagents of DCC (1 mmol) and 4-DMAP (0.12 mmol) as catalysts. Reaction at room temperature for 10 days. Chitosan-Phthaloylate-RAFT (0.0468 g) and dry DMF (5 mL) were mixed by stirring under nitrogen atmosphere. After complete dissolution, initiator 4,4-Azobis(4-cyanovaleric acid) (0.01 mmol) and NIPAAM monomer (0.50 g; 4.4 mmol) were incorporated. Polymerization conducted at 70 °C for 48 h.	[165]
Cerium (IV) attack	Chitin + acrylic monomers.	Grafting onto	Cerium (IV); Chitin + AM → Chitin-g-PAM; Chitin + AA → Chitin-g-PAA; Chitin + MA → Chitin-g-PMA; Chitin + MMA → Chitin-g-PMMA	[58]
γ-Radiation	Chitin + STY (radiated)	Grafting onto	Chitin + STY → Chitin-g-PS	[58]

Table 4. *Cont.*

Functionalization Method	Grafted Chains	Grafting Technique	Grafting Conditions	Refs.
ROP	Cyclic and vinyl monomers. Modified Chitin	Grafting from	Surface-initiated ROP graft copolymerization of L-LA and CL with initiation by hydroxy groups on chitin nanofiber, catalyzed by tin(II) 2-ethylhexanoate was performed, obtaining chitin nanofiber-graft-PLA-co-PL, which is a known biodegradable polyester. Chitin was soaked in a solution of LA/CL in toluene; tin(II) 2-ethylhexanoate (ca. 5 mol%) was incorporated. Heating at 80 °C was maintained during 48 h.	[166]
γ-Radiation	Chitosan + AA +TiO$_2$ → Chitosan-g-PAA-TiO$_2$	Grafting from	A solution of CS-PAAc (1:1 mass/vol) was prepared with different contents of TiO$_2$; 0.0, 1.0, 2.0 and 3.0 wt.% of the total polymer concentration was incorporated. Each solution was sonicated in a bath for 15 min. The solutions were transferred into small glass vials and were subjected to 60 Co-gamma rays at irradiation dose of 30 kGy.	[167]
ROP	Chitosan + CL → Chitoan-g-PCL	Grafting from	0.5 g chitosan and 15 mL MeSO$_3$H were placed in a round bottom flask under nitrogen atmosphere at 45 °C for 0.5 h. 4.19 g CL monomers were fed to the reactor, followed by stirring and reaction during 4.5 h.	[168]

Polysaccharides are abundant in nature, no matter whether in plants or animals, and are important for in vivo functions. As observed in Figure 6, there is a wide variety of polysaccharides, but the most abundant ones are cellulose and chitin. Cellulose provides structure to the cell walls of plants. Chitin, on the other side, is part of the exoskeletons of crustaceans, shellfish, and insects [169].

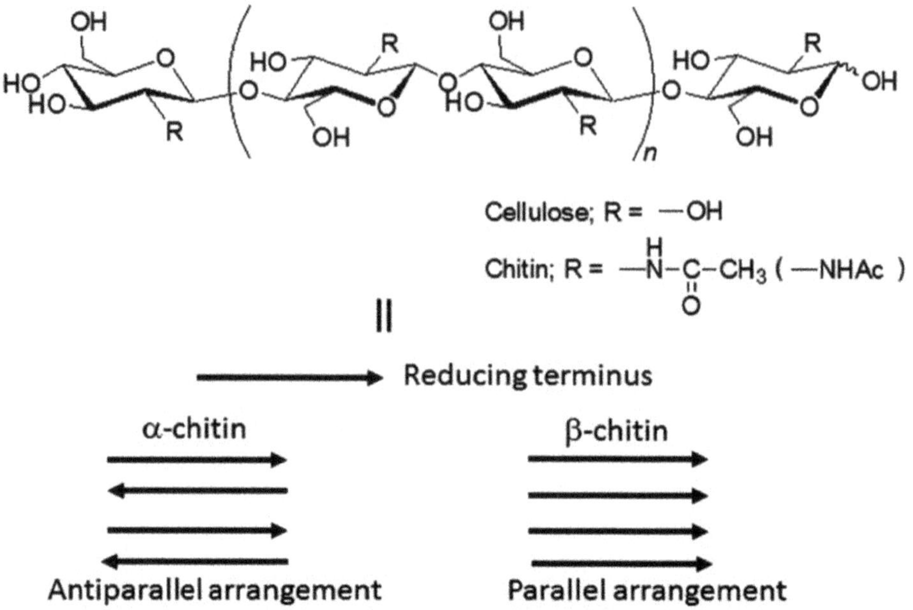

Figure 6. Cellulose and chitin structures (top). II: schematic diagram of crystalline structures for different forms of chitin. Source: Adapted with permission from Jun-ichi Kadokawa (2015). Fabrication of nanostructured and microstructured chitin materials through gelation with suitable dispersion media: RSC Adv. 5 12736 [169]. Copyright © 2021 The Royal Society of Chemistry.

Chitin is an abundant but only marginally used biomass. There are several reasons why not many practical applications for chitin have been developed [169]: (a) bulky structure; (b) insolubility in water and typical organic solvents; (c) it is harmful to recover chitin, since the available procedures require the use of strong acids and bases; (d) native chitin from crustaceans, which have exoskeletons that protect animals from their predators, has fibrous structures rich in proteins and minerals; and (e) there is a need to remove mineral and protein constituents in order to isolate chitin.

Chitosan was developed to overcome the drawbacks of chitin. Chitosan is commercially attractive for production of biocompatible polymers for environmental and biomedical applications [30]. It is basically a copolymer of N-acetyl-D-glucosamine and D-glucosamine. It is obtained from the hydration of chitin. This hydration takes place in alkaline solutions in a temperature range of 80–140 °C during 10 h [170,171].

Chitosan is a cationic polysaccharide, produced by deacetylation of chitin. Deacetylation proceeds to different levels depending on the intended uses. The physical properties of chitosan, particularly solubility, depend on molecular weight and degree of deacetylation of the material [172–174].

Modification of chitosan leads to a diversity of derivates, with differentiated properties. As shown in Figure 1 of Deng et al. [174], different chitosan moieties are possible. Each one of them is produced from grafting or other chemical or enzymatic modification forms.

They differ in antimicrobial activity. Further studies on chitin-chitosan modification are available in the literature [58,175–178].

3.2. Polymer Backbones

Polymer backbones are the most common substrates for grafting modification. Several techniques can be used to create many possible combinations. Most of these developments are focused on property improvement for industrial applications. They include synthesis of adhesives; reinforcement of mechanical properties; improvement of chemical resistance; synthesis of electro, optical, thermo responsive polymers; health care applications; and self-healing polymers, among others [179].

Polymer grafting techniques have been reported since the late 1950s [8,9,11]. Polymers based on acrylamide and acrylonitrile using polymer grafting techniques are included in the early reports. However, the huge increase in research related to synthesis of materials with controlled microstructures using polymer grafting techniques has been possible due to the advances in reversible deactivation radical polymerization (RDRP) techniques over the last three decades [180–184]. The main RDRP routes are nitroxide mediated polymerization (NMP) [185], reversible addition fragmentation transfer polymerization (RAFT) [180,186,187], and atom transfer radical polymerization (ATRP) [71,188], although there are other polymer synthesis techniques, such as ring-opening polymerization (ROP) [181] and click chemistry, among others, that can be used. Another important aspect in polymer grafting is the solvent used for the reaction, particularly when grafting proceeds as a heterogenous process. As observed in Table 19, solvents such as supercritical fluids, mainly supercritical carbon dioxide, water, DMF, or combinations of solvents are typically used for polymer grafting. These techniques have improved our skills to produce molecularly well-defined, chain-end tethered polymer brush films. The assets of RDRP have substantially impacted the synthesis and properties of surface-grafted polymers. Although vinyl monomers are widely use in graft polymerization for backbones or side-arms, other monomers coming from natural sources are increasingly being used. That is the case, for instance, of ε-caprolactone, lactic acid [189], L-lactide, and butyrolactone. It is also observed in Table 19 that other nontraditional monomers such as acrylamide (AM), N-isopropylacrylamide (NIPAAM), and acrylates and methyl acrylates, such as methyl methacrylate (MMA) and acrylic acid (AA), are being increasingly used in polymer grafting applications. To get a glimpse of the focus of research papers that involve polymer grafting as the chemical route for polymer modification, Table 5 summarizes journal reports on polymer grafting from the current period (2020–2021). As expected, an increasing trend toward the improvement of natural biopolymers using synthetic polymer arms is observable.

Table 5. Recent reports on the production of materials using polymer grafting (2020–2021).

Title [a]	Reference
Synthesis of diallyl dimethyl ammonium chloride grafted polyvinyl pyrrolidone (PVP-g-DADMAC) and its applications.	[190]
Synthesis and characterization of biocompatible hydrogel based on hydroxyethyl cellulose-g-poly(hydroxyethyl methacrylate).	[191]
Poly(vinylidene fluoride) (PVDF)/PVDF-g-polyvinylpyrrolidone (PVP)/TiO2 mixed matrix nanofiltration membranes: preparation and characterization.	[192]
Synthesis and characterization of poly (acrylonitrile-g-lignin) by semi-batch solution polymerization and evaluation of their potential application as carbon materials	[193]
High crosslinked sodium carboxyl methylstarch-g-poly (acrylic acid-co-acrylamide) resin for heavy metal adsorption: its characteristics and mechanisms.	[194]
Preparation and characterization of PLLA/chitosan-graft-poly (ε-caprolactone) (CS-g-PCL) composite fibrous mats: The microstructure, performance and proliferation assessment.	[168]
Synthesis of partly debranched starch-g-poly(2-acryloyloxyethyl trimethyl ammonium chloride) catalyzed by horseradish peroxidase and the effect on adhesion to polyester/cotton yarn.	[195]

Table 5. *Cont.*

Title [a]	Reference
Cellulose Acetate thermoplastics with high modulus, dimensional stability and anti-migration properties by using CA-g-PLA as macromolecular plasticizer.	[196]
Superwettable PVDF/PVDF- g-PEGMA ultrafiltration membranes.	[197]
Removal of malachite green using carboxymethyl cellulose-g-polyacrylamide/montmorillonite nanocomposite hydrogel.	[198]
Ultraviolet illumination responsivity of the Au/n-Si diodes with and without poly (linolenic acid)-g-poly (caprolactone)-g-poly (t-butyl acrylate) interfacial layer.	[199]
Synthesis of CO2-based polycarbonate-: G -polystyrene copolymers via NMRP.	[200]
Compatibilization of iPP/HDPE blends with PE-g-iPP graft copolymers.	[201]
Preparation, characterization of feather protein-g-poly(sodium allyl sulfonate) and its application as a low-temperature adhesive to cotton and viscose fibers for warp sizing.	[202]
Semi-Natural superabsorbents based on starch-g-poly(acrylic acid): Modification, synthesis and application.	[203]
Design and development of polymethylmethacrylate-grafted gellan gum (PMMA-g-GG)-based pH-sensitive novel drug delivery system for antidiabetic therapy.	[204]
Preparation and characterization of bioinert amphiphilic P(VDF-co-CTFE)-g-POEM graft copolymer.	[205]
Preparation, characterization of poly(acrylic acid)-g-feather protein-g-poly(methyl acrylate) and application in improving adhesion of protein to PLA fibers for sizing.	[206]
Antibacterial and pH-responsive quaternized hydroxypropyl cellulose-g-poly(THF-co-epichlorohydrin) graft copolymer: Synthesis, characterization and properties.	[207]
Thermally self-assembled biodegradable poly(casein-g-N-isopropylacrylamide) unimers and their application in drug delivery for cancer therapy.	[208]
Swelling capacity of sugarcane bagasse-g-poly(acrylamide)/attapulgite superabsorbent composites and their application as slow release fertilizer	[209]
Structure formation in pH-sensitive micro porous membrane from well-defined ethyl cellulose-g-PDEAEMA via non-solvent-induced phase separation process	[210]
Micelles with a loose core self-assembled from coil-g-rod graft copolymers displaying high drug loading capacity.	[211]
Synthesis and water absorbing properties of KGM-g-P(AA-AM-(DMAEA-EB)) via grafting polymerization method.	[212]
Preparation of hydrophilic woven fabrics: Surface modification of poly(ethylene terephthalate) by grafting of poly(vinyl alcohol) and poly(vinyl alcohol)-g-(N-vinyl-2-pyrrolidone).	[213]
The preparation, physicochemical and thermal properties of the high moisture, solvent and chemical resistant starch-g-poly(geranyl methacrylate) copolymers.	[214]
Reverse poly(ε-caprolactone)-g-dextran graft copolymers. Nano-carriers for intracellular uptake of anticancer drugs.	[215]
High-performance solid-state bendable supercapacitors based on PEGBEM-g-PAEMA graft copolymer electrolyte.	[216]
Design of well-defined polyethylene-g-poly-methyltrifluorosiloxane graft copolymers via direct copolymerization of ethylene with polyfluorosiloxane macromonomers.	[217]
Synthesis and characterization of poly(vinyl chloride-g-ε-caprolactone) brush type graft copolymers by ring-opening polymerization and "click" chemistry.	[218]
Adhesion of cornstarch-g-poly (2-hydroxyethyl acrylate) to cotton fibers in sizing.	[219]
Polynorbornene-g-poly(ethylene oxide) through the combination of ROMP and nitroxide radical coupling reactions.	[220]
Potent bioactive bone cements impregnated with polystyrene-g-soybean oil-AgNPs for advanced bone tissue applications.	[221]
Fabrication of cellulose nanocrystal-g-poly(acrylic acid-co-acrylamide) aerogels for efficient Pb(II) removal.	[222]
The compatibilization of PLA-g-TPU graft copolymer on polylactide/thermoplastic polyurethane blends.	[223]
Cellulose-g-poly-(acrylamide-co-acrylic acid) polymeric bioadsorbent for the removal of toxic inorganic pollutants from wastewaters.	[224]
Performance improvement for thin-film composite nanofiltration membranes prepared on PSf/PSf-g-PEG blended substrates.	[225]

Table 5. Cont.

Title [a]	Reference
Synthesis and characterization of poly(vinyl chloride-g-methyl methacrylate) graft copolymer by redox polymerization and Cu catalyzed azide-alkyne cycloaddition reaction.	[226]
Encapsulation of NPK fertilizer for slow release using sodium carboxymethyl cellulose-g-poly (AA-C0-AM-C0-AMPS)/ Montmorillonite clay-based nanocomposite hydrogels for sustainable agricultural applications.	[227]
Engineering a "PEG-g-PEI/DNA nanoparticle-in- PLGA microsphere" hybrid-controlled release system to enhance immunogenicity of DNA vaccine.	[228]
Chitosan-g-oligo(L,L-lactide) copolymer hydrogel for nervous tissue regeneration in glutamate excitotoxicity: In vitro feasibility evaluation	[229]

[a] Searching criteria used in Scopus: (TITLE ("-g-") AND TITLE-ABS-KEY ("poly") AND TITLE-ABS-KEY ("graft")).

Table 6. Common characterization methods.

Method	Property Measured
Nuclear magnetic resonance (NMR)	NMR identifies the nature of the group and the site where the graft is attached to the polymer backbone. ^{1}H NMR and ^{13}C NMR are utilized depending on the chemical nature of the material [73].
Infrared spectroscopy (IR)	Identification of specific functional groups grafted to backbone. It can also be used to quantify grafting functionalities in modified polyolefins by determining the intensity of the characteristic reference bands in the sample, and compare them to a calibration curve of known concentrations of the same functional group [73].
Scanning electron microscopy (SEM)	SEM is useful for the study of surface morphologies of grafted materials; also used as a tool to confirm that grafting has taken place [59].
X-ray diffraction (XRD)	XRD is useful for the study of crystallinity of chemical substances. Polymer grafting is associated with changes in XRD patterns [59].
Size exclusion chromatography (SEC)	It allows determination of molecular weight averages (MW) and molecular weight distributions (MWD) of grafted chains [230], which must be separated from the backbone before carrying out the analyses to focus on the grafts only.
Differential scanning calorimetry (DSC)	This is utilized to evaluate changes in crystallinity through the thermal behavior between backbones and grafted materials.
Thermogravimetric analysis (TGA)	This is used to study changes in thermal decomposition profiles between backbones and grafted materials.

Table 7. Summary of thermal characterization techniques employed in grafted materials.

Backbone	Grafts	Technique	Property Measured	Application	Refs.
Lignin-RAFT macrocontroller	Poly(soybean oil methacrylate) derivatives. PSBMA, PSBMAH, PSBMAEO	DSC	Tg of lignin-g-polymerized soybean oil derived methacrylate. Measured Tg: Lignin-g-PSBMA −5.6 °C, Lignin-g-PSBMAH 27.7 °C, Lignin-g-PSBMAEO −17.4 °C. DSC2000, TA Instruments. 10 mg samples were used with heating from 25 °C to 200 °C at 10 °C/min, and cooling to −70 °C at the same rate. Data was gathered from the second heating scan to 200 °C. Nitrogen flow: 50 mL/min.		[134]
		TGA	Thermal stability of biocomposites. Two MDT values were found for all samples 380 °C and 430 °C. Q5000 TGA system (TA Instruments); 25 to 600 °C, at 10 °C/min, 25 mL/min nitrogen flow.		
Lignin	Imidazole and POCl$_3$ react in to (1H-imidazol-1-yl)phosphonic group which reacts with lignin	TGA	MDT values for lignin derivatives and PP blends. Lignin MDT = 398 °C. Lignin-g-IPG MDT > 600 °C. PP =418°C. PP/Lignin = 472 °C, PP/20 % Lignin-g-IPG = 479 °C, PP/30 % Lignin-g-IPG = 483 °C. SDTQ600 thermal analyzer; 20 °C/min under nitrogen, from ambient to 600 °C.		[125,135]
		FR	PHRR (kW/m^2) / THR (MJ/m^2) PP =1350 kW/m^2//87.3 MJ/m^2, PP/Lignin = 382/333 kW/m^2//76 MJ/m^2, PP/20%Lignin-g-IPG = 380 kW/m^2//74.2 MJ/m^2, PP/30%Lignin-g-IPG = 360 kW/m^2//69.7 MJ/m^2. Heat release rate measured using an FTT UK cone calorimeter following ISO 5660. Incident flux: 35 kW/m^2.		

Table 7. Cont.

Backbone	Grafts	Technique	Property Measured	Application	Refs.
Hydroxypropyl cellulose (HPC)	Steglich esterification of PABTC over HPC using DCC and DMAP and grafting of PABTC and further polymerization of EA NIPAAM.	TGA	TGA 2950 analyzer; nitrogen atmosphere (60 cm^3/min) Heating of samples (5–10 mg) from ambient temperature to 600 °C, at 10 °C/min. Calculations of the onset, the end decomposition temperature, and the residual mass.		[108]
		DSC	Modulated DSC 2920 in nitrogen environment (60 cm^3/min). The determination of glass transition temperature (Tg) required 5–10 mg samples. Heating from ambient temperature to 200 °C, followed by cooling to −100 °C, and reheating to to 200 °C, using 10 °C/min for both, heating and cooling was used. This sequence occured three times.		
CellClAc	ATRP of 4NFA and MMA using Cu(I)Cl/2'2'BIPI macro initiator.	TGA	Shimadzu TGA-50; heating rate: 10 °C/min; nitrogen flow: 10 mL/min.		[109,112]
Cellulosic *Grewia optiva* fibers	Redox initiator FAS-H2O2 for grafting MA and KPS for polymerization of MA.	TGA	Thermo gravimetric analyses of plain and grafted cellulosic fibers carried out in presence of nitrogen on a Perkin Elmer thermal analyzer. Heating from ambient temperature to 900 °C/min, at 10 °C/min.		[110]
Cotton linter cellulose	APS with MMA.	TGA-DTA	Samples were analyzed using a Pyris Diamond TG-DTA (STA449C/3/F, Germany). Sample weight: 14.2 mg. Heating rate: 10 °C/min. Flow of nitrogen 20 mL/min.		[115]

Table 7. Cont.

Backbone	Grafts	Technique	Property Measured	Application	Refs.
Cellulose (DP = 1130)	APS and MMA embedded.	TGA-DTA	Samples analyzed in a Pyris Diamond TG/DTA equipment (STA449C/3/F, German). Sample weight: 14.2 mg. Heating/Cooling rate: 10 °C/min. Flow of nitrogen: 20 mL/min.		[118]
Cellulose nanocrystal (CNC)	Macromolecular initiator obtained from the reaction between Br-iBuBr and CNC with TEA as catalyst via SI-ATRP with sTY. Material was casted in PMMA.	TGA	Thermal degradation of studied materials analyzed in a TGA/SDTA 851e in nitrogen atmosphere. Heating from 30 to 550 °C, at 10 °C/min.		[119]
		DSC	Tg determined in a Pyris 1 DSC equipment. Heating from 30 to 200 °C at 10 °C/min.		
Cellulose nanocrystal (CNC)	Macromolecular initiator obtained from the reaction between Br-iBuBr and CNC with TEA as catalyst via SI-ATRP with sTY. Material was casted in PMMA.	TGA	Analyses carried out in a TGA Q50 (TA Instruments), under nitrogen atmosphere (20 mL/min). Heating from 30 to 500 °C at 5, 10, 15, 20, 25 and 30 °C/min for determination of degradation activation energy.		
		DSC	Analyses performed in a Star 1 (Mettler- Toledo). Heating from 25 °C to 200 °C at 5 °C/min, remaining at 200 °C for 5 min. Cooling to −50 °C at 20 °C/min, remaining there for 5 min, followed by heating to 200 °C at 5 °C/min. Tg, Tm, Tc were measured. Crystallinity (v) calculated from melting enthalpy of fully crystalline PLLA.		
Cellulose cotton fiber pulp	Esterification of maleic anhydride grafted over PHA (by double bond). PHA-g-MA grafted over cellulose cotton fiber pulp (by the anhydride group).	TGA	Analyses carried out on a Q500 TGA (TA Instruments), in nitrogen environment. Heating from ambient to 600 °C at 20 °C/min.		[120]

Table 7. Cont.

Backbone	Grafts	Technique	Property Measured	Application	Refs.
Cellulosic filter papers	Radiation-induced graft (RIGCP) copolymerization of acrylonitrile.	TGA	Analyses obtained in TGA-50 (Shimadzu). Heating rate: 10 °C/min. N_2 environment.		[121]
Hemicellulose	Hemicellulose grafting using TBD and e-caprolactona monomer	TGA	TGA/DSC 1 STARe System (Mettler Toledo). 5–15 mg of a sample. Nitrogen environment (85 mL/min). Heating from 30–500 °C at 10 °C/min. Compositions of produced gases measured in an FTIR (TA, Nicolet, iS); detector coupled to TGA. FTIR analyses from 4000 to 650 cm^{-1} with a 4 cm^{-1} resolution.		[37]
		DSC	DSCQ10 (TA Instruments). Hermetic pan (T 090127). Heating from 25 to 165 °C at 10 °C/min. Cooling to 15 °C followed by heating to 165 °C at 10 °C/min. Tg obtained from second heating ramp.		

Table 7. Cont.

Backbone	Grafts	Technique	Property Measured	Application	Refs.
CellClAc	NCHA, 4VP, DA and DAAM grafted by atom transfer radical polymerization (ATRP) using CuCl, 2'2'BIPI as catalyst.	TGA	Thermal analyses carried out in a Shimadzu TGA-50 at 10 °C/min in nitrogen environment (10 mL/min). Cell-g-DA and Cell-g-4VP exhibited higher decomposition temperatures.		[122]
Chitosan	Embedded polymerization N-VP, 4VP. Crosslinking agent: N'N'-MBA	TGA	TGA analyses carried out using a NETZSCH, STA 409 PG.4.G Luxx. 50 mL/min nitrogen environment. Heating from 20 °C to 400 °C at 10 °C/min.		[164]

Table 8. Summary of spectroscopic characterization techniques employed in grafted materials.

Backbone	Grafts	Technique	Property Measured	Application	Refs.
Lignin-RAFT macrocontroller	Poly(soybean oil methacrylate) derivatives. PSBMA, PSBMAH, PSBMAEO.	^{1}H-NMR	Amount of RAFT controller reacted (shifts 7.3, 7.6, 7.9 ppm) and nonattached monomer (shifts 5.3, 5.7 ppm). Varian Mercury 300 spectrometer, 300 MHz; tetramethyl silane as an internal reference.		[134]
		FTIR	Amount of RAFT controller and polymer attached to lignin. Area of peak at 1760 cm^{-1}. Perkin Elmer spectrum 100. ATR method, recorded at 4 cm^{-1}, 32 scans.		
Lignin	Imidazole and POCl$_3$ react in to (1H-imidazol-1-yl) phosphonic group which reacts with lignin.	FTIR	Lignin, lignin-OH, imidazole and lignin-g-IPG were compared at relevant bands (1662, 1260, 1080, 1050, 878, and 791 cm^{-1}) for determination of C=N, P=O, P-O-C, P-O, -OH, and P-N groups, respectively. A Bruker VECTOR 22 spectrometer was used.		[125,135]
		XPS	Surface elementary composition of samples (wt.%), were compared to determine the amount of functionalized or grafted groups in the surface. Elements analyzed: carbon, oxygen, phosphorous, nitrogen. Lignin, C: 68.2, O: 31.9; Modified lignin-OH, C: 67.5, O: 32.5; Lignin-g-IPG, C: 53.0, O: 31.7, P: 8.1, N: 7.2. A Thermo ESCALAB 250 spectrometer was used. Analyses were carried out from 50 up to 800 eV of binding energy.		

Table 8. Cont.

Backbone	Grafts	Technique	Property Measured	Application	Refs.
Hydroxypropyl cellulose (HPC)	Steglich esterification of PABTC over HPC using DCC and DMAP; grafting of PABTC and further polymerization of EA NIPAAM.	^{1}H NMR	NMR spectra obtained using a Bruker 300 or 200 MHz in deuterated chloroform or deuterated dimethyl sulfoxide (d6-DMSO).		[108]
CellClAc	Macro initiator, Cu(I)Cl/2'2'BIPI catalytic system via ATRP of 4NPA and MMA	^{1}H-NMR	NMR spectra obtained using a Bruker 400 MHz spectrometer at room temperature in DMSO-d$_6$.		[109]
		FTIR	FTIR data using solid samples as KBr pellets, from 4000 cm^{-1} to 450 cm^{-1}.		
Cellulosic *Grewia optiva* fibers	Redox initiator FAS-H$_2$O$_2$ for grafting of MA, and KPS for polymerization of MA.	FTIR	Chemical structures of fibers before and after grafting were studied using FTIR (PERKIN ELMER RX1). Spectrum recorded from 4000 to 400 cm^{-1}.		[110]
Cellulose acetate	Solvents: DMSO, PDX, DMAc, C$_3$H$_6$O. Initiators for grafting and polymerization: CAN, Sn(Oct)$_2$ and BPO.	^{1}H-NMR	Spectra of studied materials obtained using a Bruker WM-400 apparatus, at 300 MHz. Tetramethyl silane (TMS) used as internal standard and DMSO-d6 as solvent.		[111]
		FTIR	Grafted and ungrafted materials analyzed using a Spectrum RX1 PerkinElmer apparatus. Chloroform used as a solvent. KBr pellet technique used for grafted materials.		

Table 8. Cont.

Backbone	Grafts	Technique	Property Measured	Refs.
CellClAc	Macro initiator, Cu(I)Cl/2'2'BIPI catalytic system via ATRP controller.	FTIR	FTIR spectrophotometer on solid samples as KBr pellets, from 4000 cm^{-1} to 450 cm^{-1}.	[112]
Cellulose cotton fibers	Na$_2$CO$_3$ and thermal activation with MTC-b-CD and silver nitrate.	FTIR	MCT-β-CD grafted and ungrafted cotton fabrics were studied with attenuated total reflection (FTIR-ATR), which permits the study of solid surfaces using a diamond tipped device (SPECAC), connected to a Bruker Vertex 70 spectrophotometer.	[113]
Cotton linter cellulose	APS with MMA.	FTIR	A TENSOR27 FTIR apparatus (4500–400 cm^{-1}) was used to analyze materials.	[115]
		XRD	X-ray diffractometer (D/max-2500/PC, Rigaku Co) based on a reflection method using a Cu Kα target (40 kV and 60 mA). Diffraction angle: 5° to 60°.	

Table 8. Cont.

Backbone	Grafts	Technique	Property Measured	Application	Refs.
Microcrystalline cellulose	Ring-opening polymerization (ROP) of L-LA with DMAP in an ionic liquid AmimCl to form Cellulose-g-PLLA	^{1}H-NMR	Materials analyzed using ^{1}H-NMR (Bruker AV400-MHz). Solvent: DMSO-d6 with a drop of trifluoroacetic acid-d; internal standard: tetramethyl silane (TMS).		[116]
		XRD WAXD	WAXD carried out using an XRD-6000 X-ray diffractometer (Shimadzu, Japan); Ni-filtered Cu Ka radiation (40 kV, 30 mA); 4°/min at ambient temperature.		
		UV spectroscopy	UV analyses performed in a UV2000 equipment (UNICO, China), to follow the load and controlled release of vitamin C.		
Cellulose (DP = 1130)	APS and MMA embedded.	FTIR	A TENSOR27 FTIR apparatus (4500–400 cm^{-1}) was used to analyze materials.		[117]
		WAXD-XRD	WAXD carried out in an X-ray diffractometer (D/max-2500/PC, Rigaku Co. Ltd., Tokyo); Cu Ka radiation (40 kV, 60 mA); 2q = 5°–60°.		
Cellulose nanocrystal (CNC)	Macromolecular initiator obtained from the reaction between Br-iBuBr and CNC with TEA as catalyst via SI-ATRP with STy. Material was casted in PMMA.	FTIR	Spectra obtained at ambient temperature on a NICOLET spectrometer 10 to study surface modification of studied materials. KBr pellet method used for samples. Resolution: 4 cm^{-1}; measurement range; 4000–400 cm^{-1}, and 30 scans.		[118]
		UV-Vis	Transmittance of samples measured using a UV-vis spectrophotometer at wavelengths of 400 to 800 nm.		
		XRD	XRD patterns of studied materials were obtained using a Bruker Siemens D8 X-ray apparatus with operation at 3 kW; CuKa radiation (l = 0.154 nm); 2q = 3–60°, 0.02° step.		
		^{13}C-NMR	Chemical structures of studied materials obtained with the help of a Bruker 400 M solid-state ^{13}C-NMR apparatus.		

Table 8. *Cont.*

Backbone	Grafts	Technique	Property Measured	Application	Refs.
Cellulose nanocrystal (CNC)	Redox agent: MgH$_2$. L-LA polymerization in situ. PLLA-g-CNC particles were casted in PLLA.	FTIR	A Perkin Elmer Spectrum 100 apparatus in ATR mode, equipped with a Zn-Se crystal, was used. Modified and unmodified CNC samples. Range: 4000 to 650 cm^{-1}. Accuracy: 4 cm^{-1} for 64 scans.		[119]
		^{1}H-NMR ^{13}C-NMR	A Bruker-300 apparatus with DMSO-d6 as solvent was used. Small amounts of CDCl$_3$ were used to improve solubility of grafted product in DMSO-d6.		
		XRD	XRD analyses carried out in a PAN analytical apparatus; Cu Kα radiation (l = 1.541 Å); incidence angle: 10 to 40°, 0.07° steps.		
		XPS	XPS analyses performed in an ESCALAB 250 equipment (Thermo Scientific). Analyzed surface: 400 mm^2. Atomic concentrations estimated from areas of photoelectron peaks considering atomic sensitivity factors. All binding energies were charge-corrected to 284.8 eV.		
Cellulose cotton fiber pulp	Esterification of maleic anhydride grafted onto PHA (through double bond). PHA-g-MA grafted onto cellulose cotton fiber pulp (through the anhydride group).	FTIR	ATR-FTIR analyses of ungrafted and grafted materials acquired using KBr discs on a Vector 33 Bruker apparatus.		[120]
		XRD	XRD analyses carried out in a Bruker D8 ADVANCE X-ray apparatus with operation at 40 mA and 40 kV. Cu Kα filtered radiation (λ = 0.15418 nm). Scattering angle (2θ): 5° to 50°, 0.02° step.		

Table 8. Cont.

Backbone	Grafts	Technique	Property Measured	Application	Refs.
Cellulosic filter papers	Radiation-induced graft (RIGCP) copolymerization of acrylonitrile.	FTIR	FTIR analyses of studied materials (using KBr disks) were obtained using a Perkin Elmer-1650 FTIR apparatus.		[121]
		XRD	XRD analyses of studied materials in an angle range $2\theta = 5$–$50°$ were obtained in a Shimadzu XRD-610 equipment, with Cu Kα radiation (1.5418° A). Operating conditions: 8 °/min and 1.0 s; 30 kV and 20 mA. Peak height methods were used to estimate crystallinity index (CI).		
		XRF	XRF analyses performed using a Philips wavelength dispersive equipment X' Unique II, including flow and scintillation (fs) detectors. Qualitative analyses of chelating filter paper-metal complexes were obtained by X-ray fluorescence.		
Hemicellulose	Hemicellulose grafting using TBD and ε-caprolactone monomers	FTIR	A Nicolet Nexus spectrophotometer with single reflection ATR was used. Range: 4000 to 700 cm^{-1}. Resolution: 4 cm^{-1}; 60 scans.		[37]
		^{1}H-NMR ^{13}C-NMR HSQC	Apparatus: Bruker Advance III 400 MHz, containing a 5 mm multinuclear broad band probe (BBFO+) and z-gradient coil. 30 mg of sample dissolved in 1 mL of d6-DMSO. T = 60 °C. 128 scans for ^{1}D ^{1}H; 32 scans and 128 increments for 2D ^{1}H-^{1}H COSY. 16 scans and 256 increments for 2D ^{1}H-^{13}C HSQC.		

Table 8. Cont.

Backbone	Grafts	Technique	Property Measured	Application	Refs.
CellClAc	NCHA, 4VP, DA and DAAM grafted by ATRP using CuCl, 2'2'BIPI as catalyst.	FTIR	Apparatus: Perkin Elmer Spectrum One. Solid samples as KBr pellets.		[122]
		UV-Visible	Apparatus: Shimadzu 3600 UV-VIR-NIR.		
Chitosan	Embedded polymerizations of N-VP and 4VP. Crosslinking agent: N'N'-MBA	FTIR	Apparatus: AVATAR 370 Thermo Nicolet. Range: 4000 to 500 cm^{-1}, using the KBr pellet method. Spectral resolution: 4 cm^{-1}.		[164]
		XRD	Apparatus: Bruker Advance D8 equipment (Germany). Diffractograms comprised (2θ) 0.020 imaging; scattering rates: 3 to 800; scanning speeds: 2.0 min^{-1}; accelerated tension of 40 kV; intensity of 35 mA.		
		XRD	XRD using a Bruker D8 Advance spectrometer. Analyses carried out from 5° to 40° at 4°/min, with a current of 40 mA. Measured property: CI.		
Cellulose nanofibrils	Nitroxide TEMPO insertion and nitroxide mediated polymerization of HEMA	FTIR	Apparatus: Bruker spectrometer, accumulation of 128 scans. Resolution: 4 cm^{-1}. Range: 4000–400 cm^{-1}; absorbance mode.		[124]
		XPS	XPS analyses using Thermo Electron Scientific Instruments were performed using a 1486.6 eV Al Kα X-ray source.		

Table 8. Cont.

Backbone	Grafts	Technique	Property Measured	Application	Refs.
Lignin	Insertion of ACX onto lignin to get a RAFT macrocontroller able to polymerize AM and AA.	Dynamic light scattering (Particle size distribution)	Particle size distributions (PSD) measured in an aqueous solution (1 mg/mL) using a DLS Zeta-sizer (Malvern Instruments).		[139]
		^{1}H-NMR	Apparatus: Bruker 300 Advance; PfB used as internal standard to transform proton intensities into initiator concentration (µmol/(g lignin)).		

Table 9. Summary of digital imaging and microscopy characterization techniques employed in grafted materials.

Backbone	Grafts	Technique	Property Measured	Application	Refs.
Lignin	Imidazole and POCl$_3$ react to produce (1H-imidazol-1-yl) phosphonic, which reacts with lignin.	SEM with EDAX	S4800 FEI SEM; 5 kV accelerating voltage. Comparison of morphologies of char residues after cone calorimeter analyses. PP/Lignin displayed loosely spheroidal structures with C: 86.5 wt.%, O: 13.5 wt.%. PP/30 °% Lignin-g-IPG formed a continuous and compact spheroidal structure.		[125,135]
		Digital photo	Digital photos comparing PP composite sample test probes after cone calorimeter measurements were obtained.		
Cellulosic *Grewia optiva* fibers	FAS-H$_2$O$_2$ redox initiation for grafting of MA and KPS, for polymerization of MA.	SEM.	The morphologies of ungrafted and grafted fibers were studied using a SEM apparatus (LEO 435 VP).		[110]
Cotton linter cellulose	APS with MMA.	SEM	Apparatus: JSM-6700F scanning microscope. Samples coated with gold prior to study. Morphologies and differences between ungrafted and grafted cellulose materials were analyzed.		[115]
Microcrystalline cellulose	Ring-opening polymerization (ROP) of L-LA using DMAP in an ionic liquid (AmimCl) to produce Cellulose-g-PLLA	TEM	Apparatus: JEM-100CXa TEM at an acceleration voltage of 100 kV. Sample preparation involved dissolving them in DMSO 0.01% (w/v) and placement on a copper grid with formvar film.		[116]

Table 9. Cont.

Backbone	Grafts	Technique	Property Measured	Application	Refs.
Cellulose (DP = 1130)	Embedded APS and MMA.	SEM	Apparatus: JSM-6700F scanning microscope. Coating with a thin layer of gold required for samples.		[117]
Cellulose nanocrystal (CNC)	Macromolecular initiator obtained from Br-iBuBr and CNC with TEA as catalyst via SI-ATRP with Sty. Material was casted in PMMA.	SEM	Apparatus: su1510 (Hitachi Zosen Corporation) operating at 30 kV. Focus: measurement of homogeneity of PMMA composites.		[118]
		TEM	Apparatus: JEM-2100 electron microscope. Operation involves acceleration voltage of 200 kV. Focus: Analysis of morphological features and distribution of cellulose nanocrystals.		
Cellulose cotton fiber pulp	Esterification of maleic anhydride grafted onto PHA (through double bond). PHA-g-MA grafted onto cellulose cotton fiber pulp (through the anhydride group).	SEM	Apparatus: Nova Nano SEM 430 (FEI Company); high-resolution field emission with operation at an acceleration voltage of 15 kV. Focus: Morphological analysis of PHA/CF composite films.		[108]
Cellulosic filter papers	Radiation-induced graft (RIGCP) copolymerization of acrylonitrile.	SEM	SEM micrographs of samples were obtained on a JEOL-SEM 5400 microscope.		[109]
Chitosan	Embedded polymerization of N-VP and 4VP; Crosslinking agent: N,N'-MBA	SEM	Morphological surface images of ungrafted and grafted chitosan with or without doxocyline were obtained using a JSM-6390 SEM microscope.		[153]

Table 9. *Cont.*

Backbone	Grafts	Technique	Property Measured	Application	Refs.
Regenerated cellulose fibres (rayon)	Photo-chemical grafting of PETA without photoinitiator.	SEM	Apparatus: JSM-6510 instrument (Joel GmbH, Freising, DE). Samples prepared following standard procedures. Micrographs obtained in SE mode; acceleration voltage: 5 and 10 kV. Purpose: Analyses of qualitative fracture.		[111]
Cellulose nanofibrils	Nitroxide TEMPO insertion and NMP of HEMA	SEM	Apparatus: Hitachi S-4800 SEM microscope. Purpose: Analysis of morphologies of materials at working distances of 15 mm. Operating voltage of 2 kV.		[112]

Table 10. Summary of rheological characterization techniques used for grafted materials.

Backbone	Grafts	Technique	Property Measured	Application	Refs.
Cellulose nanocrystal (CNC)	Redox agent: MgH$_2$; in situ L-LA polymerization. PLLA-g-CNC particles were casted in PLLA.	DMA. Casted PLLA-g-CNC particles in PLLA matrix.	Apparatus: DMA 50, Metravib, tensile mode. Dimensions of sample specimens: 6.5 × 10.0 × 0.2 mm. Storage E' and loss E'' moduli were registered. Heating rate: 5 °C/min; −100–200 °C; 1 Hz.		[119]
		Molten Rheometry; Casted PLLA-g-CNC particles in PLLA matrix.	Dynamic Anton Paar MCR 102 rheometer. Materials analyzed in molten state; T = 180 °C; Parallel plates geometry, 25-mm diameter; G' and G'' moduli were registered; 100 to10^{-3} rad/s.		
Cellulose nanofibrils	Nitroxide TEMPO insertion and NMP of HEMA	Stress-strain compression	Apparatus: Instron 5848; 50 N load cell. Test: cyclic compressive stress and strain; 5 mm/min for 70% strain, followed by the same releasing rate until zero loading, for 2 min.	ELECTRONICS	[124]
Lignin	Insertion of ACX onto lignin to produce a RAFT macrocontroller to polymerize AM or AA.	Viscosity determination of an aqueous solution of grafted material.	Apparatus: Brookfield cone-and-plate viscometer. Steady shear measurements using 100 rpm.		[139]

Table 11. Summary of chromatographic characterization techniques used for grafted materials.

Backbone	Grafts	Technique	Property Measured	Application	Refs.
Lignin-RAFT macrocontroller	Poly(soybean oil methacrylate) derivatives. PSBMA, PSBMAH, PSBMAEO.	GPC	M_n and MWD of different lignin-g-poly(soybean) synthesized materials: Lignin-g-PSBMA (31.9 M g/mol/4.4); Lignin-g-PSBMAH (57.1 M g/mol/4.4); Lignin-g-PSBMAEO (64.3 M g/mol/4.0). $T = 35\ °C$; refractive index detector. Columns: HR1, HR3 and HR5E. Solvent: THF at 1 mL/min. Calibration with polystyrene standards. Samples prepared in THF (4 mg/mL) and filtered using a 0.2 mm mesh.		[134]
Hydroxypropyl cellulose (HPC)	Steglich esterification of PABTC over HPC using DCC and DMAP, and grafting of PABTC, with further polymerization of EA and NIPAAM	SEC	MWD by SEC; two PL polar gel M and one PL polar gel 5 mm guard columns; refractive index detector; Eluent: N,N-dimethylformamide/ 0.3 M LiBr, 0.5 mL/min.		[108]
Cellulose acetate	Solvents: DMSO, PDX, DMAc, C_3H_6O; initiators for grafting and polymerization: CAN, $Sn(Oct)_2$ and BPO.	GPC	Number average molecular weights (M_n) and dispersity values (Đ) of grafted PMMA extracted from samples of graft copolymer. Apparatus: Agilent 1100 with 3 PSS GPC 8 300 mm, 5 mm, 10^5, 10^5, 10^3 A columns; eluent: THF, 0.8 mL/min, at 20°C. Calibration using polystyrene standards with molecular weights ranging $200-10^6$ g/mol.		[111]
Microcrystalline cellulose	ROP of L-LA with DMAP in an ionic liquid (AmimCl) to produce Cellulose-g-PLLA.	HPLC	Apparatus: Agilent 1200 with an XDB-C18 phase column. Eluent: H_2O—methanol (20:80 by vol.), 0.8 mL/min.		[116]

Table 11. Cont.

Backbone	Grafts	Technique	Property Measured	Application	Refs.
Hemicellulose	Hemicellulose grafting using TBD and e-caprolactone monomers.	GPC	Apparatus: Malvern Viscotek HT GPC 350; refractive index (RI) and viscometer detectors. PSS-GRAM columns covering a range of 100–1,000,000 g/mol. Eluent: DMSO at 80 °C. Universal calibration using pullulan polysaccharide standards.		[37]
Lignin	Insertion of ACX onto lignin to produce a RAFT macro-controller which polymerizes AM and AA.	GPC	Apparatus: Waters Alliance 2695; eluent: aqueous solution of 0.1 M sodium phosphate buffer and 0.01% NaN$_3$, at ambient temperature, 1 mL/min.		[139]

Table 12. Summary of characterization techniques based on mechanical properties used for grafted materials.

Backbone	Grafts	Technique	Property Measured	Application	Refs.
Lignin-RAFT macro-controller	Poly(soybean oil methacrylate) derivatives. PSBMA, PSBMAH, PSBMAEO.	Tensile strength	TT s/e values for Lignin-g-PSBMAH (1.5 MPa/220 %), PSBMAH (3.0 MPa/120 %) and epoxy resin prepared from grafted materials (17 MPa/22.5 %). Tensile tests were carried out using a crosshead speed of 20 mm/min at ambient temperature. Samples were casted to obtain films. THF polymer solution samples were dried for 12 h under vacuum at ambient temperature and for 6 h, at 60 °C. Films were cut into dog-bone tensile samples of 20 mm length and 5 mm width.		[134]
Cellulosic *Grewia optiva* fibers	FAS-H_2O_2 redox initiation for grafting of MA and KPS, followed by polymerization of MA.	Swell index	Swelling of raw and grafted fibers in different solvents (DMF, water, methanol, and isobutyl alcohol) was measured.		[110]
Cellulose nanocrystal (CNC)	Macromolecular initiator obtained from the reaction between Br-iBuBr and CNC using TEA as catalyst, via SI-ATRP with Sty. Material was casted in PMMA.	Tensile strength	The mechanical properties of nanocomposites were investigated through breaking strength and elongation at break tests.		[118]
Cellulose cotton fiber pulp	Esterification of maleic anhydride grafted over PHA (through double bonds). PHA-g-MA grafted onto cellulose cotton fiber pulp (through the anhydride group).	Tensile strength	Apparatus: Lloyd Instruments, Model LR5 K; 100 mm/min (ASTM D638); pieces sized 150 × 15 mm (length × width). Mean values were determined from measurements from four specimens.		[120]

Table 12. *Cont.*

Backbone	Grafts	Technique	Property Measured	Application	Refs.
Hemicellulose	Hemicellulose grafting using TBD and ε-caprolactone monomer	Tensile strength	Apparatus: Shimadzu Autograph AG-500A. Method: ISO 527-2. Specimen size: 35 × 4 × 0.3 mm. Tests carried out at ambient conditions; 10 mm/min, 12 mm distance between grips. Properties measured: ultimate strength, Young's modulus, and elongation at break (5 repeats).		[37]
Regenerated cellulose fibers (rayon)	Photo-chemical grafting of PETA without photoinitiator.	Tensile strength	Apparatus: Z020 universal testing device with 20 kN load cell. Clamping length: 25 mm, 1 mm/min, pre-load stress of 5 N, and a clamping pressure of 2.5 bar. Specimen: 100 mm × 10 mm × 2 mm.		[123]
		Fatigue	Apparatus: 858 Mini Bionix II, MTS testing machine, Systems Corporation. Cyclic tests under tensile stresses (R = 0.1) were performed. Experiments reaching total cycle numbers of 1 million were considered as runouts. Fatigue tests performed uniformly at 10 Hz. Experiments proceeded until breakage or reaching runout conditions.		

Table 13. Summary of biological, functional, and compositional characterization techniques employed in grafted materials.

Backbone	Grafts	Technique	Property Measured	Application	Refs.
CellClAc	Macro initiator, Cu(I)Cl/2'2'BIPI catalytic system via ATRP of 4NPA and MMA	Elemental analysis	Apparatus: Leco CHNS-932. Elements determined: C, N and H elements.		[109]
CellClAc	Macro initiator, Cu(I)Cl/2'2'BIPI catalytic system via ATRP controller.	Elemental analysis	Apparatus: Leco CHNS-932. Elements determined: C, N and H elements.		[112]
Cellulose cotton fibers	Na_2CO_3 and thermal activation using MTC-b-CD and silver nitrate	Microbiological investigations	Materials treated with Ag^0 and $Ag+$ were evaluated for antibacterial activity. Gram-positive coccus (Staphylococcus aureus ATCC 25923) and Gram-negative bacillus (Escherichia coli ATCC 25922) microorganisms were used. The Kirby-Bauer diffusimetrical method was used to achieve this purpose. Bacterial cultures were standardized according to McFarland scale for 18 h, yielding 10^7–10^8 CFU/mL. After inoculation, sample discs were placed on the surface of the medium. Antibacterial activity was assessed after 24 h of incubation, at 37 °C.		[113]
Microcrystalline cellulose	ROP of L-LA with DMAP in an ionic liquid (AmimCl) to produce Cellulose-g-PLLA	Load and controlled release	A solution of grafted material (60.0 mg) and vitamin C (60.0 mg) in 2 mL of PBS was prepared and then transferred to a dialysis bag. Dialysis against 1 L distilled water for 24 h (water refreshment after 12 h) followed. Afterwards, the dialysis bag was immersed into a 400 mL phosphate buffer solution, at conditions similar to those of intestinal fluid (pH 7.40) at 37 °C. 2 mL samples were taken at predetermined times, and analyzed at l_{max} = 245 nm for vitamin C.		[116]

Table 13. *Cont.*

Backbone	Grafts	Technique	Property Measured	Application	Refs.
Cellulose cotton fiber pulp	Esterification of maleic anhydride grafted onto PHA (through double bonds).	Surface roughness	Measurement of surface roughness accomplished using an L&M CE165 PPS tester.		[120]
	PHA-g-MA grafted onto cellulose cotton fiber pulp (through the anhydride group).	Contact angle (hydrophobicity)	Surface hydrophilicity was analyzed by contact angle measurements. Tests carried out at ambient conditions using a Dataphysics OCA40 Micro instrument. Surface free energy parameters were estimated from these data.		
Cellulosic filter papers	Radiation-induced graft (RIGCP) copolymerization of acrylonitrile.	Chelating rare elements	Chelation of uranium, thorium, and lanthanides by the batch procedure took place.		[121]
Hemicellulose	Hemicellulose grafting using TBD and ε-caprolactone monomers.	Contact angle	Measurements carried out using an NRL Contact Angle Goniometer by Rame-Hart, model 100-00.		[37]
		Biodegradation test	Method: ISO 14851. Biochemical Oxygen Demand (BOD) measurements were obtained under aerobic conditions.		

Table 13. Cont.

Backbone	Grafts	Technique	Property Measured	Application	Refs.
CellClAc	NCHA, 4VP, DA and DAAM grafted by atom transfer radical polymerization (ATRP) using CuCl, 2'2'BIPI as catalyst.	Elemental analysis	Elemental analyses carried out in a Leco CHNS-932 apparatus.		[122]
		Electrical conductivity	Electrical conductivity determined with a Keithley 6517A electrometer. Cell-g-4VP was the only material that behaved as semiconductor.		
Chitosan	Embedded polymerization of N-VP and 4VP. Crosslinking agent: N'N'-MBA	Swell index	Swelling measurements were carried out in distilled water, in simulated intestinal buffer of pH 6.8 and in simulated gastric buffer of pH 1.4 (7 mL of HCl, 2 g of NaCl and 0.1 L of distilled water) at ambient conditions. A known weight of dry sample was placed into a teabag and immersed into aqueous medium. After a while, the bag was taken out and hung for 5–10 min to eliminate excess unabsorbed water, and then weighed.		[164]
		Load and controlled release	The loading capacity (milligrams of doxocyline entrapped per 100 milligrams of dried drug-loaded gels, %) of CS graft gels (CS-g) was determined as follows: the drug loaded CS-g gel was ground to fine power followed by immersion of a known amount of it into 0.1 L of HCl aqueous solution for 24 h under magnetic stirring. Then, the solution was filtered and used to determine the absorbance (A) of the drug contained in it using a Perkin Elmer Lambda 35 apparatus at 280 nm. Release profiles of the model drug (doxocyline) from the drug-loaded polymers were determined in distilled water (pH = 1.5) and in buffer solutions (pH = 6.8). All the studies were carried out in triplicate.		

67

Table 13. Cont.

Backbone	Grafts	Technique	Property Measured	Application	Refs.
Regenerated cellulose fibres (rayon)	Photo-chemical grafting of PETA without photoinitiator.	Fiber volume content	Volume of materials was determined by measuring the dimensions of samples (calliper gauge 500-171-20, Mitutoyo), and from grammage and density of fibers, referred to the total mass determined using a type 440 35N fine-scale.		[123]
Cellulose nanofibrils	Nitroxide TEMPO insertion and NMP of HEMA	BET	Apparatus: NOVA 1200e Quantachrome. Method: Brunauer–Emmett–Teller (BET) method for analysis of N_2 gas adsorption isotherms. Estimated properties: specific surface area (SBET) and average pore volumes. BET surface area improved from 9.87 m^2/g for TEMPO-Cellulose to 19.7283 m^2/g for HEMA–TEMPO-Cellulose. The microporous volume (Vtotal) increased from 0.0484 cm^3/g to 0.0705 cm^3/g, for HEMA–TEMPO-Cellulose.		[124]
		Electrical conductivity	Electrical resistivity determined using a four-point probe system (ST 2253, Suzhou Jingge Electronic Co.). The direct freezing method was compared against the unidirectional gradient freezing method. Samples produced using the direct freezing method exhibited higher resistivity values.		

Table 13. Cont.

Backbone	Grafts	Technique	Property Measured	Application	Refs.
Lignin	Insertion of ACX onto lignin to produce a RAFT macro-controller that polymerizes AM and AA	Surface tension	Apparatus: De Nouy ring-type tensiometer (Krüss), at 25 °C.		[139]
		Emulsification test	Absorbance of solutions of dissolved grafted material in deionized water and mixtures of hexanes (ultrasonication required for disolution) was measured using a Cary 300 spectrophotometer. Comparison of this value with that of the starting aqueous solution allowed calculation of lignin content in the emulsion phase.		

Table 14. Overview of studies focused on the modeling of polymer grafting.

Backbone	Functionalization Method	Graft Chains	Modeling Approach	Highlights/Comments	Ref.
Pre-polymer containing highly active chain transfer sites (pendant mercaptan groups).	CTP between polymer radicals and pendant mercaptan groups.	Vinyl polymer	Kinetic model for FRP including CTP; model equations solved numerically and approximate closed form equations were also used.	Calculated properties/variables: grafting efficiency (Equation (1)) and number average molecular weights of the different polymer molecules (Equations (2)–(6)).	[39]
Solid polymeric substrates.	Irradiation of polymer/monomer mixture.	Vinyl polymer	Calculation of r_n, r_w and number full chain length distributions (n_x) from kinetic equations.	Two limiting cases were considered in the calculation of n_x: (a) radicals terminating by disproportion, no significant chain transfer and complete conversion; and (b) no chain transfer and incomplete conversion (see Equation (7)).	[40]
Polyolefins	High temperature CTP.	Vinyl polymer	Kinetic modeling from a given polymerization scheme.	Review paper considering chemical modification of polyolefins by FRP in extruders.	[231]
Silica substrate	Formation of a growing grafted polymer chain from reaction between an initiator derived primary free radical and a "surface site."	Poly(vinylpyrrolidone)	Use of a "conservational polymerization and molecular-weight distribution" (CPMWD) numerical method to solve the kinetic equations.	The proposed numerical approach provides monomer conversion, grafting yield, and full MWDs of active and inactive homopolymer and graft polymer species.	[232]
Porous silica beads	Chain transfer reaction to fixed mercapto groups present in the pre-functionalized silica beads.	Polystyrene	Kinetic model that allows calculation of full MWD and Đ; no details on numerical approach or software used were provided.	MWDs were approximated assuming them, as a first approach, to follow a normal distribution function.	[233]
PE	CTP by FRP in melt, in a twin-screw extruder (TSE).	Poly(glycidyl methacrylate) (PGMA)	Coupling of flow characteristics from a TSE simulation program and a reactor model consisting of plug flow and "axial dispersion" reactor cells.	Overall GMA conversion was predicted, showing general good agreement with experimental data.	[234]
PE	CTP by FRP in melt, in a twin-screw extruder (TSE).	Poly(maleic anhydride) (PMAH)	Coupling of reaction kinetics to flow equations in a twin-screw extruder.	Satisfactory agreement between calculated and experimental data.	[235]

Table 14. Cont.

Backbone	Functionalization Method	Graft Chains	Modeling Approach	Highlights/Comments	Ref.
Polypropylene (PP)	CTP by FRP in melt	PMAH	A kinetic model which captures the effect of MAH and initiator concentrations on grafting yield and MWDs during the grafting process was derived.	Model performance assessed by comparing calculated and experimental data from different sources, for batch (internal mixer, static film) and continuous systems (twin-screw extruder), under a wide window of operating conditions.	[236]
Cellulose or cellulose containing fibers	Not specified (empirical approach)	Monochlorotriazinyl-β-cyclodextrin	Neural network modeling (semi-empirical approach).	A good approximation of the studied grafting process was obtained.	[237–239]
PP	CTP by FRP in solid state, in presence and absence of supercritical CO_2 ($scCO_2$).	PMAH	A kinetic model focused on polymerization and grafting rates was used.	The effect of presence or absence of $scCO_2$ on grafting rate was analyzed.	[240]
PP	CTP by FRP in solid state, with and without $scCO_2$.	Not specified	The kinetic model and grafting mechanism of the $scCO_2$ assisted grafting of PP in the solid state were reviewed.	Grafting rate controlled by diffusion in absence of $scCO_2$ and by the chemical grafting step in presence of $scCO_2$.	[241]
Polybutadiene (PB)	CTP by FRP in production of high-impact PSty (HIPS).	PSty	An existent heterogeneous model is generalized to consider bivariate distributions of graft copolymer topologies; each topology is determined by the number of trifunctional branching points per molecule.	Polymerization in two phases is assumed. Adjustment of a single kinetic parameter is required. Grafting efficiencies obtained from solvent extraction/gravimetry are contrasted against estimates obtained by the deconvolution of size exclusion chromatograms for total polymer.	[242]
Poly[ethylene-co-(1-octene)]	CTP by FRP in molten state	PMMA	A model compound approach was used.	The focus was on assessing the effect of temperature, polymerization time and initiator content on MMA polymerization rate and number average chain length, in the presence of alkoxyl radicals and alkanes.	[243]

Table 14. *Cont.*

Backbone	Functionalization Method	Graft Chains	Modeling Approach	Highlights/Comments	Ref.
PSty	Use of a Friedel-Crafts reaction, which gives rise to a mixture of unmodified PE, modified PS and a PE-g-PS copolymer.	PE	A model for predicting the complete MWD of the homopolymers, and the chemical composition distribution (CCD) of the graft copolymer was presented. Probability generating functions were used to carry out the calculations.	The model takes into account the opposing effects of rival reactions that occur in the studied grafting reaction.	[244,245]
PE	CTP by FRP	PSty	A kinetic mechanism for the whole process and application of the method of moments to solve the mass balance equations are proposed.	The model is able to calculate average molecular weights and the amount of grafted PS.	[246]
PP	CTP by FRP	PMAH	A kinetic model for the free-radical grafting of maleic anhydride MAH onto PP, considering that the reaction medium is heterogeneous due to the incomplete solubility of MAH in the polymer melt was developed.	Two different approaches are considered that derive from the limiting cases of very fast mass transfer (equilibrium) and no mass transfer between the two phases. The physical parameters considered in the model include the solubility of MAH in PP and the partition coefficient of the initiator between the two phases.	[247]
Carboxyl functionalized hydrophilic Sephadex (a cross-linked dextran gel used for gel filtration) derivatives.	Covalent amine conjugation method analyzed using XPS.	Poly(ethylene glycol) (PEG)	Langmuir and Langmuir-Freundlich isotherm models are used to study PEG-grafting kinetics. Fractional C-O intensities obtained from high-resolution C 1s scans are correlated with grafting.	The models are assessed by comparing calculated profiles and experimental data.	[248]

Table 14. Cont.

Backbone	Functionalization Method	Graft Chains	Modeling Approach	Highlights/Comments	Ref.
Various surfaces	Surface-initiated controlled radical polymerization.	Polymers from 2-methacryloyloxyethyl phosphorylcholine (MPC), methyl acrylate, acrylamide, and N-isopropylacrylamide.	A kinetic model for surface-initiated atom transfer radical polymerization (SI-ATRP) with addition of excess deactivator in solution was presented.	Polymer layer thickness, and concentrations of radical, dormant and dead polymer molecules can be calculated. A simple but reliable analytical solution is proposed for estimation of the evolution of polymer layer thickness.	[249]
Partially fluorinated polymers such as poly(vinylidene fluoride) and poly(ethylene-co-tetrafluoro ethene).	Radiation induced grafting	Poly(4-vinylpyridine)	Box-Behnken factorial designs and response surface method (RSM) were used.	A quadratic model was used to obtain the optimal set of conditions (absorbed dose, monomer concentration, grafting time and reaction temperature) to maximize the degree of grafting (G%), using a statistical software. Adequate agreement between model predictions and experimental data supported the approach proposed.	[250,251]
Poly(propylene glycol) (PPG)	CTP by FRP	Poly(styrene-co-acrylonitrile) (SAN)	A batch kinetic model for grafting copolymerization of Sty and acrylonitrile (AN) in the presence of PPG is modified to include semi-continuous and continuous processes.	The semi-continuous model is validated using experimental data, although molecular weight agreement is poor. The continuous process is simulated but not contrasted against experimental data. According to the model, graft efficiency increases drastically, and Mw decreases with decreasing St/AN weight ratio.	[252]
Starch	(Mechanically activated) CTP by FRP	Polyacrylamide	A kinetic model for an inverse emulsion copolymerization system was used. The effects of initiator, monomer, starch, and emulsifier content on reaction rate of graft co-polymerization (Rg) were determined. The kinetic model was modified to provide acceptable agreement with experimental data.	It was observed that the relationship between Rg and the concentration of components in the polymerization system can be expressed for all of the four components by the following equation: Rgα[mSt]1.24 [I]0.76[M]1.54[E]0.33, which is consistent with the kinetic equation from theoretical studies: Rpα[mSt]0.5~1 [I]0.5~1[M]1~1.5[E]0.6.	[253]

73

Table 14. Cont.

Backbone	Functionalization Method	Graft Chains	Modeling Approach	Highlights/Comments	Ref.
PB seed latex particles	CTP by emulsion FRP	SAN	Emulsion grafting copolymerization is model using a two-phase kinetic model. Monomer concentrations in the different phases is calculated.	Partition of components between phases calculated as in previous models. The model is extended to account for calculation of grafting efficiency and copolymer composition of free and grafted molecules.	[254]
PE	CTP by FRP in REX	PAA	Monomer conversion to homo- and grated-polymers are calculated separately using an incremental theory. The approach allows calculation of degree of grafting, mass of homopolymer and grafting efficiency.	The focus was on analyzing the effect of temperature and initial concentrations of monomer and initiator on degree of grafting. Both grafting and amount of homopolymer increase with temperature and monomer concentration.	[255]
Natural rubber latex (core-shell particles)	CTP by emulsion FRP	PMMA	The model consists on rate expressions of polymer chain formation. A decrease of CHPO by TEPA and a population event of radicals between core/shell phases are considered in the model.	Grafting efficiency and polymer composition for free and grafted polymer populations can be calculated in terms of initial conditions (monomer and initiator content, and rubber weight ratio and temperature).	[256]
Vinyl/divinyl copolymer synthesized by RAFT chemistry.	CTP by FRP	Non-specified RAFT synthesized graft	A RAFT crosslinking model was used to address the molecular imprinting step. The size of grafted brushes was estimated from MWD calculations for a RAFT homo-polymerization.	The model includes the two steps involved in the production of MIP responsive particles. Limitations and ways to improve the model are discussed.	[257]
High-density PE (HDPE)	Melt free-radical grafting (semi-empirical approach)	GMA	Response surface, desirability function, and artificial intelligence approaches were combined to account for polymer grafting. Input variables: monomer content, initiator concentration, and melt-processing time.	An in-house code that learns and mimics processing torque and grafting of GMA onto HDPE was developed. Optimization of the process and quantification of the competition between grafting and crosslinking was carried out.	[258]

Table 14. *Cont.*

Backbone	Functionalization Method	Graft Chains	Modeling Approach	Highlights/Comments	Ref.
PE	CTP by FRP in post-polymerization modification	Vinyl polymers	A kinetic Monte-Carlo (kMC) modeling approach for free radical induced grafting of vinyl monomers onto polyethylene (PE), assuming isothermicity, perfect macromixing and diffusion-controlled effects was developed.	Calculated variables: monomer conversion, grafting selectivity and yield, MWDs of all macromolecular species involved, average grafting (from/to) and crosslink densities, number of grafts and crosslinks per individual polymer molecule, and chain length of every graft.	[259]
PE	CTP by FRP (batch operation)	Vinyl polymers	A kMC model was used. Reactions included: initiator dissociation, hydrogen abstraction, graft chain initiation, graft (de)propagation, crosslinking, and homo (de)polymerization.	Assumptions: isothermicity and single-phase polymerization. It is found that the grafting kinetics is affected by hydrogen abstraction and depropagation kinetic rate constants, equilibrium coefficient Keq, and initial content of monomer, initiator, and polyolefin.	[260]
PE	CTP by FR? (tubular reactor)	Vinyl polymers	kMC, as in [260]	They found that functionalization, selectivity and grafting density are able to reach remarkably better values (e.g., 100%) if multiple injection points are considered for monomer, initiator and/or a temperature profile.	[261]
PE	CTP by FRP (two-phase system)	Vinyl polymers	kMC; the model of [260,261] is further extended to include two phases.	The partition of monomer and initiator between the two phases is addressed by introducing overall mass transfer coefficients. The model is tested with a monomer with low to negligible homo-propagation. Positive agreement with industrial experimental data was obtained.	[41]

Table 15. Polymerization scheme for free-radical polymerization including CTP and crosslinking.

Reaction Step	Kinetic Expression	Remarks
Initiator decomposition	$I \xrightarrow{k_d} 2R$	
First propagation	$R + M \xrightarrow{k_i} P_1$	
Propagation	$P_n + M \xrightarrow{k_p} P_{n+1}$	$n = 1, \ldots \infty$
Termination by disproportionation	$P_n + P_m \xrightarrow{k_{td}} D_n + D_m$	$n, m = 1, \ldots \infty$
Propagation through intermediate free radicals	$P_{n,b} + M \xrightarrow{k_p} P_{n+1,b}$	$n = 1, \ldots \infty$ $b = 0, \ldots \infty$
Chain transfer to polymer	$P_{n,b} + D_{m,c} \xrightarrow{k_{trp}} D_{n,b} + P_{m,c+1}$	$n, m = 1, \ldots \infty$ $b, c = 0, \ldots \infty$
Propagation through pendant double bonds (crosslinking), assuming a pseudo-kinetic rate constants method (pseudo-homopolymer) approach	$P_r + D_s \xrightarrow{k_p^{*0}} P_{r+s}$	$r, s = 1, \ldots \infty$

Table 16. Symbols of application fields.

Symbol of Corresponding Application	Meaning
	Biocomposites and biomaterials.
	Green chemistry and innovative processes.
	Electronic materials, electrical properties, and conjugated polymers.
	Surface modification, hydrophilic or hydrophobic surfaces.

Table 16. *Cont.*

Symbol of Corresponding Application	Meaning
	Flame retardancy, additives for flame retardancy in polymer blends, thermal resistance materials.
	Antimicrobial applications.
	Adhesives, polymer networks, crosslinked polymers, gels.
	Controlled release of drugs and chemicals.
	Polymer composites and blends, by extrusion, casting or co-precipitation.
	Chelating properties, membranes, effluent remediation.

Table 16. *Cont.*

Symbol of Corresponding Application	Meaning
	Mechanical properties improvement, micro and nano reinforcement.
	Application in polyolefins, polyethylene, polypropylene.
	Aerogels, light, or porous materials.

Table 17. Abbreviations used for characterization techniques.

Abbreviation	Meaning
FTIR	Fourier transformed mid-infrared spectroscopy.
ATR	Attenuated total reflection.
^{1}H-NMR	Proton nuclear magnetic resonance.
^{13}C-NMR	Carbon nuclear magnetic resonance.
TT	Tensile test.
DMA	Dynamical mechanical analysis.
GPC	Gel permeation chromatography.
TGA	Thermogravimetric analysis.
DSC	Differential scanning calorimetry.
FR	Flame retardancy.
XPS	X-ray photoelectron spectroscopy.
EDAX	Energy-dispersive X-ray analysis/spectroscopy.
WAXD	Wide Angle X-ray Scattering.
XRF	X-ray fluorescence.

Table 18. Abbreviations of properties and variables measured by characterization techniques.

Technique	Abbreviation	Explanation	Units
Chromatography	MWD	Molecular weight distribution of polymers.	-
	M_n	Number average molecular weight.	g/gmol
Thermal	T_g	Glass transition temperature.	°C
	MDT	Maximum degradation temperature.	°C
	MFI	Melt flow index.	g/10 min
	THR	Total heat release.	MJ/m^2
	PHRR	Peak heat release rate.	kW/m^2
Mechanical	σ	Maximum stress measured in tensile test.	MPa
	ε	Maximum elongation observed in tensile test.	%

Table 19. Abbreviations and formulae of some chemical compounds and materials used.

Abbreviation	Compound	Chemical Structure	Ref.
"P"	Before the name or abbreviation of a monomer means polymer of that monomer.		
"B-g-C"	Means "C" chains grafted to "B" backbone.		
THF	Tetrahydrofuran.		[17,18]
DMF	Dimethylformamide.		[121,125,139,141,144, 146,147,151,152,165]
DMSO	Dimethyl Sulfoxide.		[37,111]
PDX	1,4-dioxane.		[111,140,142,150]
PP	Polypropylene.		[135]
PE	Polyethylene.		[135]

Table 19. Cont.

Abbreviation	Compound	Chemical Structure	Ref.
DCC	Dicyclohexylcarbodiimide.		[108,165]
DMAP	4-Dimethylaminopyridine.		[108,116,165]
SBMA	(E)-2-(N-methyloctadec-9-enamido)ethyl methacrylate.		[134]
SBMAH	(E)-3-(octadec-9-enamido)propyl methacrylate.		[134]
SBMAEO	2-(N-methyl-8-(3-octyloxiran-2-yl)octanamido)ethyl methacrylate.		[134]
	4-cyano-4-(phenylcarbonothioylthio)pentanoic acid.		[134]
$POCl_3$	Phosphoryl trichloride.		[125,135]
P_4O_{10}	Phosphorus(V) oxide.		[125,136,137]

Table 19. Cont.

Abbreviation	Compound	Chemical Structure	Ref.
	Imidazole.		[125,135]
IPG	(1H-imidazol-1-yl) phosphonic group.		[125,135]
N'N'-MBA	N'N'-methylenebisacrylamide.		[98,164]
N-VP	N-vinyl pyrrolidone.		[99,164]
Co(acac)$_3$	Cobaltacetylacetonate complex.		[99,103]
MMA	Methyl methacrylate.		[58,107,111,117,140, 144,146,149]
PMMA	Poly(methyl methacrylate).		[58,107,111,117,140, 144,146,149]
AcN	Acrylonitrile.		[107,121]
EA	Ethyl acrylate.		[107,108]

Table 19. Cont.

Abbreviation	Compound	Chemical Structure	Ref.
NIPAAM	N-isopropylacrylamide.		[108,138,140,141,143,145,165]
PNIPAAM	Poly(N-isopropylacrylamide).		[108,138,140,141,143,145,165]
PABTC	Propionic acidyl butyl trithiocarbonate.		[108]
HPC	Hydroxypropyl cellulose.		[108]
DMAc	Dimethylacetamide.		[108]
CellClAc	Cellulose chloroacetate.		[109,122]
2'2'BIPI	2,2'-bipyridine.		[109,122,140,146,147]
4NPA	N-(4-nitrophenyl) acrylamide.		[109]
MA	Methyl acrylate.		[110]

Table 19. Cont.

Abbreviation	Compound	Chemical Structure	Ref.
FAS	Ferrous ammonium sulphate.	$(NH_4)_2Fe(SO_4)_2\ 6H_2O$	[110]
KPS	Potassium persulphate.		[110]
CAN	Ceric ammonium nitrate.		[111]
BPO	Benzoyl peroxide.		[111]
Sn(Oct)$_2$	Tin(II) 2-ethyl hexanoate.		[111]
NCA	N-cyclohexylacrylamide.		[112]
AgNPs	Silver nanoparticles.	Ag	[113]
b-CD	β-cyclodextrin.		[113,114]

Table 19. Cont.

Abbreviation	Compound	Chemical Structure	Ref.
MTC-b-CD	Monochlorotriazinyl-β-cyclodextrin.	CD = beta Ciclodextrina	[113,114]
APS	Ammonium persulfate.		[115,117]
AmimCl	1-allyl-3-methylimidazolium chloride.		[116]
PBS	Phosphate-buffered saline solutions.		[116]
DMAc	N,N-dimethyl acetamide.		[111,117]
Br-iBuBr	2-Bromoisobutyryl bromide.		[118]
L-LA D-LA	L-lactide. D-lactide.		[116,119,157,159]
PLLA	Poly(L-lactide).		[116,119,157,159]
PDLA	Poly(D-lactide).		

Table 19. Cont.

Abbreviation	Compound	Chemical Structure	Ref.
STY or Sty	Styrene.		[58,118,140,146,147,149]
PS or PSty	Polystyrene.		[58,118,140,146,147,149]
TEA	Triethylamine.		[118]
PMDETA	N, N, N′, N′, N″-Pentamethyldiethylenetriamine.		[118,140,146]
PHA	Polyhydroxyalkanoate.		[120]
PHA-g-MA	Poly(hydroxyalkanoate-grafted-maleic anhydride).		[120]
TBD	1,5,7- triazabicyclodecene [4.4.0].		[37]
NCHA	N-cyclohexylacrylamide.		[122]

Table 19. Cont.

Abbreviation	Compound	Chemical Structure	Ref.
CL	ε-Caprolactone..		[37,158,159,161,162]
PCL	Poly-(ε-caprolactone).		[37,158,159,161,162]
4VP	4-vinylpyridine.		[122,164]
DAAM	Diacetone acrylamide.		[122]
DA	Diallylamine.		[122]
PETA	Pentaerythritol triacrylate.		[123]
TEMPO	TEMPO nitroxide.		[124]
HEMA	2-hydroxyethyl methacrylate.		[124]
DMAEMA	2-dimethylaminoethyl methacrylate.		[138,150]

Table 19. Cont.

Abbreviation	Compound	Chemical Structure	Ref.
BA	Butyl acrylate.		[138]
EG	Ethylene glycol.		[138]
AM	Acrylamide.		[58,138,139,151,153]
AA	Acrylic Acid.		[58,138,139]
ACX	Acyl chloride xanthate.		[139,153]
LMA	Lauryl methacrylate.		[140]
	Guaiacol.		[140]
	Vainillin.		[140]

87

Table 19. Cont.

Abbreviation	Compound	Chemical Structure	Ref.
	Syringyl methacrylate.		[140]
	4-propylsyringol.		[140]
	4-propylguaiacol.		[140]
	Ferulic acid.		[140]
	Isorbide.		[140]
CuBr/HMTETA	1,1,4,7,10,10-Hexamethyltriethylenetetramine Cu(I)Br complex.		[140,143,150]

Table 19. *Cont.*

Abbreviation	Compound	Chemical Structure	Ref.
CuBr/PMDETA	N,N,N′,N″,N″-Pentamethyldiethylenetriamine Cu(I)Br complex.		[140,141,146,147]
PEG-A	Poly(ethylene glycol) acrylate. n = 9		[140,143]
PPG-A	Poly(propylen glycol) acrylate. n = 5		[140,143]
DAEA	Dehydroabietic ethyl acrylate.		[140,142]
PDAEA	Poly(dehydroabietic ethyl acrylate).		[140,142]
BMA	Butyl methacrylate.		[140,144]

Table 19. Cont.

Abbreviation	Compound	Chemical Structure	Ref.
PBMA	Poly(butyl methacrylate).		[140,144]
PEGMA	Poly(ethylene glycol) methacrylate.		[140,148]
PPh$_3$	Triphenyl phosphine.		[149]
KEX	Potassium ethyl xanthate.		[151,152]
	2-Bromopropionic acid.		[151,152]
GMA	Glycidyl methacrylate.		[153]
XCA	Xanthate carboxylic acid.		[154]

Table 19. Cont.

Abbreviation	Compound	Chemical Structure	Ref.
DMC	[2-(Methacryloyloxy)ethyl] trimethyl-ammonium chloride.		[154]
HONB	N-Hydroxy-5-norbornene-2,3-dicarboxylic.		[154]
AMBN	2,2′-azobis (2-methylbutyronitrile).		[154]
EOX	2-ethyl-2-oxazoline.		[155]
PEOX	Poly(2-ethyl-2-oxazoline).		[155]
MOX	2-methyl-2-oxazoline.		[156]
PMOX	Poly(2-methyl-2-oxazoline).		[156]
B-BL	β-Butyrolactone.		[160]
PHB	Poly(3-hydroxybutyrate).		[160]

Table 19. Cont.

Abbreviation	Compound	Chemical Structure	Ref.
EOX	Ethylene oxide.		[163]
PEO	Poly(ethylene oxide).		[161]
P4-t-Bu	1-tert-butyl-4,4,4-tris (dimethylamino)-2,2-bis[tris(dimethylamino)-phosphoranylidenamino]-2l5,4l5-catenadi (phosphazene) solution in hexane.		[163]

4. Characterization Techniques Used for Polymer Grafted Materials

The characterization of polymer grafted materials requires the use of a variety of methods due to the many possible combinations of backbones and polymer grafts [59]. The characterization methods can be classified as direct or indirect.

Direct methods are those used to identify changes in the chemical structure of grafted materials, such as the bonds between backbone and grafts. These methods include proton and carbon magnetic nuclear resonance, ^{1}H-NMR, and ^{13}C-NMR, respectively.

Indirect methods are based on differences in properties between the starting and grafted materials, relating these changes to the modified structures. Microscopy and thermal analysis are examples of indirect methods. A summary of the main characterization methods used for grafted materials is presented in Table 6.

The relevant information contained in selected articles is also gathered in this review to show how the characterization techniques were used to provide evidence of polymer grafting onto the corresponding backbones. Tables 7–12 summarize the use of thermal, spectroscopic, imaging and microscopy, rheological, chromatographical, and mechanical characterization techniques, respectively, in the analysis of polymer grafted materials. Finally, a summary of biological, functional, and composition characterization techniques used for grafted materials is provided in Table 13.

5. Modeling of Polymer Grafting

5.1. Literature review on Modeling of Polymer Grafting

An overview of the literature on the modeling of polymer grafting is summarized in Table 14. The backbones considered, the functionalization methods, the polymer chains grafted, and summary comments on the modeling approaches used to carry out the simulations are included in the table.

5.2. Modeling of Polymer Branching and Crosslinking

As observed in Table 14, most reports on the modeling of polymer grafting are related to cases where grafting involves free-radical growth of the grafts, and the generation of active sites proceeds through chain transfer to polymer reactions. In that sense, the growth of polymer grafts resembles the formation and growth of branches in polymer branching. The difference would be that the branches and primary polymer chains contain the same monomers, whereas grafts and backbones contain different monomers in polymer grafting. There are several papers focused on the modeling of polymer branching [262–268]. In some cases, as in the grafting of monochlorotriazinyl-β-cyclodextrin onto cellulose, the activation mechanism is not specified, and the modeling approach is fully empirical (neural network modeling) [237–239].

In general terms, the polymerization scheme of FRP including chain transfer to polymer (CTP) is given by the reactions shown in Table 15. The specific mathematical expressions containing CTP terms are given by Equations (1)–(7). I, R, and M in Table 15 are initiator, primary free radical, and monomer molecules, respectively; P_n and D_m denote live and dead polymer molecules, respectively, of sizes n and m. k_i, k_p, k_{td}, and k_{trp} (also denoted as k_{fp} in Equations (8) and (9)) denote initiation, propagation, termination by disproportionation, and chain transfer to polymer kinetic rate constants, respectively.

Polymer branching can be modeled using a bivariate distribution of chain length and number of branches resulting from polymerizations involving branched polymers [269]. $P_{n,b}$ in Table 15 accounts for a bivariate distribution of live polymer of length n and number of branches b. The moment equations shown below consider only the kinetic steps of propagation and chain transfer to polymer, for illustrative purposes. For a batch reactor, the application of the mass action law considering only these two kinetic steps results in Equation (1) [269]. It should be noticed that in the transfer to polymer reaction there are as many possible sites of reaction as monomeric units in the dead polymer chain participating in the reaction.

$$\frac{dP_{n,b}}{dt} = \ldots - k_p(P_{n,b}M + P_{n-1,b}M) - k_{trp}P_{n,b}\left(\sum_{m=1}^{\infty}\sum_{c=0}^{\infty} mD_{m,c}\right) + k_{trp}nD_{n,b-1}\sum_{h=1}^{\infty}\sum_{e=0}^{\infty} P_{h,c} + \ldots \quad (1)$$

$n = 1, \ldots, \infty; b = 0, \ldots, \infty$

The bivariate moments for active and inactive polymer are defined respectively as shown in Equations (2) and (3). Number and weight-averaged molecular weights, and the average number of branches, are given by Equations (4)–(6) [269].

$$\mu_{G,H} = \sum_{n=1}^{\infty}\sum_{b=0}^{\infty} n^G b^H P_{n,b} \quad (2)$$

$$\lambda_{G,H} = \sum_{n=1}^{\infty}\sum_{b=0}^{\infty} n^G b^H D_{n,b} \quad (3)$$

$$M_n = \frac{\mu_{1,0} + \lambda_{1,0}}{\mu_{0,0} + \lambda_{0,0}} W_m \quad (4)$$

$$M_w = \frac{\mu_{2,0} + \lambda_{2,0}}{\mu_{1,0} + \lambda_{1,0}} W_m \quad (5)$$

$$B_n = \frac{\mu_{0,1} + \lambda_{0,1}}{\mu_{0,0} + \lambda_{0,0}} \quad (6)$$

The moment equations for live polymer are given by Equation (7) [269].

$$\frac{d\mu_{G,H}}{dt} \ldots - k_p M \mu_{G,H} + k_p M \sum_{R=0}^{G}\binom{G}{R}\mu_{G-R,H} - k_{trp}\mu_{G,H}\lambda_{1,0} + k_{trp}\mu_{0,0}\sum_{K=0}^{H}\binom{H}{K}\lambda_{G+1,H} + \ldots \quad (7)$$

Another approach with which to address the modeling of polymer branching in FRP is to use the concept of branching density, denoted as ρ, which is given by the ratio of the number of branching points to that of monomeric units, and it can be estimated using Equation (8), which when solved leads to Equation (9) [270]. k_{fp} and k_p in Equations (8) and (9) are chain transfer to polymer and propagation kinetic rate constants, respectively, and x is monomer conversion.

$$\frac{d(x\rho)}{dx} = \frac{k_{fp}x}{k_p(1-x)} \quad (8)$$

$$\beta = -\frac{k_{fp}}{k_p}\left[1 + \frac{\ln(1-x)}{x}\right] \quad (9)$$

CTP and terminal double bond (TDB) polymerization produce tri-functional (long) branches, in addition to increasing the weight-averaged molecular weight and broadening the MWD. A reaction "similar" to TDB polymerization is the polymerization with internal (pendant) double bonds (double bonds "internal" in dead polymer chains, appearing therein due to (co)polymerization of di-functional (divinyl) monomers (e.g., butadiene). Internal double bond (IDB) polymerization produces tetra-functional (long) branches and leads eventually to the formation of crosslinked polymer (gel). Both molecular weight averages increase due to IDB polymerization and the MWD broadens considerably [43]. Crosslinking can be considered as interconnected branching, and in that sense, its growth by CTP and its modeling in terms of a crosslink density, denoted as ρ_a, can also be taken as a useful basis for the modeling of polymer grating by CTP and propagation through the intermediate free radicals. Balance equations for polymer radicals of size r (R^*_r) and ρ_a for a case of copolymerization with crosslinking of vinyl/divinyl monomers, using the pseudo-kinetic rate constants method, are given by Equations (10) and (11), respectively, where P_s is a dead polymer of size r; Q_1 is first-order moment for the dead polymer; and k_{cp} and k_{cs} are primary and secondary cyclization rate constants, respectively [271].

$$\frac{1}{V}\frac{d(V[R^*_r])}{dt} = k_p[M][R^*_{r-1}] + k_{fp}r[P_r][R^*] + k_p^* \sum_{s=1}^{r-1} s[R^*_{r-s}][P_s]$$
$$- \left(k_p^*\right)[M][R^*_r] - (k_{td} + k_{tc})[R^*][R^*_r] - \left(k_p^* + k_{fp}\right)Q_1[R^*_r] \tag{10}$$

$$\frac{d[x\overline{\rho_a}]}{dt} = \frac{k_p^*\left[\overline{F_2}(1-k_{cp}) - \overline{\rho_a}(1+k_{cs})\right]}{k_p(1-x)} \frac{x \, dx}{dt} \tag{11}$$

5.3. Main Modeling Equations for Polymer Grafting

Grafting efficiency (ε) and number average molecular weights for the different polymer populations (W_{IH}, W_{SH}, W_{II}, W_{IS}, and W_{SS}) for the grafting of vinyl polymers onto pre-formed polymer with highly active chain transfer sites of (pendant mercaptan groups) are given by Equations (12)–(17) [39]. Subscripts IH and SH in the molecular weights shown in Equations (12)–(17) account for primary chains formed by chain transfer or by disproportionation termination (without distinguishing between terminally saturated and unsaturated chains) of polymer radicals starting with I and S fragments, respectively. Subscripts II, IS, and SS, on the other hand, account for primary chains produced by combination of the appropriate pair of polymer radicals starting with I and S fragments, respectively. r in Equations (12)–(18) is the ratio of propagation to termination kinetic rate constants, namely, $r = k_p/2k_t$.

$$\varepsilon = \frac{W_{SH} + W_{SS} + W_{IS}}{W_{SH} + W_{SS} + W_{IS} + W_{IH} + W_{II}} \tag{12}$$

$$W_{IH} = m[M]_0 \int_0^\infty \left\{ \frac{y_0}{y_0 + (1-\alpha)^{C_S}} - \frac{(1-r)y_0^2}{\left[y_0 + (1-\alpha)^{C_S}\right]^2} \right\} d\alpha \tag{13}$$

$$W_{SH} = m[M]_0 \int_0^\infty \left\{ \frac{(1-\alpha)^{C_S}}{y_0 + (1-\alpha)^{C_S}} - \frac{(1-r)y_0(1-\alpha)^{C_S}}{\left[y_0 + (1-\alpha)^{C_S}\right]^2} \right\} d\alpha \tag{14}$$

$$W_{II} = m[M]_0 \int_0^\infty \left\{ \frac{(1-r)y_0^3}{\left[y_0 + (1-\alpha)^{C_S}\right]^3} \right\} d\alpha \tag{15}$$

$$W_{IS} = 2m[M]_0 \int_0^\infty \left\{ \frac{(1-r)y_0^2(1-\alpha)^{C_S}}{\left[y_0 + (1-\alpha)^{C_S}\right]^3} \right\} d\alpha \tag{16}$$

$$W_{SS} = m[M]_0 \int_0^\alpha \left\{ \frac{(1-r)y_0(1-\alpha)^{2C_S}}{\left[y_0 + (1-\alpha)^{C_S}\right]^3} \right\} d\alpha \qquad (17)$$

An example of calculation of the mole fraction chain length distribution (number distribution) (n_x) for the case of polymer grafting of vinyl polymers onto solid polymeric substrates, considering no chain transfer and incomplete conversion, is shown in Equation (18) [40].

$$n_{x,1} = \frac{[R_0]}{r[M]_0 \left\{ 1 - \left(\frac{[M]}{[M]_0}\right)^{1/r} \right\}} \left\{ 1 + \frac{[R_0](1-r)x}{r[M]_0} \right\}^{\frac{r-2}{1-r}} \qquad (18)$$

When addressing the modeling of polymer grafting of vinyl polymer onto polyolefins in extruders by CTP, Hamielec et al. [231] proposed a polymerization scheme and the corresponding kinetic equations where the prepolymer molecule bears abstractable hydrogens on its backbone and a second compound, denoted as additive (A), is bound to the prepolymer backbone via reaction with a free radical. The backbone radical then transfers its radical center to the active molecules. The radical is finally terminated with other radicals. Proper kinetic equations were written down for the participating species, and a degree of grafting, g, which is the average of number of grafted molecules per monomer unit on the prepolymer backbone, was defined as shown in Equation (19), where Q_1 is the concentration of monomer units on prepolymer backbones which remains constant during branching, $K_{a,3}$ is the kinetic coefficient for the grafting (additive addition) reaction, and R_{03} designates a primary radical with radical center on backbone R_0. Further mathematical treatment by the authors leads to Equation (20) for calculation of the full chain length distribution of the polymer population, w(r,s), where $w_0(r)$ is the initial chain length distribution of the prepolymer [231].

$$\frac{dg}{dt} = \frac{K_{a,3} R_{0,3} A}{Q_1} \qquad (19)$$

$$w(r,s) = \left(1 + \frac{s}{r}\right) \frac{w_0(r)}{(1+g)s!} (gr)^s e^{-gr} \qquad (20)$$

As stated above in Table 14, Gianoglio Pantano et al. [245] developed a very detailed model for grafting of PSty onto PE. Among the concentrations of species calculated by the model, the concentration of poly(ethylene-g-styrene) is calculated using Equation (21), and the concentration of grafted PS, denoted as Gr, is obtained from Equation (22). G(t) is a matrix array of "infinite" size whose elements contain the molar concentrations of the individual species with degree of polymerization indicated by its subscripts. **VG** is a vector that contains the corresponding reaction rate terms, $I_{1,0}$ is a bivariate moment of order 1 for PS and order 0 for PE, which represents the mass of PS grafted onto PE, and M^{PS}_1 is the molar mass of PS.

$$\frac{d\mathbf{G}(t)}{dt} = \mathbf{VG}; \; \mathbf{G}(0) = \mathbf{0} \qquad (21)$$

$$Gr = 100 \frac{I_{1,0}}{M_1^{PS}(t=0)} \qquad (22)$$

Very detailed simulation studies for the grafting of vinyl polymers onto PE using kMC were presented by Hernández-Ortiz et al. [41,259–261]. Some of the key reactions considered in this study are shown in Figure 7 and the modeling strategy is summarized in Figure 8 [259].

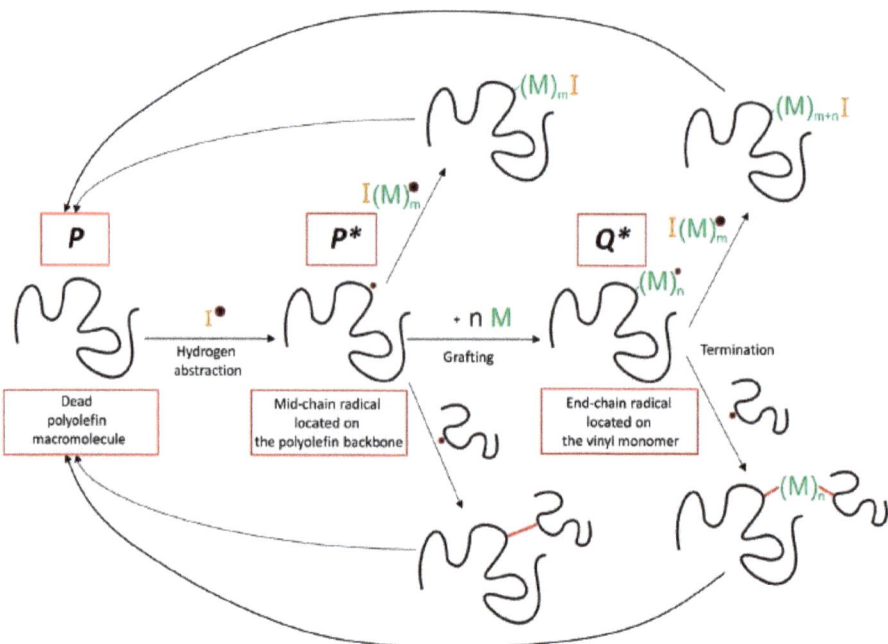

Figure 7. Some of the key reactions present in the grafting of polyolefins with vinyl monomer M. Reprinted with permission from Hernández-Ortiz et al., AIChE J., 63(11), 4944–4961 [259]. Copyright 2017 John Wiley and Sons, New York.

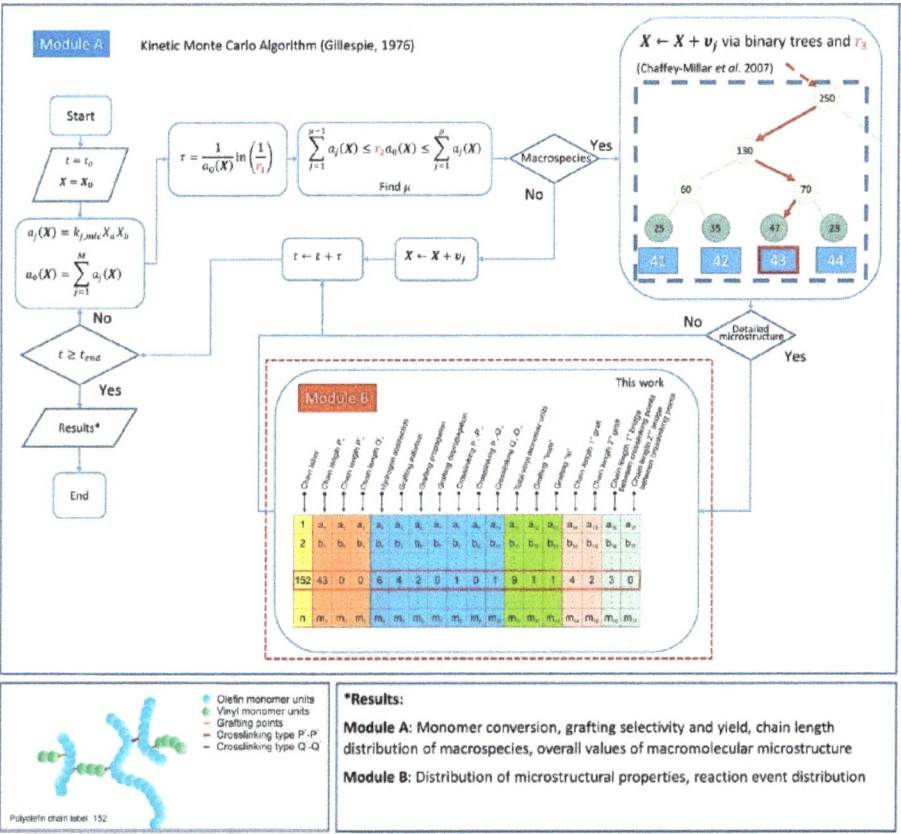

Figure 8. Simulation approach using kMC for the grafting of polyolefins with vinyl monomer M. Reprinted with permission from Hernández-Ortiz et al., AIChE J., 63(11), 4944–4961 [259]. Copyright 2017 John Wiley and Sons, New York.

6. Nomenclature, Symbols, Abbreviations, and Chemical Structures

As mentioned earlier, symbols and abbreviations used in this contribution are defined and summarized in Tables 16–19.

7. Conclusions

Polymer grafting is a useful route for the synthesis of materials with interesting mechanical, thermal, dilute solution, and melt properties, and the ability to be compatible with otherwise incompatible mixtures. Systematic studies on polymer chemistry and characterization, and even modeling studies of polymer grafting started in about the 1950s. The interest in the grafting of natural biopolymers has escalated in the last 20 or so years due to environmental and sustainability issues.

Most of the studies on modeling of polymer grafting have focused on insertion of active sites on the backbone by CTP and growth of grafts by FRP or variants of FRP, such as RDRP. The monomers of major use for polymer grafting purposes are acrylic monomers (MMA, NIPAAM, AM, AA, BA), styrene (STY) and vinyl ethers [140,184]. When fine control of grafted structures is not required, polymer grafting by FRP may be enough. However, when precise control of polymer grafts (size and separation) is required, RDRP techniques, such as NMP, ATRP, and RAFT, are more adequate. RDRP polymer grafting techniques usually proceed by the "grafting-from" route [140,180,184]. One disadvantage of RDRP techniques is that they require longer reaction times. For instance, polymer grafting by RAFT polymerization lasts from 8 to 48 h, plus the time required to prepare the related microcontrollers, which in many cases includes an esterification step through Steglich or anhydride procedures. Polymer grafting by ATRP takes from 24 h to several days.

Polymer grafting by ROP procedures using L-lactide (L-LA) and ε-caprolactone (CL) for the synthesis of poly(l-lactic acid) and poly(ε-caprolactone) polymer grafts is gaining importance [181,182,189]. Other monomers used are 2-ethyl-2-oxazoline, to produce PEOX, and ethylene glycol, to produce PEG. These reactions are commonly conducted at temperatures higher than 80 °C, which complicates solvent selection, when using metal catalysts such as $Sn(Oct)_2$. Solvents should dissolve monomer and polymer, perform adequately at the selected temperatures, and show "green" characteristics. DMF, DMSO, p-dioxane, and toluene are some of the solvents most commonly used for polymer grafting by ROP. However, the recent advent of metal-free and organocatalyzed ROP has facilitated the polymerization at room temperature. For example, poly(lactide) materials can be made by ROP of L-LA at ambient conditions in the presence of 1,5,7-triazabicyclo[4.4.0]dec-5-ene (TBD) and 1,8-diazabicyclo[5.4.0]undec-7-ene (DBU) [272].

Polymer grafting is also important in the synthesis of "dendrigraft copolymers." A large variety of heterogeneous dendrigraft copolymer architectures with core-shell and core-shell-corona morphologies can be produced, at significantly lower costs than for conventional dendrimer syntheses [18].

As observed in Table 14, the modeling of polymer grafting has focused on CTP or site formation by irradiation with FRP chemistry, in conventional flasks, stirred-tank reactors, or extruders. Other chemical routes have been addressed using semi-empirical approaches only. Therefore, there is still much to do in and contribute to this area.

Author Contributions: E.V.-L. and A.P. conceived the original idea (with discussions with A.R.-A., J.P.-A., M.G.H.-L., and A.M.). E.V.-L., M.G.H.-L., and A.M. assured funding acquisition through a project where experimental and theoretical work on polymer grafting of biopolymers from lignocellulosic biomasses has been carried out and provided background and motivation for this contribution. M.Á.V.-H., G.S.C.-D., A.R.-A., and E.V.-L. carried out critical literature reviews on different topics of the review. M.Á.V.-H., A.R.-A., and E.V.-L. wrote the original draft of the paper; E.V.-L., A.M., A.P., and Y.M. read and corrected different versions of the manuscript and provided extra references and discussion points. All authors have read and agreed to the published version of the manuscript.

Funding: This research was funded by: (a) Consejo Nacional de Ciencia y Tecnología (CONACYT, México), PhD scholarships granted to M.A.V.-H. and G.S.C.-D.; (b) DGAPA-UNAM, Projects PAPIIT IG100718, IV100119, TA100818, and TA102120, granted to E.V.-L.—the first two—and to A.R.-A.—the last two; and PASPA sabbatical support to E.V.-L. while at the University of Waterloo, in Ontario, Canada; (c) Facultad de Química, UNAM, research funds granted to E.V.-L. (PAIP 5000-9078) and A.R.-A. (PAIP 5000-9167); (d) NSERC funding to A.P.; and (e) the Department of Chemical Engineering, University of Waterloo, Canada, partial sabbatical support to E.V.-L. with research funds from A.P. No funding was received for APC.

Institutional Review Board Statement: Not applicable.

Informed Consent Statement: Not applicable.

Data Availability Statement: Data sharing is not applicable to this article.

Conflicts of Interest: The authors declare no conflict of interest. The funders had no role in the design of the study; in the collection, analyses, or interpretation of data; in the writing of the manuscript, or in the decision to publish the results.

References

1. Hadjichristidis, N.; Pitsikalis, M.; Iatrou, H.; Driva, P.; Chatzichris, M.; Sakellariou, G.; Lohse, D. Graft copolymers. In *Encyclopedia of Polymer Science and Technology*, 2nd ed.; Matyjaszewski, K., Ed.; John Wiley & Sons: Hoboken, NJ, USA, 2010; pp. 1–38, ISBN 978-047-144-026-0.

2. Slagman, S.; Zuilhof, H.; Franssen, M.C.R. Laccase-Mediated Grafting on Biopolymers and Synthetic Polymers: A Critical Review. *ChemBioChem* **2017**, *19*, 288–311. [CrossRef]
3. Stannett, V.T. Block and graft copolymerization. In *Journal of Polymer Science: Polymer Letters*, 1st ed.; Ceresa, R.J., Ed.; John Wiley & Sons: Hoboken, NJ, USA, 1973; Volume 1, pp. 669–670. [CrossRef]
4. Meier, D.J. Theory of block copolymers. I. Domain formation in A-B block copolymers. *J. Polym. Sci. C Polym. Symp.* **1969**, *26*, 81–98. [CrossRef]
5. Helfand, E.; Block Copolymer Theory. III. Statistical Mechanics of the Microdomain Structure. *Macromolecules* **1975**, *8*, 552–556. [CrossRef]
6. Helfand, E.; Wasserman, Z.R. Block Copolymer Theory. 4. Narrow Interphase Approximation. *Macromolecules* **1976**, *9*, 879–888. [CrossRef]
7. Helfand, E. Block copolymers, polymer-polymer interfaces, and the theory of inhomogeneous polymers. *Acc. Chem. Res.* **1975**, *8*, 295–299. [CrossRef]
8. Blanchette, J.A.; Nielsen, L.E. Characterization of graft polymers. *J. Polym. Sci.* **1956**, *20*, 317–326. [CrossRef]
9. Merret, F.M. Graft polymers with preset molecular configurations. *J. Polym. Sci.* **1957**, *24*, 467–477. [CrossRef]
10. Gluckman, M.S.; Kampf, M.J.; O'brien, L.J.; Fox, T.G.; Graham, R.K. Graft copolymers from polymers having pendant mercaptan groups. II. Synthesis and characterization. *J. Polym. Sci.* **1959**, *37*, 411–423. [CrossRef]
11. Miller, M.L. Block and graft polymers I. Graft polymers from acrylamide and acrylonitrile. *Can. J. Chem.* **1957**, *36*, 303–308. [CrossRef]
12. Beevers, R.B.; White, E.F.T.; Brown, L. Physical properties of vinyl polymers. Part 3.—X-ray scattering in block, random and graft methyl methacrylate + acrylonitrile copolymers. *Trans. Faraday Soc.* **1960**, *56*, 1535–1541. [CrossRef]
13. Oster, G.; Oster, G.K.; Moroson, H. Ultraviolet induced crosslinking and grafting of solid high polymers. *J. Polym. Sci.* **1959**, *XXXIV*, 671–684. [CrossRef]
14. Kobayashi, Y. Gamma-ray–induced graft copolymerization of styrene onto cellulose and some chemical properties of the grafted polymer. *J. Polym. Sci.* **1961**, *51*, 359–372. [CrossRef]
15. Bridgeford, D.J. Catalytic Deposition and Grafting of Olefin Polymers into Cellulosic Materials. *Ind. Eng. Chem. Prod. Res. Dev.* **1962**, *1*, 45–52. [CrossRef]
16. Huang, R.Y.-M.; Immergut, B.; Immergut, E.H.; Rapson, W.H. Grafting vinyl polymers onto cellulose by high energy radiation. I. High energy radiation-induced graft copolymerization of styrene onto cellulose. *J. Polym. Sci. A Gen. Pap.* **1963**, *1*, 1257–1270. [CrossRef]
17. McManus, N.; Zhu, S.-H.; Tzoganakis, C.; Penlidis, A. Grafting of ethylene-ethyl acrylate-maleic anhydride terpolymer with amino-terminated polydimethylsiloxane during reactive processing. *J. Appl. Polym. Sci.* **2006**, *101*, 4230–4237. [CrossRef]
18. Cadena, L.-E.; Gauthier, M. Phase-segregated dendrigraft copolymer architectures. *Polymers* **2010**, *2*, 596–622. [CrossRef]
19. Aridi, T.; Gauthier, M. Chapter 6. Arborescent polymers with a mesoscopic scale. In *Complex Macromolecular Architectures: Synthesis, Characterization, and Self-Assembly*, 1st ed.; Hadjichristidis, N., Hirao, A., Tezuka, Y., Du Prez, F., Eds.; John Wiley & Sons: Hoboken, NJ, USA, 2011; pp. 169–194. ISBN 978-047-082-514-3.
20. Moingeon, F.; Wu, Y.; Cadena-Sánchez, L.; Gauthier, M. Synthesis of arborescent styrene homopolymers and copolymers from epoxidized substrates. *J. Polym. Sci. A Polym. Chem.* **2012**, *50*, 1819–1826. [CrossRef]
21. Whitton, G.; Gauthier, M. Arborescent polypeptides from γ-benzyl l-glutamic acid. *J. Polym. Sci. A Polym. Chem.* **2013**, *51*, 5270–5279. [CrossRef]
22. Aridi, T.; Gauthier, M. Synthesis of arborescent polymers by click grafting. *Mater. Res. Soc. Symp. Proc.* **2014**, *1613*, 23–31. [CrossRef]
23. Dockendorff, J.; Gauthier, M. Synthesis of arborescent polystyrene-g-[poly(2-vinylpyridine)-b- polystyrene] core-shell-corona copolymers. *J. Polym. Sci. A Polym. Chem.* **2014**, *52*, 1075–1085. [CrossRef]
24. Whitton, G.; Gauthier, M. Arborescent micelles: Dendritic poly(γ-benzyl l-glutamate) cores grafted with hydrophilic chain segments. *J. Polym. Sci. A Polym. Chem.* **2016**, *54*, 1197–1209. [CrossRef]
25. Gauthier, M.; Whitton, G. Arborescent unimolecular micelles: Poly(γ-benzyl L-glutamate) core grafted with a hydrophilic shell by copper(I)-catalyzed azide-alkyne cycloaddition coupling. *Polymers* **2017**, *9*, 540. [CrossRef]
26. Gauthier, M.; Aridi, T. Synthesis of arborescent polystyrene by "click" grafting. *J. Polym. Sci. A Polym. Chem.* **2019**, *57*, 1730–1740. [CrossRef]
27. Roy, D.; Semsarilar, M.; Guthrie, J.T.; Perrier, S. Cellulose modification by polymer grafting: A review. *Chem. Soc. Rev.* **2009**, *38*, 2046–2064. [CrossRef] [PubMed]
28. Wohlhauser, S.; Delepierre, G.; Labet, M.; Morandi, G.; Thielemans, W.; Weder, C.; Zoppe, J.O. Grafting Polymers from Cellulose Nanocrystals: Synthesis, Properties, and Applications. *Macromolecules* **2018**, *51*, 6157–6189. [CrossRef]
29. Jenkins, D.W.; Hudson, S.M. Review of vinyl graft copolymerization featuring recent advances toward controlled radical-based reactions and illustrated with chitin/chitosan trunk polymers. *Chem. Rev.* **2001**, *101*, 3245–3274. [CrossRef] [PubMed]
30. Thakur, V.K.; Thakur, M.K. Recent Advances in Graft Copolymerization and Applications of Chitosan: A Review. *ACS Sustain. Chem. Eng.* **2014**, *2*, 2637–2652. [CrossRef]
31. Kaur, L.; Gupta, G.D. A review on microwave assisted grafting of polymers. *Int. J. Pharm. Sci. Res.* **2017**, *8*, 422–426. [CrossRef]

32. Brodin, M.; Vallejos, M.; Opedal, M.T.; Area, M.C.; Chinga-Carrasco, G. Lignocellulosics as sustainable resources for production of bioplastics—A review. *J. Clean. Prod.* **2017**, *162*, 646–664. [CrossRef]
33. Niphadkar, S.; Bagade, P.; Ahmed, S. Bioethanol production: Insight into past, present and future perspectives. *Biofuels* **2018**, *9*, 229–238. [CrossRef]
34. Banerjee, J.; Singh, R.; Vijayaraghavan, R.; MacFarlane, D.; Patti, A.F.; Arora, A. Bioactives from fruit processing wastes: Green approaches to valuable chemicals. *Food Chem.* **2017**, *225*, 10–22. [CrossRef]
35. Neuling, U.; Kaltschmitt, M. Review of Biofuel Production—Feedstock, Processes and Markets. *J. Oil Palm Res.* **2019**, *29*, 137–167. [CrossRef]
36. Vega-Hernández, M.Á.; Rosas-Aburto, A.; Vivaldo-Lima, E.; Vázquez-Torres, H.; Cano-Díaz, G.S.; Pérez-Salinas, P.; Hernández-Luna, M.G.; Alcaraz-Cienfuegos, J.; Zolotukhin, M.G. Development of polystyrene composites based on blue agave bagasse by in situ RAFT polymerization. *J. Appl. Polym. Sci.* **2019**, *136*, 47089. [CrossRef]
37. Farhat, W.; Venditti, R.; Ayoub, A.; Prochazka, F.; Fernández-de-Alba, C.; Mignard, N.; Taha, M.; Becquart, F. Towards thermoplastic hemicellulose: Chemistry and characteristics of poly-(ε-caprolactone) grafting onto hemicellulose backbones. *Mater. Des.* **2018**, *153*, 298–307. [CrossRef]
38. Sun, Y.; Ma, Z.; Xu, X.; Liu, X.; Liu, L.; Huang, G.; Liu, L.; Wang, H.; Song, P. Grafting Lignin with Bioderived Polyacrylates for Low-Cost, Ductile, and Fully Biobased Poly(lactic acid) Composites. *ACS Sustain. Chem. Eng.* **2020**, *8*, 2267–2276. [CrossRef]
39. Fox, T.G.; Gluckman, M.S.; Gornick, F.; Graham, R.K.; Gratch, S. Graft copolymers from polymers having pendant mercaptan groups. I. Kinetic considerations. *J. Polym. Sci.* **1959**, *XXXVII*, 397–409. [CrossRef]
40. Zimmerman, J. Molecular weight distributions of vinyl polymers grafted to a solid polymeric substrate by irradiation (theoretical). *J. Polym. Sci.* **1960**, *XLIV*, 107–116. [CrossRef]
41. Hernández-Ortiz, J.C.; Van Steenberge, P.H.M.; Duchateau, J.N.E.; Toloza, C.; Schreurs, F.; Reyniers, M.-F.; Marin, G.B.; D'hooge, D.R. A two-phase stochastic model to describe mass transport and kinetics during reactive processing of polyolefins. *Chem. Eng. J.* **2019**, *377*, 119980. [CrossRef]
42. Gandhi, A.; Verma, S.; Imam, S.S.; Vyas, M. A review on techniques for grafting of natural polymers and their applications. *Plant Arch* **2019**, *19*, 972–978.
43. Wei, L.; McDonald, A.G. A Review on Grafting of Biofibers for Biocomposites. *Materials* **2016**, *9*, 303. [CrossRef]
44. De Jesús Muñoz Prieto, E.; Rivas, B.; Sánchez, J. Natural polymer grafted with syntethic monomer by microwave for water treatment—A review. *Cienc. Desarro.* **2012**, *4*, 219–240. [CrossRef]
45. Barsbay, M.; Güven, O. A short review of radiation-induced raft-mediated graft copolymerization: A powerful combination for modifying the surface properties of polymers in a controlled manner. *Radiat. Phys. Chem.* **2009**, *78*, 1054–1059. [CrossRef]
46. Francis, R.; Joy, N.; Aparna, E.P.; Vijayan, R. Polymer Grafted Inorganic Nanoparticles, Preparation, Properties, and Applications: A Review. *Polym. Rev.* **2014**, *54*, 268–347. [CrossRef]
47. Garcia-Valdez, O.; Champagne, P.; Cunningham, M.F. Graft modification of natural polysaccharides via reversible deactivation radical polymerization. *Prog. Polym. Sci.* **2018**, *76*, 151–173. [CrossRef]
48. Sun, H.; Yang, L.; Thompson, M.P.; Schara, S.; Cao, W.; Choi, W.C.; Hu, Z.; Zang, N.; Tan, W.; Gianneschi, N.C. Recent Advances in Amphiphilic Polymer–Oligonucleotide Nanomaterials via Living/Controlled Polymerization Technologies. *Bioconjugate Chem.* **2019**, *30*, 1889–1904. [CrossRef] [PubMed]
49. Chung, T.C. Synthesis of functional polyolefin copolymers with graft and block structures. *Prog. Polym. Sci.* **2002**, *27*, 39–85. [CrossRef]
50. Zhou, T.; Zhu, Y.; Li, X.; Liu, X.; Yeung, K.W.K.; Wu, S.; Wang, X.; Cui, Z.; Yang, X.; Chu, P.K. Surface functionalization of biomaterials by radical polymerization. *Prog. Mater. Sci.* **2016**, *83*, 191–235. [CrossRef]
51. Ayyavoo, J.; Nguyen, T.P.N.; Jun, B.-M.; Kim, I.-C.; Kwon, Y.N. Protection of polymeric membranes with antifouling surfacing via surface modifications. *Colloids Surf. A Physicochem. Eng. Asp.* **2016**, *506*, 190–201. [CrossRef]
52. Weber, C.; Hoogenboom, R.; Schubert, U.S. Temperature responsive bio-compatible polymers based on poly(ethylene oxide) and poly(2-oxazoline)s. *Prog. Polym. Sci.* **2012**, *37*, 686–714. [CrossRef]
53. Sun, H.; Choi, W.; Zang, N.; Battistella, C.; Thompson, M.P.; Cao, W.; Zhou, X.; Forman, C.; Gianneschi, N.C. Bioactive Peptide Brush Polymers via Photoinduced Reversible-Deactivation Radical Polymerization. *Angew. Chem. Int. Ed.* **2019**, *58*, 17359–17364. [CrossRef]
54. Kumar, R.; Sharma, R.K.; Singh, A.P. Grafted cellulose: A bio-based polymer for durable applications. *Polym. Bull.* **2018**, *75*, 2213–2242. [CrossRef]
55. Sharma, S.; Kumar, A. *Lignin. Biosynthesis and Transformation for Industrial Applications*; Springer Series on Polymer and Composite Materials; Springer Nature: Cham, Switzerland, 2020; pp. 1–252. ISBN 978-303-040-663-9.
56. Mourya, V.K.; Inamdar, N.N. Chitosan-modifications and applications. *React. Funct. Polym.* **2008**, *68*, 1013–1051. [CrossRef]
57. Argüelles-Monal, W.M.; Lizardi-Mendoza, J.; Fernández-Quiroz, D.; Recillas-Mota, M.T.; Montiel-Herrera, M. Chitosan Derivatives: Inducing new functionalities with a controlled molecular architecture for innovative materials. *Polymers* **2018**, *10*, 342. [CrossRef]
58. Kurita, K. Controlled functionalization of the polysaccharide chitin. *Prog. Polym. Sci.* **2001**, *26*, 1921–1971. [CrossRef]
59. Lele, V.V.; Kumari, S.; Niju, H. Syntheses, characterization and applications of graft copolymers of sago starch. *Starch* **2018**, *70*, 1700133. [CrossRef]

60. Radhakrishnan, B.; Ranjan, R.; Brittain, W.J. Surface initiated polymerization from silica nanoparticles. *Soft Matter* **2006**, *2*, 386–396. [CrossRef] [PubMed]
61. Foster, J.C.; Radzinsky, S.C.; Matson, J.B. Graft polymer synthesis by RAFT transfer-to. *J. Polym. Sci. A Polym. Chem.* **2017**, *55*, 2865–2876. [CrossRef]
62. Bhattacharya, A. Radiation and industrial polymers. *Prog. Polym. Sci.* **2000**, *25*, 371–401. [CrossRef]
63. Desmet, T.; Morent, R.; De Geyter, N.; Leys, C.; Schacht, E.; Dubruel, P. Nonthermal plasma technology as a versatile strategy for polymeric biomaterials surface modification: A review. *Biomacromolecules* **2009**, *10*, 2351–2378. [CrossRef]
64. Ngo, T.H.A.; Tran, D.T.; Dinh, C.H. Surface photochemical graft polymerization of acrylic acid onto polyamide thin film composite membranes. *J. Appl. Polym. Sci.* **2017**, *134*, 44418. [CrossRef]
65. Singh, V.; Tiwari, A.; Tripathi, D.N.; Sanghi, R. Microwave assisted synthesis of guar-g-polyacrylamide. *Carbohydr. Polym.* **2004**, *58*, 1–6. [CrossRef]
66. Bhattacharya, A.; Misra, B.N. Grafting: A versatile means to modify polymers technics, factors and applications. *Prog. Polym. Sci.* **2004**, *29*, 767–814. [CrossRef]
67. Sosnik, A.; Gotelli, G.; Abraham, G.A. Microwave-assisted polymer synthesis (MAPS) as a tool in biomaterials science: How new and how powerful. *Prog. Polym. Sci.* **2011**, *36*, 1050–1078. [CrossRef]
68. Kumar, D.; Pandey, J.; Raj, V.; Kumar, P. A review on the modification of polysaccharide through graft copolymerization for various potential applications. *Open Med. Chem. J.* **2017**, *11*, 109–126. [CrossRef]
69. Fan, G.; Zhao, J.; Zhang, Y.; Guo, Z. Grafting modification of kevlar fiber using horseradish peroxidase. *Polym. Bull.* **2006**, *56*, 507–515. [CrossRef]
70. Cannatelli, M.D.; Ragauskas, A.J. Conversion of lignin into value-added materials and chemicals via laccase-assisted copolymerization. *Appl. Microbiol. Biotechnol.* **2016**, *100*, 8685–8691. [CrossRef] [PubMed]
71. Ran, J.; Wu, L.; Zhang, Z.; Xu, T. Atom transfer radical polymerization (ATRP): A versatile and forceful tool for functional membranes. *Prog. Polym. Sci.* **2014**, *39*, 124–144. [CrossRef]
72. Crawford, D.E. Extrusion-back to the future: Using an established technique to reform automated chemical synthesis. *Beilstein J. Org. Chem.* **2017**, *13*, 65–75. [CrossRef]
73. Moad, G. The synthesis of polyolefin graft copolymers by reactive extrusion. *Prog. Polym. Sci.* **1999**, *24*, 81–142. [CrossRef]
74. Moad, G. Chemical modification of starch by reactive extrusion. *Prog. Polym. Sci.* **2011**, *36*, 218–237. [CrossRef]
75. Russell, K.E. Free radical graft polymerization and copolymerization at higher temperatures. *Prog. Polym. Sci.* **2002**, *27*, 1007–1038. [CrossRef]
76. Monties, B. *Les Polymères Végétaux: Polymères Pariétaux et Alimentaires non Azotés*, 1st ed.; Gauthier-Villars: Paris, France, 1980; ISBN 978-204-010-480-1.
77. Casarrubias-Cervantes, R.A. Análisis Fisicoquímico de Procesos de Pretratamiento de Materiales Lignocelulósicos para su Uso en Polímeros Conductores. Bachelor Degree, Facultad de Química—Universidad Nacional Autónoma de México, Ciudad Universitaria, CDMX, 2019, Biblioteca Digital UNAM. Available online: http://132.248.9.195/ptd2019/abril/0788572/Index.html (accessed on 25 May 2020).
78. Koshijima, T.; Muraki, E. Radiation Grafting of Methyl Methacrylate onto Lignin. *J. Jpn. Wood Res. Soc.* **1964**, *10*, 110–115.
79. Koshijima, T.; Muraki, E. Degradation of Lignin-Methyl Metacrylate Graft Copolymer by γ-Ray Irradiation. *J. Jpn. Wood Res. Soc.* **1964**, *10*, 116–119.
80. Koshijima, T.; Timell, T.E. Factors Affecting Number Average Molecular Weights Determination of Hardwood Xylan. *J. Jpn. Wood Res. Soc.* **1966**, *12*, 166–172.
81. Koshijima, T.; Muraki, E. Solvent Effects upon Radiation-Induced Graft-copolymerization of Styrene onto Lignin. *J. Jpn. Wood Res. Soc.* **1966**, *12*, 139.
82. Koshijima, T. Oxidation of Lignin-Styrene Graft polymer. *J. Jpn. Wood Res. Soc.* **1966**, *12*, 114.
83. Koshijima, T.; Muraki, E. Radial Grafting on Lignin (II). Grafting of styrene into Lignin by Initiators. *J. Jpn. Wood Res. Soc.* **1967**, *13*, 355–358.
84. Koshijima, T.; Muraki, E.; Naito, K.; Adachi, K. Radical Grafting on Lignin. IV. Semi-Conductive Properties of Lignin-Styrene Graftpolymer. *J. Jpn. Wood Res. Soc.* **1968**, *14*, 52–54.
85. Meister, J.J. Modification of Lignin. *J. Macromol. Sci. Polymer Rev.* **2002**, *42*, 235–289. [CrossRef]
86. Hon, D.N.S. *Chemical Modification of Lignocelulosic Materials*, 1st ed.; Marcel Dekker, Inc.: New York, NY, USA, 1996; ISBN 978-082-479-472-9.
87. McDowall, D.J.; Gupta, B.S.; Stannett, V.T. Grafting of vinyl monomers to cellulose by ceric ion initiation. *Prog. Polym. Sci.* **1984**, *10*, 1–50. [CrossRef]
88. Bhattacharyya, S.N.; Maldas, D. Graft copolymerization onto cellulosics. *Prog. Polym. Sci.* **1984**, *10*, 171–270. [CrossRef]
89. Hon, D.N.S. *Graft Copolymerization of Lignocellulosic Fibers*; ACS Symposium Series 187; American Chemical Society: Washington, DC, USA, 1982; ISBN 978-084-120-721-9.
90. Mansour, O.Y.; Nagaty, A. Grafting of synthetic polymers to natural polymers by chemical processes. *Prog. Polym. Sci.* **1985**, *11*, 91–165. [CrossRef]
91. Feldman, D.; Lacasse, M.; Bernaczuk, L.M. Lignin-polymer sytems and some applications. *Prog. Polym. Sci.* **1986**, *12*, 271–299. [CrossRef]

92. Matyjaszewski, K.; Möller, M. Celluloses and polyoses/hemicelluloses. In *Polymer Science: A Comprehensive Reference*; Elsevier: Amsterdam, The Netherlands, 2012; ISBN 978-008-087-862-1.
93. Pantelakis, S.; Tserpes, K. *Revolutionizing Aircraft Materials and Processes*; Springer Nature AG: Cham, Switzerland, 2020; ISBN 978-303-035-346-9.
94. Rol, F.; Belgacem, M.N.; Gandinia, A.; Bras, J. Recent advances in surface-modified cellulose nanofibrils. *Prog. Polym. Sci.* **2019**, *88*, 241–264. [CrossRef]
95. Zahran, M.K.; Morsy, M.; Mahmoud, R. Grafting of acrylic monomers onto cotton fabric using an activated cellulose thiocarbonate–azobisisobutyronitrile redox system. *J. Appl. Polym. Sci.* **2003**, *91*, 1261–1274. [CrossRef]
96. Chauhan, G.S.; Lal, H.; Sharma, R.; Sarwade, B.D. Grafting of a styrene–acrylonitrile binary monomer mixture onto cellulose extracted from pine needles. *J. Appl. Polym. Sci.* **2001**, *83*, 2000–2007. [CrossRef]
97. Sabaa, M.W.; Mokhtar, S.M. Chemically induced graft copolymerization of itaconic acid onto cellulose fibers. *Polym. Test.* **2002**, *21*, 337–343. [CrossRef]
98. Gupta, K.C.; Sahoo, S. Grafting of N,N'-methylenebisacrylamide onto cellulose using Co(III)-acetylacetonate complex in aqueous medium. *J. Appl. Polym. Sci.* **2000**, *76*, 906–912. [CrossRef]
99. Gupta, K.C.; Sahoo, S. Co(III) acetylacetonate-complex-initiated grafting of N-vinyl pyrrolidone on cellulose in aqueous media. *J. Appl. Polym. Sci.* **2001**, *81*, 2286–2296. [CrossRef]
100. Gupta, K.C.; Khandekar, K. Temperature-Responsive Cellulose by Ceric(IV) Ion-Initiated Graft Copolymerization of N-Isopropylacrylamide. *Biomacromolecules* **2003**, *4*, 758–765. [CrossRef]
101. Gupta, K.C.; Khandekar, K. Graft copolymerization of acrylamide–methylacrylate comonomers onto cellulose using ceric ammonium nitrate. *J. Appl. Polym. Sci.* **2002**, *86*, 2631–2642. [CrossRef]
102. Gupta, K.C.; Khandekar, K. Graft copolymerization of acrylamide onto cellulose in presence of comonomer using ceric ammonium nitrate as initiator. *J. Appl. Polym. Sci.* **2006**, *101*, 2546–2558. [CrossRef]
103. Gupta, K.C.; Sahoo, S. Graft Copolymerization of Acrylonitrile and Ethyl Methacrylate Comonomers on Cellulose Using Ceric Ions. *Biomacromolecules* **2001**, *2*, 239–247. [CrossRef]
104. Gupta, K.C.; Sahoo, S.; Khandekar, K. Graft Copolymerization of Ethyl Acrylate onto Cellulose Using Ceric Ammonium Nitrate as Initiator in Aqueous Medium. *Biomacromolecules* **2002**, *3*, 1087–1094. [CrossRef] [PubMed]
105. Gupta, K.C.; Khandekar, K. Ceric(IV) ion-induced graft copolymerization of acrylamide and ethyl acrylate onto cellulose. *Polym. Int.* **2005**, *55*, 139–150. [CrossRef]
106. Toledano-Thompson, T.; Loría-Bastarrachea, M.I.; Aguilar-Vega, M.J. Characterization of henequen cellulose microfibers treated with an epoxide and grafted with poly(acrylic acid). *Carbohydr. Polym.* **2005**, *62*, 67–73. [CrossRef]
107. Mansour, O.Y.; Nagieb, Z.A.; Basta, A.H. Graft polymerization of some vinyl monomers onto alkali-treated cellulose. *J. Appl. Polym. Sci.* **1991**, *43*, 1147–1158. [CrossRef]
108. Semsarilar, M.; Ladmiral, V.; Perrier, S. Synthesis of a cellulose supported chain transfer agent and its application to RAFT polymerization. *J. Polym. Sci. A Polym. Chem.* **2010**, *48*, 4361–4365. [CrossRef]
109. Cankaya, N.; Temüz, M. Characterization and monomer reactivity ratios of grafted cellulose with n-(4-nitrophenyl)acrylamide and methyl methacrylate by atom transfer radical polymerization. *Cell. Chem. Technol.* **2012**, *46*, 551–558.
110. Thakur, V.K.; Thakur, M.K.; Gupta, R.K. Rapid synthesis of graft copolymers from natural cellulose fibers. *Carbohydr. Polym.* **2013**, *98*, 820–828. [CrossRef]
111. Routray, C.; Tosh, B. Graft copolymerization of methyl methacrylate (mma) onto cellulose acetate in homogeneous medium: Effect of solvent, initiator and homopolymer inhibitor. *Cell. Chem. Technol.* **2013**, *47*, 171–190.
112. Cankaya, N.; Temüz, M.M. Monomer reactivity ratios of cellulose grafted with N-cyclohexylacrylamide and methyl methacrylate by atom transfer radical polymerization. *Cell. Chem. Technol.* **2014**, *48*, 209–215.
113. Popescu, O.; Dunca, S.; Grigoriu, A. Antibacterial action of silver applied on cellulose fibers grafted with monochlorotriazinyl-β-cyclodextrin. *Cell. Chem. Technol.* **2013**, *47*, 247–255.
114. Popescu, O.; Grigoriu, A.; Diaconescu, R.M.; Vasluianu, E. Optimization of the cellulosic materials functionalization with monochlorotriazinyl-β-cyclodextrin in basic medium. *Ind. Textilá* **2012**, *63*, 68–75.
115. Sun, Z.; Chen, F. Homogeneous grafting copolymerization of methylmethacrylate onto cellulose using ammonium persulfate. *Cell. Chem. Technol.* **2014**, *48*, 217–223.
116. Dai, L.; Shen, Y.; Li, D.; Xiao, S.; He, J. Cellulose-graft-poly(l-lactide) as a degradable drugdelivery system: Synthesis, degradation and drug release. *Cell. Chem. Technol.* **2014**, *48*, 237–245.
117. Xiaoming, S.; Songlin, W.; Shanshan, G.; Fushan, C.; Fusheng, L. Study on grafting copolymerization of methyl methacrylate onto cellulose under heterogeneous conditions. *Cell. Chem. Technol.* **2016**, *50*, 65–70.
118. Yin, Y.; Jiang, T.X.; Wang, H.; Gao, W. Modification of cellulose nanocrystal via SI-ATRP of styrene and themechanism of its reinforcement of polymethylthacrylate. *Carbohydr. Polym.* **2016**, *142*, 206–212. [CrossRef] [PubMed]
119. Paula, E.L.; Roig, F.; Mas, A.; Habas, J.P.; Mano, V.; Vargas Pereira, F.; Robin, J.J. Effect of surface-grafted cellulose nanocrystals on the thermal and mechanical properties of PLLA based nanocomposites. *Eur. Polym. J.* **2016**, *84*, 173–187. [CrossRef]
120. Zhao, C.; Li, J.; He, B.; Zhao, L. Fabrication of hydrophobic biocomposite by combining cellulosic fibers with polyhydroxyalkanoate. *Cellulose* **2017**, *24*, 2265–2274. [CrossRef]

121. Badawy, S.M. Functional cellulosic filter papers prepared by radiation-induced graft copolymerization for chelation of rare earth elements. *Cell. Chem. Technol.* **2017**, *51*, 551–558.
122. Çankaya, N.; Temüz, M.M.; Yakuphanoglu, F. Grafting of some monomers onto cellulose by atom transfer radical polymerization. Electrical conductivity and thermal properties of resulting copolymers. *Cell. Chem. Technol.* **2018**, *52*, 19–26.
123. Müssig, J.; Kelch, M.; Gebert, B.; Hohe, J.; Luke, M.; Bahners, T. Improvement of the fatigue behaviour of cellulose/polyolefin composites using photo-chemical fibre surface modification bio-inspired by natural role models. *Cellulose* **2020**, *27*, 5815–5827. [CrossRef]
124. Chen, Y.; Yu, Z.; Han, Y.; Yang, S.; Fan, D.; Li, G.; Wang, S. Combination of water soluble chemical grafting and gradient freezing to fabricate elasticity enhanced and anisotropic nanocellulose aerogels. *Appl. Nanosci.* **2020**, *10*, 411–419. [CrossRef]
125. Eraghi Kazzaz, A.; Hosseinpour Feizi, Z.; Fatehi, P. Grafting strategies for hydroxy groups of lignin for producing materials. *Green Chem.* **2019**, *21*, 5714–5752. [CrossRef]
126. Abe, A.; Dusek, K.; Kobayashi, S. *Biopolymers. Lignin, Proteins, Bioactive Nanocomposites*, 1st ed.; Springer: Heidelberg/Berlin, Germany, 2010; Volume 232, ISBN 978-364-213-630-6.
127. Huang, J.; Fu, S.; Gan, L. *Lignin Chemistry and Applications*, 1st ed.; Elsevier: Amsterdam, The Netherlands, 2019; ISBN 978-012-813-963-9.
128. Marton, J. *Lignin. Structure and Reactions*, 1st ed.; Advances in Chemistry Series 59; American Chemical Society: Washington, DC, USA, 1966; ISBN 978-084-122-239-7.
129. Glasser, W.G.; Sarkanen, S. *Lignin. Properties and Materials*, 1st ed.; ACS Symposium Series 397; American Chemical Society: Washington, DC, USA, 1989; ISBN 978-084-121-248-0.
130. Lewis, N.G.; Sarkanen, S. *Lignin and Lignan Biosynthesis*; ACS Symposium Series 697; American Chemical Society: Washington, DC, USA, 1998; ISBN 084-123-566-X.
131. Katahira, R.; Elder, T.J.; Beckham, G.T. Chapter 1 A brief introduction to lignin structure. In *Lignin Valorization. Emerging Approaches*, 1st ed.; Beckham, G.T., Ed.; The Royal Society of Chemistry: Croydon, London, UK, 2018; pp. 1–20, ISBN 978-178-801-035-1.
132. Laurichesse, S.; Avérous, L. Chemical modification of lignins: Towards biobased polymers. *Prog. Polym. Sci.* **2014**, *39*, 1266–1290. [CrossRef]
133. Figueiredo, P.; Lintinen, K.; Hirvonen, J.T.; Kostiainen, M.A.; Santos, H.A. Properties and chemical modifications of lignin: Towards lignin-based nanomaterials for biomedical applications. *Prog. Mater. Sci.* **2018**, *93*, 233–269. [CrossRef]
134. Xu, Y.; Yuan, L.; Wang, Z.; Wilbon, P.A.; Wang, C.; Chu, F.; Tang, C. Lignin and soy oil-derived polymeric biocomposites by "grafting from" RAFT polymerization. *Green Chem.* **2016**, *18*, 4974–4981. [CrossRef]
135. Yu, Y.; Fu, S.; Song, P.; Lou, X.; Jin, Y.; Lu, F.; Wu, Q.; Ye, J. Functionalized lignin by grafting phosphorus-nitrogen improves the thermal stability and flame retardancy of polypropylene. *Polym. Degrad. Stabil.* **2012**, *97*, 541–546. [CrossRef]
136. Prieur, B.; Meub, M.; Wittermann, M.; Klein, R.; Bellayer, S.; Fontaine, G.; Bourbigot, S. Phosphorylation of lignin: Characterization and investigation of the thermal decomposition. *RSC Adv.* **2017**, *7*, 16866–16877. [CrossRef]
137. Prieur, B.; Meub, M.; Wittemann, M.; Klein, R.; Bellayer, S.; Fontaine, G.; Bourbigot, S. Phosphorylation of lignin to flame retard acrylonitrile butadiene styrene (ABS). *Polym. Degrad. Stabil.* **2016**, *127*, 32–43. [CrossRef]
138. Liu, H.; Chung, H. Lignin-Based Polymers via Graft Copolymerization. *J. Polym. Sci. A Polym. Chem.* **2017**, *55*, 3515–3528. [CrossRef]
139. Gupta, C.; Washburn, N.R. Polymer-grafted lignin surfactants prepared via Reversible Addition−Fragmentation Chain-Transfer polymerization. *Langmuir* **2014**, *30*, 9303–9312. [CrossRef]
140. Ganewatta, M.S.; Lokupitiya, H.N.; Tang, C. Lignin biopolymers in the age of controlled polymerization. *Polymers* **2019**, *11*, 1176. [CrossRef]
141. Kim, Y.S.; Kadla, J.F. Preparation of a thermoresponsive lignin-based biomaterial through atom transfer radical polymerization. *Biomacromolecules* **2010**, *11*, 981–988. [CrossRef] [PubMed]
142. Wang, J.; Yao, J.; Korich, K.; Li, S.; Ma, S.; Ploehn, H.J.; Iovine, P.M.; Wang, C.; Tang, C. Combining renewable gum rosin and lignin: Towards hydrophobic polymer composites by controlled polymerization. *J. Polym. Sci. A Polym. Chem.* **2011**, *49*, 3728–3738. [CrossRef]
143. Diao, B.; Zhang, Z.; Zhu, J.; Li, J. Biomass-based thermogelling copolymers consisting of lignin and grafted poly (N-isopropylacrylamide), poly (ethylene glycol), and poly (propylene glycol). *RSC Adv.* **2014**, *4*, 42996–43003. [CrossRef]
144. Yu, J.; Wang, J.; Wang, C.; Liu, Y.; Xu, Y.; Tang, C.; Chu, F. UV-Absorbent Lignin-Based Multi-Arm Star Thermoplastic Elastomers. *Macromol. Rapid Commun.* **2015**, *36*, 398–404. [CrossRef]
145. Gao, G.; Dallmeyer, J.I.; Kadla, J.F. Synthesis of lignin nanofibers with ionic-responsive shells: Water-expandable lignin-based nanofibrous mats. *Biomacromolecules* **2012**, *13*, 3602–3610. [CrossRef] [PubMed]
146. Hilburg, S.L.; Elder, A.N.; Chung, H.; Ferebee, R.L.; Bockstaller, M.R.; Washburn, N.R. A universal route towards thermoplastic lignin composites with improved mechanical properties. *Polymer* **2014**, *55*, 995–1003. [CrossRef]
147. Shah, T.; Gupta, C.; Ferebee, R.L.; Bockstaller, M.R.; Washburn, N.R. Extraordinary toughening and strengthening effect in polymer nanocomposites using lignin-based fillers synthesized by ATRP. *Polymer* **2015**, *72*, 406–412. [CrossRef]
148. Kai, D.; Low, Z.W.; Liow, S.S.; Abdul Karim, A.; Ye, H.; Jin, G.; Li, K.; Loh, X.J. Development of lignin supramolecular hydrogels with mechanically responsive and self-healing properties. *ACS Sustain. Chem. Eng.* **2015**, *3*, 2160–2169. [CrossRef]

149. Li, H.; Pang, Z.; Gao, P.; Wang, L. Fe (III)-catalyzed grafting copolymerization of lignin with styrene and methyl methacrylate through AGET ATRP using triphenyl phosphine as a ligand. *RSC Adv.* **2015**, *5*, 54387–54394. [CrossRef]
150. Liu, X.; Yin, H.; Zhang, Z.; Diao, B.; Li, G. Functionalization of lignin through ATRP grafting of poly(2-dimethylaminoethyl methacrylate) for gene delivery. *Colloids Surf. B Biointerfaces* **2015**, *125*, 230–237. [CrossRef] [PubMed]
151. Silmore, K.S.; Gupta, C.; Washburn, N.R. Tunable Pickering emulsions with polymer-grafted lignin nanoparticles (PGLNs). *J. Colloid Interface Sci.* **2016**, *466*, 91–100. [CrossRef] [PubMed]
152. Gupta, C.; Nadelman, E.; Washburn, N.R.; Kurtis, K.E. Lignopolymer Superplasticizers for Low-CO2 Cements. *ACS Sustain. Chem. Eng.* **2017**, *5*, 4041–4049. [CrossRef]
153. Gupta, C.; Sverdlove, M.J.; Washburn, N.R. Molecular architecture requirements for polymer-grafted lignin superplasticizers. *Soft Matter* **2015**, *11*, 2691–2699. [CrossRef]
154. Liu, Z.; Lu, X.; Xie, J.; Feng, B.; Han, Q. Synthesis of a novel tunable lignin-based star copolymer and its flocculation performance in the treatment of kaolin suspension. *Sep. Purif. Technol.* **2019**, *210*, 355–363. [CrossRef]
155. Nemoto, T.; Konishi, G.-I.; Tojo, Y.; An, Y.C.; Funaoka, M. Functionalization of lignin: Synthesis of lignophenol–graft–poly (2-ethyl-2-oxazoline) and its application to polymer blends with commodity polymer. *J. Appl. Polym. Sci.* **2012**, *123*, 2636–2642. [CrossRef]
156. Mahata, D.; Jana, M.; Jana, A.; Mukherjee, A.; Mondal, N.; Saha, T.; Sen, S.; Nando, G.B.; Mukhopadhyay, C.K.; Chakraborty, R.; et al. Lignin-graft-polyoxazoline conjugated triazole a novel anti-infective ointment to control persistent inflammation. *Sci. Rep.* **2017**, *7*, 46412. [CrossRef] [PubMed]
157. Chung, Y.L.; Olsson, J.V.; Li, R.J.; Frank, C.W.; Waymouth, R.M.; Billington, S.L.; Sattely, E.S. A renewable lignin-lactide copolymer and application in biobased composite. *ACS Sustain. Chem. Eng.* **2013**, *1*, 1231–1238. [CrossRef]
158. Liu, X.; Zong, E.; Jiang, J.; Fu, S.; Wang, J.; Xu, B.; Li, W.; Lin, X.; Xu, Y.; Wang, C.; et al. Preparation and characterization of Lignin–graft–poly (ε-caprolactone) copolymers based on lignocellulosic butanol residue. *Int. J. Biol. Macromol.* **2015**, *81*, 521–529. [CrossRef]
159. Sun, Y.; Yang, L.; Lu, X.; He, C. Biodegradable and renewable poly (lactide)-lignin composites: Synthesis, interface and toughening mechanism. *J. Mater. Chem. A* **2015**, *3*, 3699–3709. [CrossRef]
160. Kai, D.; Zhang, K.; Liow, S.S.; Loh, X.J. New dual functional phb-grafted lignin copolymer: Synthesis, mechanical properties, and biocompatibility studies. *ACS Appl. Bio Mater.* **2018**, *2*, 127–134. [CrossRef]
161. Pérez–Camargo, R.A.; Saenz, G.; Laurichesse, S.; Casas, M.T.; Puiggalí, J.; Avérous, L.; Müller, A.J. Nucleation crystallization, and thermal fractionation of poly (ε-caprolactone)-grafted-lignin: Effect of grafted chains length and lignin content. *J. Polym. Sci. Part B Polym. Phys.* **2015**, *53*, 1736–1750. [CrossRef]
162. Laurichesse, S.; Avérous, L. Synthesis, thermal properties, rheological and mechanical behaviors of lignins-grafted-poly(ε-caprolactone). *Polymer* **2013**, *54*, 3882–3890. [CrossRef]
163. Schmidt, B.V.K.J.; Molinari, V.; Esposito, D.; Tauer, K.; Antonietti, M. Lignin-based polymeric surfactants for emulsion polymerization. *Polymer* **2017**, *112*, 418–426. [CrossRef]
164. Tapdiqov, S.Z. A drug-loaded gel based on graft radical co-polymerization of n-vinylpyrrolidone and 4-vinylpyridine with chitosan. *Cell. Chem. Technol.* **2020**, *54*, 429–438. [CrossRef]
165. Cheaburu-Yilmaz, C.N. On the development of chitosan-graft-poly(n-isopropylacrylamide) by raft polymerization technique. *Cell. Chem. Technol.* **2020**, *54*, 1–10. [CrossRef]
166. Kadokawa, J.-I. Preparation and Grafting Functionalization of Self-Assembled Chitin Nanofiber Film. *Coatings* **2016**, *6*, 27. [CrossRef]
167. Mahmoud, G.A.; Sayed, A.; Thabit, M.; Safwat, G. Chitosan biopolymer based nanocomposite hydrogels for removal of methylene blue dye. *SN Appl. Sci.* **2020**, *2*, 968. [CrossRef]
168. Xu, Y.; Liu, B.; Zou, L.; Sun, C.; Li, W. Preparation and characterization of PLLA/chitosan-graft-poly(ε-caprolactone) (CS-g-PCL) composite fibrous mats: The microstructure, performance and proliferation assessment. *Int. J. Biol. Macromol.* **2020**, *162*, 320–332. [CrossRef] [PubMed]
169. Kadokawa, J.-I. Fabrication of nanostructured a nd microstructured chitin materials through gelation with suitable dispersion media. *RSC Adv.* **2015**, *5*, 12736–12746. [CrossRef]
170. Stefan, J.; Lorkowska-Zawicka, B.; Kaminski, K.; Szczubialka, K.; Nowakowska, M.; Korbut, R. The current view on biological potency of cationically modified chitosan. *J. Physiol. Pharmacol.* **2014**, *65*, 341–347. [PubMed]
171. Jiang, T.; Deng, M.; James, R.; Nair, L.S.; Laurencin, C.T. Micro- and nanofabrication of chitosan structures for regenerative engineering. *Acta Biomater.* **2014**, *10*, 1632–1645. [CrossRef]
172. Lai, G.-J.; Shalumon, K.T.; Chen, S.-H.; Chen, J.P. Composite chitosan/silk fibroin nanofibers for modulation of osteogenic differentiation and proliferation of human mesenchymal stem cells. *Carbohydr. Polym.* **2014**, *111*, 288–297. [CrossRef]
173. Jayakumar, R.; Menon, D.; Manzoor, K.; Nair, S.V.; Tamura, H. Biomedical applications of chitin and chitosan based nanomaterials—A short review. *Carbohydr. Polym.* **2010**, *82*, 227–232. [CrossRef]
174. Deng, Z.; Wang, T.; Chen, X.; Liu, Y. Applications of chitosan based biomaterials: A focus on dependent antimicrobial properties. *Mar. Life Sci. Technol.* **2020**, *2*, 398–413. [CrossRef]
175. Wen, J.; Li, Y.; Wang, L.; Chen, X.; Cao, Q.; He, N. Carbon Dioxide Smart Materials Based on Chitosan. *Prog. Chem.* **2020**, *32*, 417–422. (In Chinese)

176. Sashiwa, H.; Aiba, S.-I. Chemically modified chitin and chitosan as biomaterials. *Prog. Polym. Sci.* **2004**, *29*, 887–908. [CrossRef]
177. Crini, G.; Badot, P.-M. Application of chitosan, a natural aminopolysaccharide, for dye removal from aqueous solutions by adsorption processes using batch studies: A review of recent literature. *Prog. Polym. Sci.* **2008**, *33*, 399–447. [CrossRef]
178. Mittal, H.; Ray, S.S.; Kaith, B.S.; Bhatia, J.K.; Sharma, S.J.; Alhassan, S.M. Recent progress in the structural modification of chitosan for applications in diversified biomedical fields. *Eur. Polym. J.* **2018**, *109*, 402–434. [CrossRef]
179. Iyer, B.V.S.; Yashin, V.V.; Hamer, M.J.; Kowalewski, T.; Matyjaszewski, K.; Balazsa, A.C. Ductility, toughness and strain recovery in self-healing dualcross-linked nanoparticle networks studied by computer simulations. *Prog. Polym. Sci.* **2015**, *40*, 121–137. [CrossRef]
180. Derry, M.J.; Fielding, L.A.; Armes, S.P. Polymerization-induced self-assembly of block copolymer nanoparticles via RAFT non-aqueous dispersion polymerization. *Prog. Polym. Sci.* **2016**, *52*, 1–18. [CrossRef]
181. Bednarek, M. Branched aliphatic polyesters by ring-opening (co)polymerization. *Prog. Polym. Sci.* **2016**, *58*, 27–58. [CrossRef]
182. Yildirim, I.; Weber, C.; Schubert, U.S. Old meets new: Combination of PLA and RDRP to obtain sophisticated macromolecular architectures. *Prog. Polym. Sci.* **2018**, *76*, 111–150. [CrossRef]
183. Wang, W.; Lu, W.; Goodwin, A.; Wang, H.; Yin, P.; Kang, N.-G.; Hong, K.; Mays, J.W. Recent advances in thermoplastic elastomers from living polymerizations: Macromolecular architectures and supramolecular chemistry. *Prog. Polym. Sci.* **2019**, *95*, 1–31. [CrossRef]
184. Mocny, P.; Klok, H.-A. Complex polymer topologies and polymer—nanoparticle hybrid films prepared via surface-initiated controlled radical polymerization. *Prog. Polym. Sci.* **2020**, *100*, 101185. [CrossRef]
185. Vivaldo-Lima, E.; Jaramillo-Soto, G.; Penlidis, A. Nitroxide-mediated polymerization (NMP). In *Encyclopedia of Polymer Science and Technology*, 1st ed.; John Wiley & Sons: New York, NY, USA, 2016; pp. 1–48. ISBN 978-047-144-026-0.
186. Olivier, A.; Meyer, F.; Raquez, J.-M.; Damman, P.; Dubois, P. Surface-initiated controlled polymerization as a convenient method for designing functional polymer brushes: From self-assembled monolayers to patterned surfaces. *Prog. Polym. Sci.* **2012**, *37*, 157–181. [CrossRef]
187. Radzevicius, P.; Krivorotova, T.; Makuska, R. Synthesis by one-pot RAFT polymerization and properties of amphiphilic pentablock copolymers with repeating blocks of poly(2-hydroxyethyl methacrylate) and poly(butyl methacrylate). *Eur. Polym. J.* **2017**, *87*, 69–83. [CrossRef]
188. Chmielarz, P.; Fantin, M.; Park, S.; Isse, A.A.; Gennaro, A.; Magenau, A.J.; Sobkowiak, A.; Matyjaszewski, K. Electrochemically mediated atom transfer radical polymerization (eATRP). *Prog. Polym. Sci.* **2017**, *69*, 47–78. [CrossRef]
189. Maharana, T.; Pattanaik, S.; Routaray, A.; Nath, N.; Sutar, A.K. Synthesis and characterization of poly(lactic acid) based graft copolymers. *React. Funct. Polym.* **2015**, *93*, 47–67. [CrossRef]
190. Mehta, A.; Pandey, J.P.; Sen, G. Synthesis of Diallyl dimethyl ammonium chloride grafted polyvinyl pyrrolidone (PVP-g-DADMAC) and its applications. *Mater. Sci. Eng. B Solid State Mater. Adv. Technol.* **2021**, *263*, 114750. [CrossRef]
191. El-Sayed, N.; Awad, H.; El-Sayed, G.M.; Nagieb, Z.A.; Kamel, S. Synthesis and characterization of biocompatible hydrogel based on hydroxyethyl cellulose-g-poly(hydroxyethyl methacrylate). *Polym. Bull.* **2020**, *77*, 6333–6347. [CrossRef]
192. Mahdavi, H.; Mazinani, N.; Heidari, A.A. Poly(vinylidene fluoride) (PVDF)/PVDF-g-polyvinylpyrrolidone (PVP)/TiO2 mixed matrix nanofiltration membranes: Preparation and characterization. *Polym. Int.* **2020**, *69*, 1187–1195. [CrossRef]
193. Oliveira, T.S.; Brazil, T.R.; Guerrini, L.M.; Rezende, M.C.; Oliveira, M.P. Synthesis and characterization of poly (acrylonitrile-g-lignin) by semi-batch solution polymerization and evaluation of their potential application as carbon materials. *J. Polym. Res.* **2020**, *27*, 340. [CrossRef]
194. Zhang, M.; Yang, P.; Lan, G.; Liu, Y.; Cai, Q.; Xi, J. High crosslinked sodium carboxyl methylstarch-g-poly (acrylic acid-co-acrylamide) resin for heavy metal adsorption: Its characteristics and mechanisms. *Environ. Sci. Pollut. Res.* **2020**, *27*, 38617–38630. [CrossRef]
195. Wang, L.; Zhang, X.; Xu, J.; Wang, Q.; Fan, X. Synthesis of partly debranched starch-g-poly(2-acryloyloxyethyl trimethyl ammonium chloride) catalyzed by horseradish peroxidase and the effect on adhesion to polyester/cotton yarn. *Process Biochem.* **2020**, *97*, 176–182. [CrossRef]
196. Xu, R.-M.; Yang, T.T.; Vidovic, E.; Jia, R.-N.; Zhang, J.-M.; Mi, Q.-Y.; Zhang, J. Cellulose Acetate Thermoplastics with High Modulus, Dimensional Stability and Anti-migration Properties by Using CA-g-PLA as Macromolecular Plasticizer. *Chin. J. Polym. Sci.* **2020**, *38*, 1141–1148. [CrossRef]
197. Wu, Q.; Tiraferri, A.; Li, T.; Xie, W.; Chang, H.; Bai, Y.; Liu, B. Superwettable PVDF/PVDF-g-PEGMA Ultrafiltration Membranes. *ACS Omega* **2020**, *5*, 23450–23459. [CrossRef] [PubMed]
198. Peighambardoust, S.J.; Aghamohammadi-Bavil, O.; Foroutan, R.; Arsalani, N. Removal of malachite green using carboxymethyl cellulose-g-polyacrylamide/montmorillonite nanocomposite hydrogel. *Int. J. Biol. Macromol.* **2020**, *159*, 1122–1131. [CrossRef]
199. Gürsel, U.; Taran, S.; Gökçen, M.; Ari, Y.; Alli, A. Ultraviolet illumination responsivity of the Au/n-Si diodes with and without poly (linolenic acid)-g-poly (caprolactone)-g-poly (t-butyl acrylate) interfacial layer. *Surf. Rev. Lett.* **2020**, *27*, 1950207. [CrossRef]
200. Song, P.; Guo, R.; Ma, W.; Wang, L.; Ma, F.; Wang, R. Synthesis of CO2-based polycarbonate-g-polystyrene copolymers via NMRP. *Chem. Commun.* **2020**, *56*, 9493–9496. [CrossRef] [PubMed]
201. Klimovica, K.; Pan, S.; Lin, T.-W.; Peng, X.; Ellison, C.J.; LaPointe, A.M.; Bates, F.S.; Coates, G.W. Compatibilization of iPP/HDPE Blends with PE-g-iPP Graft Copolymers. *ACS Macro. Lett.* **2020**, *9*, 1161–1166. [CrossRef]

202. Li, W.; Yu, Z.; Wu, Y.; Liu, Q. Preparation, characterization of feather protein-g-poly(sodium allyl sulfonate) and its application as a low-temperature adhesive to cotton and viscose fibers for warp sizing. *Eur. Polym. J.* **2020**, *136*, 109945. [CrossRef]
203. Czarnecka, E.; Nowaczyk, J. Semi-Natural superabsorbents based on Starch-g-poly(acrylic acid): Modification, synthesis and application. *Polymers* **2020**, *12*, 1794. [CrossRef]
204. Bhosale, R.R.; Gangadharappa, H.V.; Osmani, R.A.M.; Gowda, D.V. Design and development of polymethylmethacrylate-grafted gellan gum (PMMA-g-GG)-based pH-sensitive novel drug delivery system for antidiabetic therapy. *Drug Deliv. and Transl. Res.* **2020**, *10*, 1002–1018. [CrossRef] [PubMed]
205. Patel, R.; Patel, M.; Sung, J.-S.; Kim, J.H. Preparation and characterization of bioinert amphiphilic P(VDF-co-CTFE)-g-POEM graft copolymer. *Polym. Plast. Technol. Mater.* **2020**, *59*, 1077–1087. [CrossRef]
206. Li, W.; Wu, Y.; Wu, J.; Ni, Q. Preparation, characterization of poly(acrylic acid)-g-feather protein-g-poly(methyl acrylate) and application in improving adhesion of protein to PLA fibers for sizing. *React. Funct. Polym.* **2020**, *152*, 104607. [CrossRef]
207. Deng, J.-R.; Zhao, C.-L.; Wu, Y.-X. Antibacterial and pH-responsive Quaternized Hydroxypropyl Cellulose-g-Poly(THF-co-epichlorohydrin) Graft Copolymer: Synthesis, Characterization and Properties. *Chin. J. Polym. Sci.* **2020**, *38*, 704–714. [CrossRef]
208. Cuggino, J.C.; Ambrosioni, F.E.; Picchio, M.L.; Nicola, M.; Jiménez Kairuz, A.F.; Gatti, G.; Minari, R.J.; Calderon, M.; Alvarez Igarzabal, C.I.; Gugliotta, L.M. Thermally self-assembled biodegradable poly(casein-g-N-isopropylacrylamide) unimers and their application in drug delivery for cancer therapy. *Int. J. Biol. Macromol.* **2020**, *154*, 446–455. [CrossRef] [PubMed]
209. Kenawy, E.R.; Seggiani, M.; Cinelli, P.; Elnaby, H.M.H.; Azaam, M.M. Swelling capacity of sugarcane bagasse-g-poly(acrylamide)/attapulgite superabsorbent composites and their application as slow release fertilizer. *Eur. Polym. J.* **2020**, *133*, 109769. [CrossRef]
210. Jiang, P.; Ji, H.; Li, G.; Chen, S.; Lv, L. Structure formation in pH-sensitive micro porous membrane from well-defined ethyl cellulose-g-PDEAEMA via non-solvent-induced phase separation process. *J. Macromol. Sci. Pure Appl. Chem.* **2020**, *57*, 461–471. [CrossRef]
211. Huang, Q.; Xu, Z.; Cai, C.; Lin, J. Micelles with a Loose Core Self-Assembled from Coil-g-Rod Graft Copolymers Displaying High Drug Loading Capacity. *Macromol. Chem. Phys.* **2020**, *221*, 2000121. [CrossRef]
212. Wang, Z.; Wu, L.; Zhou, D.; Ji, P.; Zhou, X.; Zhang, Y.; He, P. Synthesis and Water Absorbing Properties of KGM-g-P(AA-AM-(DMAEA-EB)) via Grafting Polymerization Method. *Polym. Sci. Ser. B* **2020**, *62*, 238–244. [CrossRef]
213. Erdoğan, M.K.; Akdemir, Ö.; Hamitbeyli, A.; Karakışla, M. Preparation of hydrophilic woven fabrics: Surface modification of poly(ethylene terephthalate) by grafting of poly(vinyl alcohol) and poly(vinyl alcohol)-g-(N-vinyl-2-pyrrolidone). *J. Appl. Polym. Sci.* **2020**, *137*, 48584. [CrossRef]
214. Worzakowska, M. The preparation, physicochemical and thermal properties of the high moisture, solvent and chemical resistant starch-g-poly(geranyl methacrylate) copolymers. *J. Thermal. Anal. Calorim.* **2020**, *140*, 189–198. [CrossRef]
215. Delorme, V.; Lichon, L.; Mahindad, H.; Hunger, S.; Laroui, N.; Daurat, M.; Godefroy, A.; Coudane, J.; Gary-Bobo, M.; Van Den Berghe, H. Reverse poly(ε-caprolactone)-g-dextran graft copolymers. Nano-carriers for intracellular uptake of anticancer drugs. *Carbohydr. Polym.* **2020**, *232*, 115764. [CrossRef] [PubMed]
216. Kang, D.A.; Kim, K.; Karade, S.S.; Kim, H.; Kim, J.H. High-performance solid-state bendable supercapacitors based on PEGBEM-g-PAEMA graft copolymer electrolyte. *Chem. Eng. J.* **2020**, *384*, 123308. [CrossRef]
217. Tian, B.; Cai, Y.; Zhang, X.; Fan, H.; Li, B.-G. Design of Well-Defined Polyethylene-g-poly-methyltrifluorosiloxane Graft Copolymers via Direct Copolymerization of Ethylene with Polyfluorosiloxane Macromonomers. *Ind. Eng. Chem. Res.* **2020**, *59*, 4557–4567. [CrossRef]
218. Öztürk, T.; Meyvacı, E.; Arslan, T. Synthesis and characterization of poly(vinyl chloride-g-ε-caprolactone) brush type graft copolymers by ring-opening polymerization and "click" chemistry. *J. Macromol. Sci. Pure Appl. Chem.* **2020**, *57*, 171–180. [CrossRef]
219. Zha, X.; Sadi, M.S.; Yang, Y.; Luo, T.; Huang, N. Adhesion of cornstarch-g-poly (2-hydroxyethyl acrylate) to cotton fibers in sizing. *J. Adhes. Sci. Technol.* **2020**, *34*, 461–479. [CrossRef]
220. Nicolas, C.; Zhang, W.; Choppé, E.; Fontaine, L.; Montembault, V. Polynorbornene-g-poly(ethylene oxide) Through the Combination of ROMP and Nitroxide Radical Coupling Reactions. *J. Polym. Sci.* **2020**, *58*, 645–653. [CrossRef]
221. Ilhan, E.; Karahaliloglu, Z.; Kilicay, E.; Hazer, B.; Denkbas, E.B. Potent bioactive bone cements impregnated with polystyrene-g-soybean oil-AgNPs for advanced bone tissue applications. *Mater. Technol.* **2020**, *35*, 179–194. [CrossRef]
222. Chen, Y.; Li, Q.; Li, Y.; Zhang, Q.; Huang, J.; Wu, Q.; Wang, S. Fabrication of cellulose nanocrystal-g-poly(acrylic acid-co-acrylamide) aerogels for efficient Pb(II) removal. *Polymers* **2020**, *12*, 333. [CrossRef]
223. Mo, X.-Z.; Wei, F.-X.; Tan, D.-F.; Pang, J.-Y.; Lan, C.-B. The compatibilization of PLA-g-TPU graft copolymer on polylactide/thermoplastic polyurethane blends. *J. Polym. Res.* **2020**, *27*, 33. [CrossRef]
224. Guleria, A.; Kumari, G.; Lima, E.C. Cellulose-g-poly-(acrylamide-co-acrylic acid) polymeric bioadsorbent for the removal of toxic inorganic pollutants from wastewaters. *Carbohydr. Polym.* **2020**, *228*, 115396. [CrossRef] [PubMed]
225. He, M.; Li, T.; Hu, M.; Chen, C.; Liu, B.; Crittenden, J.; Chu, L.-Y.; Ng, H.Y. Performance improvement for thin-film composite nanofiltration membranes prepared on PSf/PSf-g-PEG blended substrates. *Sep. Purif. Technol.* **2020**, *230*, 115855. [CrossRef]
226. Savaş, B.; Öztürk, T. Synthesis and characterization of poly(vinyl chloride-g-methyl methacrylate) graft copolymer by redox polymerization and Cu catalyzed azide-alkyne cycloaddition reaction. *J. Macromol. Sci. Pure Appl. Chem.* **2020**, 1–7. [CrossRef]

227. Ahuja, D.; Rainu, A.S.; Singh, M.; Kaushik, A. Encapsulation of NPK fertilizer for slow release using sodium carboxymethyl cellulose-g-poly (AA-C0-AM-C0-AMPS)/ Montmorillonite clay-based nanocomposite hydrogels for sustainable agricultural applications. *Trends Carbohydr. Res.* **2020**, *12*, 15–23.
228. Lu, Y.; Wu, F.; Duan, W.; Zhou, X.; Kong, W. Engineering a "PEG-g-PEI/DNA nanoparticle-in- PLGA microsphere" hybrid controlled release system to enhance immunogenicity of DNA vaccine. *Mater. Sci. Eng. C* **2020**, *106*, 110394. [CrossRef]
229. Grebenik, E.A.; Surin, A.M.; Bardakova, K.N.; Dermina, T.S.; Minaev, N.V.; Veryasova, N.N.; Artyukhova, M.A.; Krasilnikova, I.A.; Bakaeva, Z.V.; Sorokina, E.G.; et al. Chitosan-g-oligo(L,L-lactide) copolymer hydrogel for nervous tissue regeneration in glutamate excitotoxicity: In vitro feasibility evaluation. *Biomed. Mater.* **2020**, *15*, 015011. [CrossRef] [PubMed]
230. Barth, H.G.; Jackson, C.; Boyes, B.E. Size Exclusion Chromatography. *Anal. Chem.* **1994**, *66*, 595–620. [CrossRef] [PubMed]
231. Hamielec, A.; Gloor, P.; Zhu, S. Kinetics of, free radical modification of polyolefins in extruders—Chain scission, crosslinking and grafting. *Can. J. Chem. Eng.* **1991**, *69*, 611–618. [CrossRef]
232. Chaimberg, M.; Cohen, Y. Kinetic Modeling of Free-Radical Graft Polymerization. *AIChE J.* **1994**, *40*, 294–311. [CrossRef]
233. Guillot, J.; Leroux, D. Modelling of size-exclusion chromatograms from molecular weight distribution calculations. Application to the grafting of polymers onto functionalized silica. *Macromol. Chem. Phys.* **1994**, *195*, 1463–1470. [CrossRef]
234. Hojabr, S.; Baker, W.; Russell, K.; McLellan, P.; Huneault, M. Melt grafting of glycidyl methacrylate onto polyethylene: An experimental and mathematical modeling study. *Int. Polym. Proc.* **1998**, *13*, 118–128. [CrossRef]
235. Machado, A.; Gaspar-Cunha, A.; Covas, J. Modelling of the grafting of maleic anhydride onto polyethylene in an extruder. *Mater. Sci. Forum* **2004**, *455–456*, 763–766. [CrossRef]
236. Giudici, R. Mathematical modeling of the crafting of maleic anhydride onto polypropylene. *Macromol. Symp.* **2007**, *259*, 354–364. [CrossRef]
237. Diaconescu, R.; Grigoriu, A.-M.; Luca, C. Neural network modeling of monochlorotriazinyl-β-cyclodextrin grafting on cellulosic supports. *Cell. Chem. Technol.* **2007**, *41*, 385–390.
238. Luca, C.; Grigoriu, A.-M.; Diaconescu, R.; Secula, M. Modeling and simulation of monochlorotriazinyl-β-cyclodextrin paper grafting by artificial neural network. *Rev. Chim.* **2011**, *62*, 1033–1038.
239. Grigoriu, A.; Racu, C.; Diaconescu, R.; Grigoriu, A.-M. Modeling of the simultaneous process of wet spinning-grafting of bast fibers using artificial neural networks. *Textile Res. J.* **2012**, *82*, 324–335. [CrossRef]
240. Tong, G.-S.; Liu, T.; Hu, G.-H.; Hoppe, S.; Zhao, L.; Yuan, W.-K. Modelling of the kinetics of the supercritical CO_2 assisted grafting of maleic anhydride onto isotactic polypropylene in the solid state. *Chem. Eng. Sci.* **2007**, *62*, 5290–5294. [CrossRef]
241. Wang, J.; Ran, Y.; Ding, L.; Wang, D. Advances in supercritical CO_2 assisted grafting of polypropylene in solid state. *Chem. React. Eng. Technol.* **2008**, *24*, 173–177.
242. Casis, N.; Estenoz, D.; Vega, J.; Meira, G. Bulk prepolymerization of styrene in the presence of polybutadiene: Determination of grafting efficiency by size exclusion chromatography combined with a new extended model. *J. Appl. Polym. Sci.* **2009**, *111*, 1508–1522. [CrossRef]
243. Badel, T.; Beyou, E.; Bounor-Legaré, V.; Chaumont, P.; Flat, J.; Michel, A. Free radical graft copolymerization of methyl methacrylate onto polyolefin backbone: Kinetics modeling through model compounds approach. *Macromol. Chem. Phys.* **2009**, *210*, 1087–1095. [CrossRef]
244. Gianoglio Pantano, I.; Asteasuain, M.; Sarmoria, C.; Brandolin, A. Graft copolymers for blend compatibilization. Mathematical modeling of the grafting process. In Proceedings of the 2010 AIChE Annual Meeting Conference Proceedings, Salt Lake City, UT, USA, 7–12 November 2010; pp. 1–14.
245. Gianoglio Pantano, I.; Asteasuain, M.; Sarmoria, C.; Brandolin, A. Graft Copolymers for Blend Compatibilization: Mathematical Modeling of the Grafting Process. *Macromol. React. Eng.* **2012**, *6*, 406–418. [CrossRef]
246. Gianoglio Pantano, I.A.; Brandolin, A.; Sarmoria, C. Mathematical modeling of the graft reaction between polystyrene and polyethylene. *Polym. Degrad. Stabil.* **2011**, *96*, 416–425. [CrossRef]
247. Aguiar, L.; Pessôa-Filho, P.; Giudici, R. Mathematical modeling of the grafting of maleic anhydride onto poly(propylene): Model considering a heterogeneous medium. *Macromol. Theory Simul.* **2011**, *20*, 837–849. [CrossRef]
248. Damodaran, V.; Fee, C.; Popat, K. Modeling of PEG grafting and prediction of interfacial force profile using x-ray photoelectron spectroscopy. *Surf. Interface Anal.* **2012**, *44*, 144–149. [CrossRef]
249. Zhou, D.; Gao, X.; Wang, W.-J.; Zhu, S. Termination of surface radicals and kinetic modeling of ATRP grafting from flat surfaces by addition of deactivator. *Macromolecules* **2012**, *45*, 1198–1208. [CrossRef]
250. Nasef, M.; Shamsaei, E.; Ghassemi, P.; Aly, A.; Yahaya, A. Modeling, prediction, and multifactorial optimization of radiation-induced grafting of 4-vinylpyridine onto poly(vinylidene fluoride) films using statistical simulator. *J. Appl. Polym. Sci.* **2013**, *127*, 1659–1666. [CrossRef]
251. Nasef, M.M.; Ali, A.; Saidi, H.; Ahmad, A. Modeling and optimization aspects of radiation induced grafting of 4-vinylpyridene onto partially fluorinated films. *Radiat. Phys. Chem.* **2014**, *94*, 123–128. [CrossRef]
252. Wu, L.L.; Bu, Z.; Gong, C.; Li, B.-G.; Hungenberg, K.-D. Graft Copolymerization of Styrene and Acrylonitrile in the Presence of Poly(propylene glycol): Modeling and Simulation of Semi-Batch and Continuous Processes. *Macromol. React. Eng.* **2012**, *6*, 384–394. [CrossRef]
253. Xie, X.-L.; Tong, Z.-F.; Huang, Z.-Q.; Zhang, Y.-Q. Kinetics model of graft co-polymerization of acrylamide onto mechanically-activated starch in inverse emulsion. *J. Chem. Eng. Chin. Univ.* **2014**, *28*, 567–573.

254. Liu, X.; Nomura, M. Kinetic modeling and simulation of emulsion grafting copolymerization of styrene and acrylonitrile in the presence of polybutadiene seed latex particles. *Ind. Eng. Chem. Res.* **2014**, *53*, 17580–17588. [CrossRef]
255. Zhang, M.; Jia, Y. Kinetic study on free radical grafting of polyethylene with acrylic acid by reactive extrusion. *J. Appl. Polym. Sci.* **2014**, *131*, 40990. [CrossRef]
256. Sirirat, T.; Vatanatham, T.; Hansupalak, N.; Rempel, G.; Arayapranee, W. Kinetics and modeling of methyl methacrylate graft copolymerization in the presence of natural rubber latex. *Korean J. Chem. Eng.* **2015**, *32*, 980–992. [CrossRef]
257. Oliveira, D.; Dias, R.; Costa, M. Modeling RAFT Gelation and Grafting of Polymer Brushes for the Production of Molecularly Imprinted Functional Particles. *Macromol. Symp.* **2016**, *370*, 52–65. [CrossRef]
258. Saeb, M.; Rezaee, B.; Shadman, A.; Formela, K.; Ahmadi, Z.; Hemmati, F.; Kermaniyan, T.; Mohammadi, Y. Controlled grafting of vinylic monomers on polyolefins: A robust mathematical modeling approach. *Des. Monomers Polym.* **2017**, *20*, 250–268. [CrossRef]
259. Hernández-Ortiz, J.; Van Steenberge, P.; Reyniers, M.-F.; Marin, G.; D'hooge, D.; Duchateau, J.; Remerie, K.; Toloza, C.; Vaz, A.; Schreurs, F. Modeling the reaction event history and microstructure of individual macrospecies in postpolymerization modification. *AIChE J.* **2017**, *63*, 4944–4961. [CrossRef]
260. Hernández-Ortiz, J.; Van Steenberge, P.; Duchateau, J.; Toloza, C.; Schreurs, F.; Reyniers, M.-F.; Marin, G.; D'hooge, D. Sensitivity Analysis of Single-Phase Isothermal Free Radical–Induced Grafting of Polyethylene. *Macromol. Theory Simul.* **2018**, *27*, 1800036. [CrossRef]
261. Hernández-Ortiz, J.; Van Steenberge, P.; Duchateau, J.; Toloza, C.; Schreurs, F.; Reyniers, M.-F.; Marin, G.; D'hooge, D. The Relevance of Multi-Injection and Temperature Profiles to Design Multi-Phase Reactive Processing of Polyolefins. *Macromol. Theory Simul.* **2019**, *28*, 1900035. [CrossRef]
262. Penlidis, A.; MacGregor, J.F.; Hamielec, A.E. Dynamic modeling of emulsion polymerization reactors. *AIChE J.* **1985**, *31*, 881–889. [CrossRef]
263. Iedema, P.D.; Grcev, S.; Hoefsloot, H.C.J. Molecular weight distribution modeling of radical polymerization in a CSTR with long chain branching through transfer to polymer and terminal double bond. *Macromolecules* **2003**, *36*, 458–476. [CrossRef]
264. Dias, R.; Costa, M. A new look at kinetic modeling of nonlinear free radical polymerizations with terminal branching and chain transfer to polymer. *Macromolecules* **2003**, *36*, 8853–8863. [CrossRef]
265. Krallis, A.; Kiparissides, C. Mathematical modeling of the bivariate molecular weight-Long chain branching distribution of highly branched polymers: A population balance approach. *Chem. Eng. Sci.* **2007**, *62*, 5304–5311. [CrossRef]
266. Kryven, I.; Iedema, P.D. A novel approach to population balance modeling of reactive polymer modification leading to branching. *Macromol. Theory Simul.* **2013**, *22*, 89–106. [CrossRef]
267. Wang, R.; Luo, Y.; Li, B.-G.; Zhu, S. Modeling of Branching and Gelation in RAFT Copolymerization of Vinyl/Divinyl Systems. *Macromolecules* **2009**, *42*, 85–94. [CrossRef]
268. Yaghini, N.; Iedema, P.D. Molecular weight and branching distribution modeling in radical polymerization with transfer to polymer and scission under gel conditions and allowing for multiradicals. *Macromolecules* **2014**, *47*, 4851–4863. [CrossRef]
269. Penlidis, A.; Vivaldo-Lima, E.; Hernández-Ortiz, J.; Saldívar-Guerra, E. Chapter 12: Polymer Reaction Engineering. In *Handbook of Polymer Synthesis, Characterization, and Processing*, 1st ed.; John Wiley & Sons: New York, NY, USA, 2013; pp. 251–271, ISBN 978-047-063-032-7.
270. Zhu, S.; Hamielec, A. Polymerization kinetic modeling and macromolecular. In *Polymer Science: A Comprehensive Reference*, 1st ed.; Matyjaszewski, K., Möller, M., Eds.; Elsevier B.V: London, UK, 2012; Chapter 4.32; Volume 4, pp. 779–831, ISBN 978-008-087-862-1.
271. Quintero-Ortega, I.; Vivaldo-Lima, E.; Luna-Bárcenas, G.; Alvarado, J.; Louvier-Hernández, J.; Sanchez, I. Modeling of the Free-Radical Copolymerization Kinetics with Cross Linking of Vinyl/Divinyl Monomers in Supercritical Carbon Dioxide. *Ind. Eng. Chem. Res.* **2005**, *44*, 2823–2844. [CrossRef]
272. Dong, P.; Sun, H.; Quan, D. Synthesis of poly(L-lactide-co-5-amino-5-methyl-1,3-dioxan-2-ones) [P(L-LA-co-TAc)] containing amino groups via organocatalysis and post-polymerization functionalization. *Polymer* **2016**, *97*, 614–622. [CrossRef]

Article

Ethylene Polymerization via Zirconocene Catalysts and Organoboron Activators: An Experimental and Kinetic Modeling Study

Luis Valencia [1], Francisco Enríquez-Medrano [2], Ricardo López-González [2], Priscila Quiñonez-Ángulo [3], Enrique Saldívar-Guerra [2], José Díaz-Elizondo [2], Iván Zapata-González [4,*] and Ramón Díaz de León [2,*]

1. Materials Technology and Chemistry, Alfa Laval Tumba AB, SE-14782 Tumba, Sweden; luisalex_val@hotmail.com
2. Research Center for Applied Chemistry, Enrique Reyna Hermosillo, No.140, Col. San José de los Cerritos, Saltillo 25294, Mexico; javier.enriquez@ciqa.edu.mx (F.E.-M.); ricardo.lopez@ciqa.edu.mx (R.L.-G.); enrique.saldivar@ciqa.edu.mx (E.S.-G.); alejandro.diaz@ciqa.edu.mx (J.D.-E.)
3. Centro de Graduados e Investigación en Química, Tecnológico Nacional de México/I.T.R. de Tijuana, A.P. 1166, Tijuana 22000, Mexico; priscila.quinonez@tectijuana.edu.mx
4. Cátedras CONACYT—Instituto Tecnológico de Tijuana, Centro de Graduados e Investigación en Química, Tijuana 22000, Mexico
* Correspondence: ivan.zapata@tectijuana.edu.mx or ijzapatago@conacyt.mx (I.Z.-G.); ramon.diazdeleon@ciqa.edu.mx (R.D.d.L.); Tel.: +52-664-623-3772 (I.Z.-G.); +52-844-438-9830 (R.D.d.L.)

Abstract: Forty years after the discovery of metallocene catalysts, there are still several aspects that remain unresolved, especially when the "conventional" alkylaluminum activators are not used. Herein, we systematically investigated the synthesis of polyethylene (PE) via three different zirconocene catalysts, with different alkyl substituents, activated via different organoboron compounds. The polymerization behavior, as well as the properties of the materials, were evaluated. The results demonstrate that the highest catalytic activity is shown by bis(cyclopentadienyl)dimethylzirconium activated by trityl tetra(pentafluorophenyl)borate. Additionally, it was found that toluene is the optimum solvent for these systems and at these reaction conditions. Moreover, to validate our experimental results, a comprehensive mathematical model was developed on the basis of thermodynamic and kinetic principles. The concentration of ethylene transferred to the solvent phase (toluene) in a liquid–vapor equilibrium (LVE) system was estimated based on Duhem's theorem. Arrhenius expressions for the kinetic rate constants of a proposed kinetic mechanism were estimated by a kinetic model, in which the rate of polymerization was fitted by a least-square optimization procedure and the molecular weight averages by the method of moments. The simulations of the coordination polymerization suggest the presence of two types of active sites, principally at low temperatures, and the reactivation of the deactivated sites via a boron-based activator. However, the effect of the temperature on the reactivation step was not clear; a deeper understanding via designed experiments is required.

Keywords: ethylene polymerization; metallocene; zirconium-based catalyst; organoboron compounds; kinetic modeling

1. Introduction

Polyethylene (PE), one of the most used and commercialized thermoplastics in the world, is produced by the polymerization of ethylene which is catalyzed via two main different routes: using heterogeneous processes with Ziegler–Natta catalysts, or via metallocene catalytic systems. Since the discovery of the catalytic activity of the homogenous catalysts based on biscyclopentadienyl titanium or zirconium dialkyl systems in the ethylene polymerization in the 1980s by Kaminsky and Sinn [1–4], metallocene systems have revolutionized the polyolefins field, because they enable the production of PE with narrow molecular weight distributions, low content of extractables, good processability, and superior properties [5]. Moreover, metallocene catalysts, in comparison to Ziegler–Natta types, show a single type of active site, which enables predictions of the properties of the resulting polymers.

Several factors play an important role in the olefin's polymerization via metallocene catalysis. For instance, the formation of weakly coordinating anions with a weak bonding to the metallocene active centers (acting as co-catalysts). The anions interact with the cationic metal species, in the reaction medium, creating active sites (ion-pairs), followed by the subsequent polymerization. Methylaluminoxane (MAO) is a popular activator due to its high efficiency; however, a large excess of MAO is usually required, and, despite extensive efforts, its detailed active-site structure has not yet been fully elucidated [6–8]. A prominent alternative to replace MAO is the use of other bulky coordinating anions such as organoboranes, e.g., tris(pentafluorophenyl)borane (B1) [9,10], and organoborates such as N,N-dimethylanilinium tetra(pentafluorophenyl)borate (B2) or trityl tetra(pentafluorophenyl)borate (B3) [11,12]. These types of activators can ionize the metallocene (pre-alkylated) catalyst, acting as Lewis acids, leading to excellent active cationic metallocene catalysts for the polymerization of olefins in quasi-equimolar amounts between the metallocene catalyst and the boron-based activator, and resulting in catalytic complexes with a definite chemical structure [13–15]. A breakthrough in this field was the introduction of the weakly coordinating tris(pentafluorophenyl)borate [B(C_6F_5)$_3$] as a counterion, which can abstract a methyl group from the alkylated metallocene catalyst, to form ionic species such as [CP_2ZrMe]$^+$[MeB(C_6F_5)$_3$]$^-$, followed by the coordination of a monomer molecule and subsequent propagation [10]. Nevertheless, residual coordinative interactions between the activated metal center and the anion, via the abstracted methyl group, can slightly decrease the catalyst reactivity. Ionic organoboron activators, such as [$C_6H_5NHMe_2$]$^+$[B(C_6F_5)$_4$]$^-$ and [(C_6H_5)$_3$C]$^+$[B(C_6F_5)$_4$]$^-$, on the other hand, avoid this form of ion–ion contact between the abstracted methyl group and the transition metal atom, while the cationic species takes the function of activating the metal active center (Figure 1).

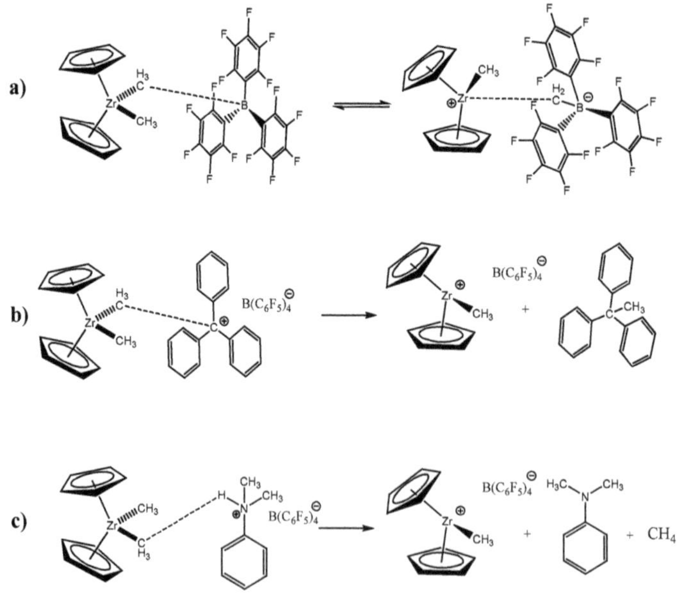

Figure 1. Schematic representation of the activation of a zirconocene catalyst (CP$_2$ZrCl$_2$) by (a) [B(C_6F_5)$_3$], (b) [(C_6H_5)$_3$C]$^+$[B(C_6F_5)$_4$]$^-$ and (c) [$C_6H_5NHMe_2$]$^+$[B(C_6F_5)$_4$]$^-$.

There are a wide variety of metallocene catalysts with different symmetries and substitutions, because the configuration of the catalyst is another factor governing polymerization behavior. For instance, the steric and electronic environment of ligand substituents of the metal catalyst, as well as the ion–ion interactions between the electrophilic metal and the counterion, are critical factors that can dramatically alter the polymerization behavior due to steric hindrance and electronic factors.

Several works have previously studied this behavior. For instance, Ewen and Chien [13,16], studied the effect of different alkyl substituents in cyclopentadienyl (CP) groups for several zirconocene (Zr) catalysts, reporting the following behavior in terms of catalyst efficiency: $(MeCP)_2ZrCl_2 > (EtCP)_2ZrCl_2 > CP_2ZrCl_2 > (Me_5CP)CPZrCl_2$. Through this, they concluded that single alkyl substituents increase the catalytic activity due to electro donating effects, while the steric hindrance of bulky substituents has a detrimental effect instead. Zr-based catalysts have been also studied in heterogeneous systems for ethylene polymerizations; for example, Charles et al. reported ethylene polymerization using catalysts derived from the activation of Zr aluminohydride complexes, supported on silica, which was previously treated with MAO. The results were compared with those using the more traditional Zr dichloro complexes, finding higher activity in the former [17]. Zeolites (ZSM-5) [18], and solid polymethylaluminoxane [19] are among the supports reported for carrying out ethylene polymerizations catalyzed by Zr-based metallocenes, achieving high catalytic activities, high molecular weights, and narrow distributions. Although these works provide general features about the influence of the alkyl groups on the ligand substituents, and the influence of using solid supports during the polymerization, they were all carried out using MAO as the activator.

Few works have studied in-detail the ethylene polymerization behavior when the metallocene is activated by the bulky, weakly-coordinating organoboron anions (B). In this sense, our research group reported the use of tris(pentafluorophenyl)borane and N,N-dimethylanilinium tetrakis(pentafluorophenyl)borate (B1 and B2 in this work, respectively) to act in conjunction with MAO as activators on ethylene polymerization by using the catalyst CP_2ZrCl_2. The addition of these organoboron compounds of ionic and nonionic nature in a molar ratio B1(or B2)/Zr = 5 promoted a partial deactivation of the catalyst, causing a reduction in the catalytic activity; however, the crystallinity degree, as well as the macromolecular, thermal, and dynamic-mechanical properties of the obtained polyethylenes were improved, especially with B1 as co-activator in this evaluated catalytic system [14]. In the same context, González-Hernández et al. [19] reported the ethylene polymerization using catalysts derived from Zr aluminohydride complexes activated with tris(pentafluorophenyl)borane (B1), although with limited utility (catalytic activity) of these catalysts systems when compared with the corresponding use of MAO as the activator. Supported zirconocene catalysts activated by boron compounds for olefin polymerizations are not as widely reported in the literature, but there are some related works such as that reported by Charoenchidet et al. who treated silica with tris(pentafluorophenyl)borane (B1 in this work) to produce borane-functionalized support, which was then used as a support and co-catalyst for the CP_2ZrCl_2, CP_2ZrCl_2/Triisobutylaluminum (TIBA), $CP_2Zr(CH_3)_2$ and $CP_2Zr(CH_3)_2$/TIBA catalyst systems for ethylene polymerizations. The activations of the catalysts were carried out in two ways: pre-activation, and in situ activation. The pre-activated and in situ-activated metallocene systems produced PE with M_w between 96 and 154 Kg mol^{-1}, and dispersity index ($Đ$) around 3. The bulk density of PE products was higher for the in situ-activated systems, but there was no significant difference between the products of both types of zirconocenes [20].

On the other hand, the kinetics of the catalyst coordination polymerization has been previously simulated, however a low number of reports can be found, compared to free-radical polymerization systems. Chien and Wang [13] reported the first kinetic model to study polymerization using zirconocene dichloride (CP_2ZrCl_2) and MAO as the catalyst and co-catalyst, respectively. The kinetic mechanism proposed the chain transfer to MAO, β-hydride chain transfer, multiple types of active sites, and deactivation step. Estrada and Hamielec [21] developed a model with two types of active sites, where the first one experienced a gradual transition (a state change) to the second type; this step was supported on the bimodal molecular weight distribution observed in the size exclusion chromatography (SEC) measurements. Both models did not provide an estimation of the ethylene concentration in the liquid phase. Moreover, Jiang et al. [22] carried out a comparative study between different models: in one of them, the reactivation of MAO was included as part of the kinetic mechanism, resulting in better agreement with the experimental polymerization rate profiles. A strategy of parameter estimation was reported by Ahmadi et al., in which a multivariable nonlinear optimization problem was solved using the Nelder–Mead simplex method [23]. The methodology combined the numerical solution of the kinetic model with the optimization algorithm, resulting in good agreement with the experimental data. Mehdiabadi and Soares [24] carried out a semi-batch reaction of a constrained geometry catalyst with MAO, and a kinetic model was proposed and then refined based on monomer uptake curves and polymer yield data. The deactivation of the catalyst/MAO system during ethylene polymerization was of the first order; the mechanism also included reversible activation and deactivation with MAO. The mechanism described the full kinetic picture. To the best knowledge of the authors, no study exists dealing with the modeling of zirconocene catalyst coordination polymerization using organoboron activators.

In this work, we aim to provide insights into the polymerization of ethylene catalyzed by Zr catalysts activated by organoboron compounds. Three Zr-based catalysts, with different ligand substituents, activated by three different organoboron compounds (B1, B2, and B3), were used for the PE synthesis. This work is focused on establishing the relationship between the catalytic system configuration with the polymerization behavior and with the final properties of the resultant polymers, in terms of molecular weight characteristics, crystallinity, and thermal behavior. Furthermore, the catalytic system leading to the highest catalytic activity was further analyzed, employing different solvents to elucidate the role over the features of the polymers. Moreover, a kinetic mechanism is proposed for the B3/Zr catalytic system, based on previous studies of MAO, and a mathematical model has been developed to estimate the kinetic rate coefficients of the two types of active species in the propagation, the chain transfer to monomer, polymer transition, spontaneous deactivation, and reactivation steps. With the knowledge of the kinetic parameters, the catalytic system is deeply studied, and some unexpected behaviors are analyzed.

2. Materials and Methods

2.1. Materials

All manipulations were carried out under an inert atmosphere using an MBraun glove box or via standard Schlenk techniques. Toluene, hexane, heptane, and isooctane were purchase from Sigma Aldrich (anhydrous grade) and were distilled twice from sodium and benzophenone before use. Bis(cyclopentadienyl)dimethyl zirconium (97%) ($CP_2Zr(CH_3)_2$), dimethylbis(*t*-butylcyclopentadienyl)zirconium (97%) ((t-butyl-CP)$_2$Zr(CH$_3$)$_2$) and dimethylbis(indenyl)zirconium (97%) ((ind)$_2$Zr(CH$_3$)$_2$) were purchased from Sigma Aldrich and used as received. Tris(pentafluorophenyl)borane (95%) ($B(C_6F_5)_3$), *N*,*N*-dimethylanilinium tetra(pentafluorophenyl)borate (98%) ($[C_6H_5N(CH_3)_2H]^+[B(C_6F_5)_4]^-$) and trityl tetrakis (pentafluorophenyl)borate (97%) ($[(C_6H_5)_3C]^+[B(C_6F_5)_4]^-$) were supplied by Strem Chemicals and were used as received. Polymer-grade ethylene was purchased from Praxair and was purified by passing it through 3–4 Å activated molecular sieves.

2.2. Polymerization Reactions

All polymerizations were performed in a 1 L stainless steel Parr reactor through the following procedure: three vacuum–argon cycles were first undertaken at 150 °C before the reaction to eliminate any traces of moisture. Then, the reactor was cooled down to room temperature and filled with 200 mL of solvent under an argon atmosphere. The reactor stirring system was set at 100 rpm and it was heated to 50 °C. The catalyst system was then fed into the reactor as follows: (i) boron compound solution (B) in 5 mL of toluene; (ii) metal catalyst solution in 5 mL of toluene. In all cases, 12.8 mmol of zirconium catalyst (Zr) was employed, and the B/Zr molar ratio was fixed to 2.5. The polymerizations were then initiated by introducing the ethylene monomer to the reactor at a continuous flow. All experiments were performed at an ethylene pressure of 1 bar and for 45 min. The reactions were terminated by the addition of acidified methanol. The resultant polymers were filtered off, washed with methanol, and vacuum dried.

2.3. Characterization

The molecular weight characteristics were determined by high-temperature size exclusion chromatography (SEC) with an Alliance chromatograph (GPC V-2000) equipped with two on-line detectors: a differential viscometer and refractometer, using three linear columns, PLgel 10 μm MIXED-B. The calibration was conducted under polystyrene standards using 1,2,4-trichlorobencene as eluent, and the measurements were carried out at a flow rate of 1 mL/min at 140 °C. The molar mass number and weight averages of the different polymers relative to polystyrene standards were corrected using the well-known principle of universal calibration employing the unique parameters for the Mark–Houwink–Sakurada equation for polyethylene: $K = 0.000323$ dL/g and $a = 0.735$. The melting temperature and crystallinity degree of the polymers was measured by differential scanning calorimetry (DSC), where the different thermograms were obtained through a TA instrument DSC 2920 at a heating rate of 10 °C/min under an inert atmosphere. Each sample was heated twice to eliminate the thermal history.

3. Mathematical Modeling

3.1. Kinetic Scheme

As the first approach to understand the mechanism, the following kinetic scheme is proposed in the ethylene polymerization in the work of Estrada and Hamielec [21] and Jiang [22].

To maintain the simplicity of the mechanism, here we considered that the complete catalyst has been instantaneously activated, producing the total concentration of the active sites (C_{Act}); other works have used the named Instantaneous Initiation Hypothesis [25]. As shown in Figure 2, the monomer addition to active sites results in polymeric chains, which are denoted as the active polymer of type 1, ($P_{r,1}$, where r is the degree of polymerization of the active polymer of type 1). In the propagation step, these active species add monomers in the chains, increasing the degree of polymerization. Estrada and Hamielec [21] firstly assumed the gradual transition of the active polymer of type 1 ($P_{r,1}$) to the active polymer of type 2 ($P_{r,2}$) in the catalyst coordination systems. The transition reaction is supported in ethylene polymerization via zirconocene/organoboron catalysts by the two polymeric populations found in the deconvolution of the SEC signal, as will be discussed later. Additionally, species $P_{r,2}$ increases their chain length by propagation. Both active polymer types can undergo a chain transfer to monomer reaction by the abstraction of a proton H from a monomer molecule to the active polymer of type 1 or 2, obtaining a dead polymer ($D_{r,i}$, where i denotes the polymer type) and an active polymer type either 1 or 2 with one monomeric unit in the chain. The deactivation of $P_{r,1}$ is negligible, and therefore only $P_{r,2}$ is spontaneously deactivated, which can present a reactivation by catalyst and monomer, similarly to MAO cocatalyst polymerization.

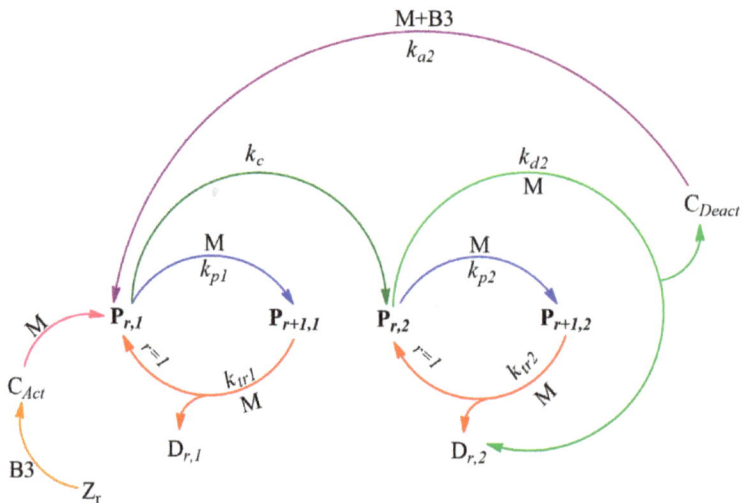

Figure 2. Proposed kinetic mechanism for the ethylene polymerization by zirconocene catalysts and organoboron activators. Kinetic parameters: k_{pi} denotes the propagation rate constant for active polymer type i; k_{tri} denotes the transfer chain to monomer rate constant for active polymer type i; k_c is the transition rate constant form $P_{r,1}$ to $P_{r,2}$; k_{d2} is the spontaneous deactivation rate constant of active polymer type 2; and k_{a2} is the re-activation of C_{Deact} species, via the reaction with M and B3. The description of each species appears in the text.

3.2. Population Balance Equations

The population mass balances for each species are derived from the kinetic scheme previously described: active sites (C_{Act}), deactivated sites (C_{Deact}), active polymer type 1 or 2 ($P_{r,1}$ and $P_{r,2}$, respectively) and dead polymer type 1 or 2 ($D_{r,1}$ and $D_{r,2}$, respectively) have been considered in this work, in Equations (S1)–(S8) shown in the Supplementary Materials.

Polymerization rate R_p, Equation (1), is principally based on the monomer consumed by the propagation, and the long chain hypothesis (LCH) is assumed; therefore, the monomer consumption in the initiation and reactivation steps is negligible. The contribution of the chain transfer to the monomer is also discarded, because $k_{pi} \gg k_{tri}$, as has been estimated in other coordination polymerization systems.

$$R_p = \sum_{i=1}^{2} \left(k_{pi}[M]_l Y_{0,i}\right) \tag{1}$$

The monomer is fed on demand to keep constant pressure; therefore, the ethylene concentration is almost constant with time (dM/dt = 0), and as a result, the course of the polymerization rate throughout the reaction can be directly tracked in the flowmeter measurements.

A polymerization rate model of the kinetic mechanism presented in this study, Figure 2, was reported by Jiang et al. [22] (Equation (2)), named herein as Model 1.

$$\begin{aligned}R_p =\ & k_{p1}^*[C_{Act}]e^{-k_c(t-t_0)} \\ & + k_{p2}^*[C_{Act}]\left(\frac{k_c k_{d2}}{(k_c - k_{d2} - k_a^*)(k_{d2} + k_a')}e^{-(k_{d2}-k_a')(t-t_0)} - \frac{k_c - k_a^*}{(k_c - k_{d2} - k_a^*)}e^{-k_c(t-t_0)}\frac{k_a^*}{k_a^* + k_{d2}}\right)\end{aligned} \tag{2}$$

where $k_{p1}^* = k_{p1}[M]_l$, $k_{p2}^* = k_{p2}[M]_l$, and $k_a^* = k_a[B][M]_l$

If the reactivation step of the deactivated sites is negligible, Equation (2) is transformed into Equation (3), as reported by Vela-Estrada et al. [21] and named as Model 2.

$$R_p = [M]_l \left(k_{p1}^* e^{-k_c(t-t_0)} + \frac{k_{p2}^* k_c}{k_c - k_{d2}} \left(e^{-k_{d2}(t-t_0)} - e^{-k_c(t-t_0)} \right) \right) \quad (3)$$

3.3. Liquid–Vapor Equilibrium (LVE)

The concentration of ethylene transferred to the liquid phase ($[M]_l$) in a liquid–vapor equilibrium (LVE) system is estimated based on Duhem's theorem, in which a T, P-flash calculation was developed. We assumed that toluene at a pressure (P) equal or lower than its bubble-point pressure (P_{Bubl}), Equation (S9), was partially evaporated because the pressure was reduced so an LVE was established between the toluene/ethylene phases, but it should have been greater than the drew-point pressure (P_{Dew})

Vapor pressure of a pure species (P_i^{sat}, where i = ethylene or toluene) is obtained by Equation (4), using the Antoine equation, and whose parameters are shown in Table 1.

$$\log_{10}(P_i^{Sat}[bar]) = A_i - \left(\frac{B_i}{T[K] + C_i} \right) \quad (4)$$

Table 1. Parameters used in the Antoine equation for toluene and ethylene.

Component i	A_i	B_i	C_i	Temp. Range (K)	Ref.
Toluene	4.08	1346.38	53.51	273.13–297.89	[26,27]
Ethylene	3.87	584.14	18.31	149.37–188.57	[28,29]

The calculations of equation for P_{Bubl} or P_{Dew} involve $\{x_i\} = \{z_i\}$ and $\{y_i\} = \{z_i\}$, respectively, with z_i being the overall composition in the components.

First, P must lie in the range of the following constraint, Equation (5):

$$P_{Dew} \geq P \geq P_{Bubl} \quad (5)$$

The K-value correlations are estimated by Equation (6):

$$K_i = \frac{P_i^{sat}}{P} \quad (6)$$

The moles in the vapor phase (v) are calculated via a nonlinear algebraic equation, Equation (7), which is solved by the Newton method. The total molar mass in the liquid is obtained by the difference with respect to the total molar amount.

$$\sum_{i=1}^{2} \frac{z_i K_i}{1 + v(K_i - 1)} = 1 \quad (7)$$

The factions y_i and x_i are calculated by Equations (8) and (9), respectively

$$y_i = \frac{z_i K_i}{1 + v(K_i - 1)} \quad (8)$$

$$x_i = \frac{y_i}{K_i} \quad (9)$$

The total moles in the liquid phase n_L, Equation (10), is obtained from the mass balances of toluene, and ethylene, in both the phases, where n_{et0} and n_{Tol0} are the initial total moles of ethylene and toluene, respectively.

$$n_L = \frac{n_{et0} - n_{Tol0}\varphi}{(x_{Et} - x_{Tol}\varphi)} \tag{10}$$

and

$$\varphi = \frac{y_{Et}}{y_{Tol}} \tag{11}$$

Finally, the ratio between the moles of ethylene and the solvent volume (the volume of the ethylene is very low, so it is negligible) produces the concentration of ethylene.

$$[M]_l = \frac{n_{Et,L}}{V_{Tol,L}} \tag{12}$$

3.4. Optimization of the Parameter Estimation

The parameters k_{p1}^*, k_{p2}^*, k_c, k_{d2}, and k_{a2}^* in Equations (2) or (3) were found by an optimization tool in Matlab 2015a, and the minimization function, named fmincon [30], in which the objective function was defined as Equation (13):

$$\begin{aligned}&\min \sum_{i=1}^{n} \left(\frac{R_p^P(t_i) - R_p^M(t_i)}{R_p^M(t_i)} \right)^2 \\ &\text{s.t.} \\ &k_{p1}^* k_{p2}^* k_a^* k_c k_{d2} \geq 0 \\ &k_{p1}^* k_{p2}^* k_a^* k_c k_{d2} \in R^n \end{aligned} \tag{13}$$

Here, $R_p^P(t_i)$ denotes the predicted polymerization rate at the time i, and $R_{p(i)}^M$ is the measured polymerization rate at the time i. Additionally, the coefficient of determination denoted R^2 was calculated (Equation (14)):

$$R^2 = 1 - \frac{SS_{Res}}{SS_{Tot}} \tag{14}$$

where

$$SS_{Res} = \sum_{i=1}^{n} \frac{\left(R_{p(i)}^P - R_{p(i)}^M\right)^2}{n-1} \tag{15}$$

$$SS_{Tot} = \sum_{i=1}^{n} \frac{\left(R_{p(i)}^P - \widehat{R}_{p(i)}^M\right)^2}{n-1} \tag{16}$$

The standard deviation (S) is defined in Equation (17)

$$S = \sqrt{\sum_{i=1}^{n} \frac{\left(R_{p(i)}^P - R_{p(i)}^M\right)^2}{n-1}} \tag{17}$$

In a nonlinear model the covariance-variance of the kinetic parameter p ($\hat{\beta}_p$) is estimated by Equation (18) [31]:

$$\text{var}\,\hat{\beta}_p = SS_{Res} \sqrt{\left\{(J^T J)^{-1}\right\}_{pp}} \tag{18}$$

where J is the matrix of derivatives of the non-linear model with respect to the parameters (similar to the Jacobian matrix of the system) and $\left\{(J^TJ)^{-1}\right\}_{pp}$ is equal to pth diagonal term of the matix $(J^TJ)^{-1}$. Each element of J is

$$J_{ij} = \frac{\partial F(x_i, \beta)}{\partial \beta_j} \quad (19)$$

where F is the non-linear model and the x_i are the experimental points.

3.5. Method of Moments

In this section, the mathematical model was developed by using the method of moments, in which an overall kinetic behavior is obtained. After writing out the mass balances of two polymer populations (Equations (S1)–(S8)), the three first moments were derived (Equations (20)–(31)), considering the moment definitions shown in Table 2.

Table 2. Definition of moments.

Species	Definition
j-th moment for the active polymer of type i, $i = 1, 2$	$Y_{j,i} = \sum_{n=1}^{N} n^j [P_{n,i}]$
j-th moment for the dead polymer of type i, $i = 1, 2$	$Z_{j,i} = \sum_{n=1}^{N} n^j [P_{n,i}]$

Zeroth moments

$$\frac{dY_{0,1}}{dt} = k_a^*([C_{Act}] - Y_{0,1} - Y_{0,2}) - k_c Y_{0,1} \quad (20)$$

$$\frac{dY_{0,2}}{dt} = k_c Y_{0,1} - k_{d2} Y_{0,2} \quad (21)$$

$$\frac{dZ_{0,1}}{dt} = k_{tr1}^* Y_{0,1} \quad (22)$$

$$\frac{dZ_{0,2}}{dt} = (k_{d2} + k_{tr2}^*) Y_{0,2} \quad (23)$$

First Moments

$$\frac{dY_{1,1}}{dt} = k_a^*([C_{Act}] - Y_{0,1} - Y_{0,2}) + k_{p1}^* Y_{0,1} - k_c Y_{1,1} - k_{tr1}^* Y_{1,1} + k_{tr1}^* Y_{0,1} \quad (24)$$

$$\frac{dY_{1,2}}{dt} = k_c Y_{1,1} + k_{tr2}^* Y_{0,2} - k_{tr2}^* Y_{1,2} + k_{tr2}^* Y_{0,2} - k_{d2} Y_{1,2} \quad (25)$$

$$\frac{dZ_{1,1}}{dt} = k_{t1}^* Y_{1,1} \quad (26)$$

$$\frac{dZ_{1,2}}{dt} = k_{tr2}^* Y_{1,2} + k_{d2} Y_{1,2} \quad (27)$$

Second Moment

$$\frac{dY_{2,1}}{dt} = k_a^*([C_{Act}] - Y_{0,1} - Y_{0,2}) + 2k_{p1}^* Y_{1,1} + k_{p1}^* Y_{0,1} - k_c Y_{2,1} - k_{tr1}^* Y_{2,1} + k_{tr1}^* Y_{0,1} \quad (28)$$

$$\frac{dY_{2,2}}{dt} = k_c Y_{2,1} + 2k_{p2}^* Y_{1,2} + k_{p2}^* Y_{0,2} - k_{tr2}^* Y_{2,2} + k_{tr2}^* Y_{0,2} - k_{d2} Y_{2,2} \quad (29)$$

$$\frac{dZ_{2,1}}{dt} = k_{tr1}^* Y_{2,1} \quad (30)$$

$$\frac{dZ_{2,2}}{dt} = k^*_{tr2} Y_{2,2} + k_{d2} Y_{2,2} \tag{31}$$

where $k^*_{tr1} = k_{tr1}[M]_l$ and $k^*_{tr2} = k_{tr2}[M]_l$

Number (M_n) and weight (M_w) average molecular weights were calculated by Equations (32) and (33), respectively.

$$M_n = \left(\frac{\sum_{i=1}^{2} Y_{1,i} + \sum_{i=1}^{2} Z_{1,i}}{\sum_{i=1}^{2} Y_{0,i} + \sum_{i=1}^{2} Z_{0,i}} \right) MW_{mon} \tag{32}$$

$$M_n = \left(\frac{\sum_{i=1}^{2} Y_{2,i} + \sum_{i=1}^{2} Z_{2,i}}{\sum_{i=1}^{2} Y_{1,i} + \sum_{i=1}^{2} Z_{1,i}} \right) MW_{mon} \tag{33}$$

4. Results and Discussion

4.1. Polymerizations

With the aim to provide insights regarding the ethylene polymerization conceived by zirconocenes and organoboron activators, we carried out a series of experiments using three different zirconocene catalysts, with different ligand substituents, activated via three different organoboron compounds. The polymerization behavior, as well as the final properties, were evaluated. Moreover, we validated our experimental results via a kinetic modeling study, which is presented in the next section of the article. The general properties of the result polymers are shown in Table 3.

The ethylene polymerization was greatly influenced by the type of organoboron compound used as the activator. Considering the catalytic activity, shown in Table 3, trityl tetrakis(pentafluorophenyl)borate (B3) was the one promoting the ethylene polymerization to the greatest extent, irrespectively of the zirconocene precursors, which presumably was due to the possible remaining interaction between the abstracted methyl group and the metal cation in the case of B1. While in B2, one of the by-products was a trisubstituted amine that might be able to trap the coordinatively unsaturated cation formed in this reaction, as has been previously reported for other systems [32]: both phenomena could potentially decrease the catalytic activity.

On the other hand, it was found that the non-substituted CP ring (Zr1) led to the highest catalytic activity, suggesting that by employing these organoboron compounds as activators, the bulkiness of the t-butyl substituent groups on the CP ring (Zr2), shows dominance towards the electro-donating effect, decreasing the polymerization activity due to steric hindrance. On the other hand, Zr1 also led to higher catalytic activity than Zr3, which has an indenyl substituent group. There is a discrepancy in the literature regarding the difference in the electronic environment in indenyl groups compared to CP. The general understanding has been that indenyl is more electron-rich than CP [33,34]. However, some authors have also suggested that indenyl is a poorer donor based on reduction potentials. Moreover, Nguyen et al. suggest that indenyl is electron-richer than CP, but an anodic shift of the reduction potential presumably occurs because an η5 to η3 haptotropic shift accompanies reduction, thus stabilizing the product [35]. Further understanding of this complex phenomenon is required, and at this point, any explanation appears to be merely speculative.

Table 3. General features of the synthesized polyethylene (PE) using different catalytic systems.

Sample [a]	Catalyst [b]	Co-Catalyst [c]	A [d]	M_w (Kg/mol)	Đ [e]	X [f] (%)	T_m (°C)
PE1	Zr1	B1	24.6	282	2.5	62.4	135.7
PE2	Zr1	B2	20.3	302	3.9	74.5	136.9
PE3	Zr1	B3	61.0	456	3.1	65.8	136.1
PE4	Zr2	B1	9.2	467	3.04	66.7	134.0
PE5	Zr2	B2	8.4	143	7.46	73.5	134.3
PE6	Zr2	B3	47.7	332	2.04	70.1	137.5
PE7	Zr3	B1	23.0	261	2.75	61.9	134.9
PE8	Zr3	B2	13.0	231	4.47	72.7	134.3
PE9	Zr3	B3	26.3	366	4.86	65.8	136.2

[a] Isothermal polymerizations (50 °C) were performed in 200 mL of toluene for 45 min: Zr = 1.28×10^{-4} mol, molar ratio B/Zr = 2.5. Pressure = 1 bar. [b] Zr1 = $CP_2Zr(CH_3)_2$; Zr2 = $(t\text{-butyl-CP})_2Zr(CH_3)_2$; Zr3 = $(ind)_2Zr(CH_3)_2$. [c] B1 = $B(C_6F_5)_3$; B2 = $[HNMe_2Ph][B(C_6F_5)_4]$; B3 = $[CPh_3][B(C_6F_5)_4]$. [d] Catalytic activity ($Kg_{PE}\ mol_{Zr}^{-1}\ h^{-1}$) [e] Dispersity index ($M_w/M_n$) was determined by SEC. [f] Crystallinity degree calculated by DSC.

Concerning the molecular weight characteristics (shown in Table 3), the lowest molecular weight was exhibited by employing B2 as the activator, accompanied by an increase in dispersity index (Đ), suggesting chain transfer reactions during the polymerization, which was also reflected in the multimodal behavior of the molecular weight distributions (MWDs), shown in Figure 3. The MWD was further deconvoluted by statistical procedures to provide an approximate notion of the number of active sites carrying out the polymerization, considering that each active site possesses different probabilities of chain transfer and termination, therefore producing polymers with individual molar mass distributions, with the observed MWD being a superposition of all products.

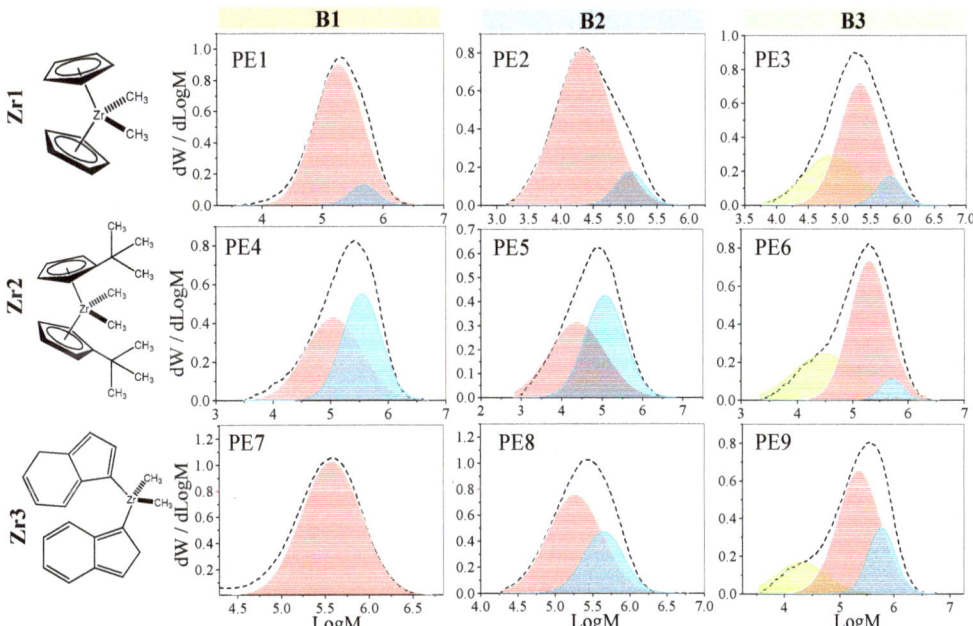

Figure 3. Deconvoluted molecular weight distributions of the synthesized PE using the different catalytic systems.

A strong influence of the activator over the MWD was observed, while a narrow bimodal molecular weight distribution was observed by using B1 as a co-catalyst; at least three kinds of active sites were observed in the case of B3. Concerning B2 as the activator, a broad bimodal MWD was exhibited, presenting the Zr2 + B2 system as the highest polydispersity (Đ = 7.5) with the lowest catalytic activity due to the poor compatibility between (t-butyl-CP)$_2$Zr(CH$_3$) and [C$_6$H$_5$N(CH$_3$)$_2$H]$^+$[B(C$_6$F$_5$)$_4$] as the catalyst system for ethylene polymerization. The crystallinity degree was determined through DSC (values are shown in Table 3). A relatively high crystallinity degree of around 60–75% was observed in all cases, as expected for high density polyethylene (HDPE). However, a slightly higher degree of crystallinity was observed with B2 (around 72%) as the activator, attributed to the relatively lower molecular weight. Nevertheless, this activator also led to the broadest MWD, suggesting the presence of chain transfer reactions during the polymerization; however, it could be assumed that β-hydride was not predominant, because branching would take place thus decreasing the crystallinity. Concerning the melting temperature, no significant differences were exhibited among the samples, remaining above 134 °C as expected.

To elucidate the influence of different solvents over the polymerization behavior and polymer properties, isooctane, heptane, and hexane were tested for the catalytic system CP$_2$Zr(CH$_3$)$_2$ + [(C$_6$H$_5$)$_3$C]$^+$[B(C$_6$F$_5$)$_4$]$^-$ (Zr1 + B3), which led to the highest catalytic activity. The results were compared with those of the sample PE3, in which toluene was used. The results are shown in Table 4. As can be observed, a significant difference in the catalytic activity and molecular weight characteristics was exhibited by varying the solvent. The highest catalytic activity was found when using toluene as solvent, followed by isooctane, heptane and finally hexane; which results correlate to the solubility parameter of the solvents, implying that they play a fundamental role in the polymerization behavior. This behavior is presumably related to the higher miscibility of the catalyst system (solubility parameter not calculated in this work); however, additionally, the solubility parameter is related to the solvency behavior of a specific solvent, and could therefore influence the nature of the equilibrium of the complexation reaction together with the solvation effect. Further investigations are required to understand this behavior. Concerning the molecular weight characteristics, the highest molecular weight was obtained by using hexane, which is attributed to the reduced concentration of active sites, implied by the low catalytic activity. On the other hand, by using heptane, the lowest molecular weight was obtained, which suggests the increase in termination reactions, and which is supported by the higher polydispersity value reported in Table 4. Similar results were observed by using toluene and isooctane as solvents.

Table 4. Influence of solvents over the ethylene polymerization using CP$_2$Zr(CH$_3$)$_2$ (Zr1) as the catalyst, activated by [CPh$_3$][B(C$_6$F$_5$)$_4$] (B3).

Sample [a]	Solvent	δ [b]	A [c]	M_w (Kg/mol)	Đ [d]
PE3	Toluene	8.8	61.0	456	3.2
PE10	Isooctane	6.9	46.0	473	2.8
PE11	Heptane	7.4	33.0	310	3.6
PE12	Hexane	7.3	10.5	613	2.0

[a] Isothermal polymerizations (50 °C) were performed in 200 mL of solvent for 45 min: Zr = 1.28 × 10^{-4} mol, molar ratio B/Zr = 2.5. Pressure = 1 bar. [b] Solubility parameter (cal/cc)$^{1/2}$. [c] Catalytic activity (Kg$_{PE}$/mol$_{Zr}$/h). [d] Dispersity index (M_w/M_n), determined by SEC.

4.2. Kinetic Parameter Estimations

The catalytic system (B + Zr) that showed the highest catalytic activity was B3/Zr1 (in toluene); therefore, in the following mathematical modeling sections, a polymerization series using the aforementioned system was synthesized. The resulting experimental data were analyzed in Table 5. The ratio B3/Zr1 was varied in two levels (1 and 2.5) and the operating temperature was changed in three levels (40, 50, and 60 °C). Additionally, the operating pressure was increased to 1.5 bar to increase the catalytic activity.

Table 5. Reaction parameters and general features of the synthesized PEs using $CP_2Zr(CH_3)_2$ (Zr1)/[CPh$_3$][B(C$_6$F$_5$)$_4$] (B3), used for kinetic modeling.

Sample [a]	Temperature (°C)	B/Zr	A [b]
PE13	40	1	12.07
PE14	40	2.5	12.75
PE15	50	1	15.59
PE16	50	2.5	16.29
PE17	60	2.5	15.70

[a] Polymerizations were performed in 200 mL of toluene for 45 min: Zr = 1.28 × 10^{-4} mol, molar ratio B/Zr = 2.5. Pressure = 1.5 bar. [b] Catalytic activity (Kg$_{PE}$ mol$_{Zr}^{-1}$ h^{-1}).

The thermodynamic state of the liquid and vapor phases was considered under VLE because the operational pressure (P = 150 bar, T = 40, 50, and 60 °C) lay between the bounds of P_{Bubl} and P_{Dew} under all the studied conditions (Equation (5)). Therefore, a T, P-flash calculation protocol was carried out, resulting in $[M]_l$ = 0.17, 0.14, and 0.11 mol L^{-1}, as shown in Table 6. As the temperature increased, a higher amount of mass of both components (ethanol and toluene) was transferred to the vapor phase, resulting in a lower concentration of ethylene in the liquid phase. According to Lee et al. [36], the solubility of the ethylene in toluene is inversely proportional to a temperature increase; the results of our predictions are in agreement with these experimental findings. Figure 4 shows the dependency of $[M]_l$ with respect to the operating temperature and pressure.

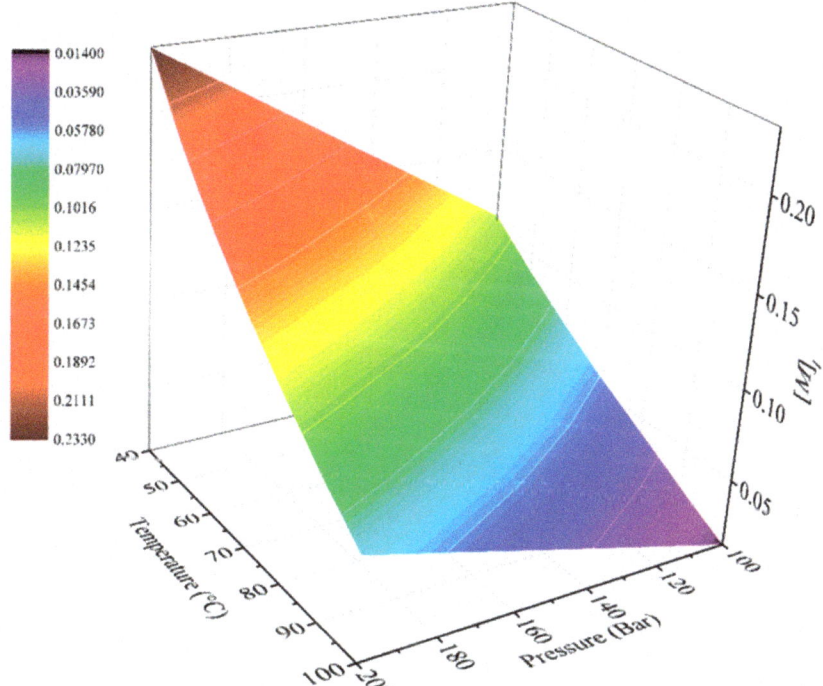

Figure 4. Mapping of the influence of the temperature (°C) and pressure (bar) on the monomer concentration in the liquid phase (toluene).

Table 6. Operating conditions, $[M]_l$, estimated kinetic rate constants, coefficients of determination (R^2), standard deviation (S), and operating pressure P = 150 bar.

Run	PE13		PE14		PE15		PE16		PE17	
Temp. (°C)	40		40		50		60		60	
[Zr] (mol L⁻¹)	0.17		0.17		0.14		0.11		0.11	
$[M]_l$ (mol L⁻¹)	5.53×10^{-4}		1.38×10^{-3}		5.51×10^{-4}		1.38×10^{-3}		1.38×10^{-3}	
[B3] (mol L⁻¹)					5.53×10^{-4}					
Model	1	2	1	2	1	2	1	2	1	2
$k_{p1} \times 10^{-5}$ (L mol⁻¹ min⁻¹)	2.52 ± 0.74	10.93 ± 0.10	0.31 ± 1.85	16.72 ± 0.24	10.33 ± 1.56	22.80 ± 0.22	5.62 ± 2.13	25.56 ± 0.46	36.15 ± 0.67	42.43 ± 1.16
$k_c \times 10^2$ (min⁻¹)	9.00 ± 0.32	7.27 ± 0.12	11.00 ± 6.65	8.26 ± 0.24	9.00 ± 1.59	7.27 ± 0.18	11.00 ± 0.52	8.26 ± 0.28	11.00 ± 0.2	8.26 ± 2.22
$k_{p2} \times 10^{-5}$ (L mol⁻¹ min⁻¹)	40.76 ± 5.03	12.92 ± 0.45	60.81 ± 13.57	18.95 ± 1.40	48.98 ± 12.21	15.53 ± 1.20	61.84 ± 10.87	24.17 ± 4.23	62.38 ± 7.44	29.19 ± 7.76
$k_{d2} \times 10^1$ (L mol⁻¹ min⁻¹)	3.72 ± 5.5	5.47 ± 0.44	3.90 ± 1.24	6.58 ± 0.67	3.88 ± 1.01	5.47 ± 0.28	4.87 ± 1.76	11.12 ± 0.50	5.64 ± 6.70	11.10 ± 5.14
$k_{a2} \times 10^{-2}$ (L² mol⁻² min⁻¹)	0.08 ± 1.34	—	0.03 ± 0.12	—	0.23 ± 0.49	—	0.15 ± 0.17	—	1.18 ± 2.52	—
$k_{tr1} \times 10^{-2}$ [a] (L mol⁻¹ min⁻¹)	0.20	1.10	0.20	1.83	1.00	2.60	2.60	4.00	5.80	6.60
$k_{tr2} \times 10^{-2}$ [a] (L mol⁻¹ min⁻¹)	4.90	3.30	6.80	4.15	8.00	6.00	8.10	5.00	10.00	8.00
R^2	0.94	0.92	0.94	0.90	0.74	0.27	[b]	[b]	0.98	—
S	10.44	9.52	24.90	18.31	16.16	27.10	37.65	43.93	10.69	64.01

[a] Values of k_{tr1} and k_{tr2} are not statistical estimates under any circumstances. [b] Values of R^2 were not calculated, due to the high values of S.

The next step was to obtain the optimal values of the kinetic rate constants k_{p1}, k_{p2}, k_{d2}, k_c, and k_{a2} by fitting the experimental polymerization rate profiles to Model 1 (Equation (2)) and Model 2 (Equation (3)), by using Matlab tools. It must be noted that the lower bound of each parameter was limited to the value previously found at a lower temperature for experiments in the same series, i.e., PE14, PE16, and PE17. It is clear that these constraints are based on the expected physical behavior, described by the Arrhenius expression.

Comparisons between the optimization results of Model 1 (dashed lines), Model 2 (dotted lines), and the experimental profiles (continuous lines) for the five selected experiments are illustrated in Figure 5, and the kinetic parameter values and the coefficients of determination, R^2, are presented in Table 6. The best fit with the experimental curves corresponds to predictions by Model 1 with the reactivation step of the deactivated catalytic system, with closer values of R^2 to the unity than those obtained by Model 2 (Table 6). Therefore, it is assumed that the reactivation has high importance in the adequate prediction of the polymerization rate profile.

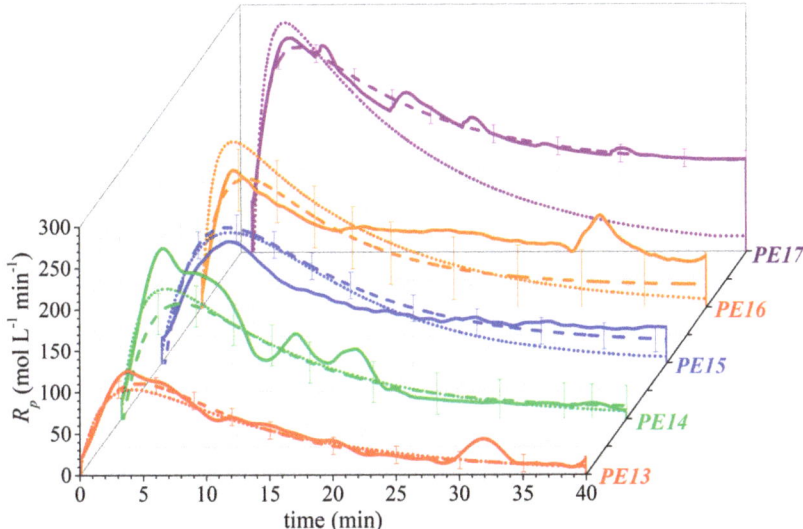

Figure 5. Polymerization rate profiles using Zr1/B3 as a catalytic system, varying the temperature and the [B]$_0$ for experimental data (continues lines): Model 1 (dashed lines), and Model 2 (dotted lines). The operating conditions are shown in Table 6. Bars denote the standard deviations (S) for Model 1.

To estimate the transfer chain kinetic rate constants (k_{tr1} and k_{tr2}), the system of ordinary differential equations (ODEs) was solved by a routine called ode23s in Matlab. The previously estimated parameters for Model 1, shown in Table 6, were used, and the M_n and M_w values, computed by Equations (41) and (42), were fitted to the corresponding experimental values. The values of k_{tr1} and k_{tr2} were assigned until they reached a good agreement with the experimental data, but the low bound in the estimation was constrained to the fitted value at a lower temperature, analogous to the procedure previously described. As presented in Table 7, the fitted M_n and M_w values showed an excellent agreement with the experimental data with low values of the relative changes, and the proper values for k_{tr1} and k_{tr2} are summarized in Table 6.

Table 7. Fitted (Model 1) and experimental molecular weights (gr mol^{-1}) of polyethylene, and the relative change (R.C.). P = 150 bar.

Run	PE13	PE14	PE15		PE16		PE17
Time (min)	40	40	30	40	30	40	40
M_n exp. (gr mol^{-1})	248,358	234,792	185,668	224,373	147,564	131,544	176,695
M_n fitted (gr mol^{-1})	249,940	230,240	211,596	211,370	152,230	152,026	174,650
R.C.	0.006	−0.020	0.123	−0.062	0.031	0.135	−0.008
M_w exp. (gr mol^{-1})	507,982	477,332	480,188	455,629	380,701	349,233	352,227
M_w fitted (gr mol^{-1})	513,070	492,690	452,210	451,740	328,990	328,590	349,270
R.C.	0.010	0.031	−0.062	−0.009	−0.157	−0.063	−0.008

All the kinetic rate constants of the series PE14, PE16, and PE17, where the temperature was increased at 40, 50, and 60 °C, respectively, with constant molar ratio B3/Zr1 = 2.5, were used in the estimation of the activation energy (E_A) and steric factor (A_0). Figure 6 depicts an Arrhenius plot ($ln\ (k_j)\ versus\ T^{-1}$) and the fitted equations for all the ks values. An excellent fit was achieved for all the kinetic rate constants ($R^2 > 0.92$), with the exception of k_c, giving $R^2 = 0.04$, attributed to the almost-null increase in the constant with respect to the temperature, as shown in Table 6. In fact, when an optimization procedure without a constrained low bound on k_c was run, a negative effect of the temperature on this rate constant was computed. This was probably due to a reversible transition reaction taking place, in which the difference between the $E_A(k_c)$ and $E_A(k_{-c})$ predominately led to the formation of the active polymer of type 2 at low temperatures, but at high temperatures, the formation of active chains of type 1 could be favored, as reported for the ethylene polymerization using MAO and CP$_2$ZrCl$_2$ [22,23]. Hence, as a first approximation of the transition reaction in this system, in the next section a value of $k_c = 0.11$ min^{-1} will be used for all the simulations.

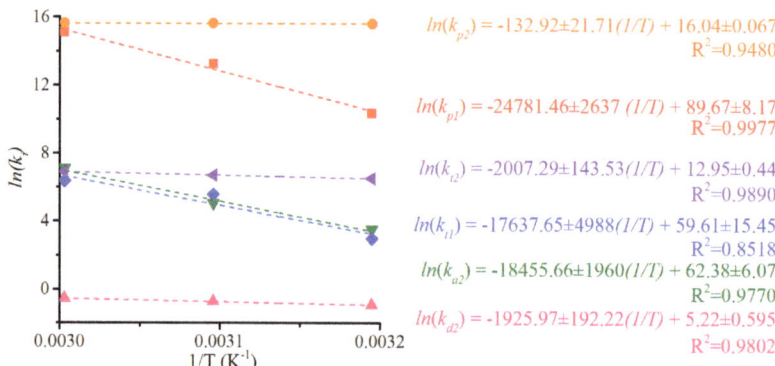

Figure 6. Arrhenius plot for a set of k_i values for the ethylene coordination polymerization using a ratio B/Zr = 2.5 and their linear regressions.

On the other hand, contrary to the findings reported by Jiang et al. [23], in which the reactivation reaction only was important at low temperatures (at 50 °C k_{a2} = 19 L^2 mol^{-2} min^{-1}, at 60 and 70 °C k_{a2} = 0), in this work the effect of the temperature on the reactivation of catalytic systems (C_{Deact}) has been well described, obtaining $R^2 = 0.9880$ in the linear regression, as shown in Figure 6: $E_a\ (k_{a2}) = 153.44 \pm 16.29$ kJ mol^{-1} and $A_0\ (k_{a2}) = 1.23 \times 10^{27}$ (exp($\pm$6.07)) L^2 mol^{-2} min^{-1}. The values estimated for the spontaneous deactivation are $E_A\ (k_{d2}) = 16.01 \pm 1.60$ kJ mol^{-1} and $A_0\ (k_{d2}) = 184.96$(exp($\pm$0.59)) L mol^{-1} min^{-1}.

Additionally, high differences in the E_A and A_0 values between k_{p1} and k_{p2} were obtained (Figure 6), attributable to more reactive active polymers of type 2 than those initially generated of type 1. While the values of $E_A\ (k_{p2}) = 1.09 \pm 0.18$ kJ mol^{-1} and

A_0 (k_{p2}) = 9.16 × 10^{27}(exp(±0.07)) L^2 mol^{-2} min^{-1} were estimated for the former type, the following values of E_A (k_{p1}) = 206.03 ± 21.92 kJ mol^{-1} and A_0 (k_{p1}) = 8.77 × 10^{38} (exp(±8.17)) L mol^{-1} min^{-1} were estimated for the propagation step of the latter. As expected, the estimated behavior for the monomer rate constants for each type was related to the k_p values with E_A (k_{tr1}) = 146.63 ± 41.47 kJ mol^{-1} and A_0 (k_{tr1}) = 7.73 × 10^{25} (exp(±15.46))L mol^{-1} min^{-1}; and E_A (k_{tr2}) = 16.68 ± 1.19 kJ mol^{-1} and A_0 (k_{tr2}) = 4.21 × 10^5 (exp(±0.44)) L mol^{-1} min^{-1}. Given such high values of E_A (k_{p1}), E_A (k_{tr1}), and E_A (k_{a2}), at high temperature (i.e., 70 °C) one would expect the active polymer of type 1 to present lower molecular weights than type 2, as well as a competitive step between the transition to type 2 and the chain transfer to monomer.

It is important to note a strong effect of the boron-based activator on the kinetic parameters, because a concentration change induces different values during the fitting procedure, as shown in Table 6. If the values of E_A and A_0 are estimated with the available experimental data at 40 and 50 °C with [B3] = 5.53 × 10^{-4} mol L^{-1}, the values in Table 8 are obtained. A considerable decrease in the frequency factor and the activation energy of k_{p1} is observed in comparison to those estimated values at a high concentration of B3 (Table 6). Additionally, the energetic barrier is increased for k_{p2} with respect to a high concentration of B, but its frequency factor is greatly decreased, ascribable to an important steric hindrance of the active site with the monomer.

Table 8. Parameter estimation results of the PE polymerization using B3 = 5.53 × 10^{-4} mol L^{-1}.

Kinetic Rate Constant	A_0	E_A (kJ mol^{-1})
k_{p1} (L mol^{-1} min^{-1})	1.54 × 10^4	15.44
k_{p2} (L mol^{-1} min^{-1})	3.68 × 10^9	41.20
k_{d2} (L mol^{-1} min^{-1})	14.64	3.56
k_{a2} (L^2 mol^{-2} min^{-1})	1.56 × 10^{20}	118.60
k_{tr1} (L mol^{-1} s^{-1})	7.54 × 10^{21}	135.27
k_{tr2} (L mol^{-1} s^{-1})	1.10 × 10^{13}	84.60

4.3. Kinetic Simulations

The estimated kinetic rate constants were used as inputs in the mathematical model, and the results are shown in Figure 7. First, the effect of the temperature was studied in the series PE10, PE12, and PE14, corresponding to polymerizations at 40, 50, and 60 °C, respectively (Figure 7a,c,e). As expected, a higher temperature leads to an increase in the polymerization rate (Figure 7a versus Figure 7e). The R_p curves show an increase in the first five minutes of the reaction, which is attributed to the gradual generation of active polymer of type 2 ($Y_{0,2}$), with a higher propagation rate than type 1 ($Y_{0,2}$). Then, the profiles reach a maximum when the highest concentration of ($Y_{0,2}$) is generated, as confirmed in Figure 7b,d,f. At longer reaction times, the deactivation of both active polymer chains is greater than their generation, resulting in a decrease in ethylene consumption.

The concentration curves at 40 °C are shown in Figure 7b, in which $Y_{0,1}$ constantly decreases, but $Y_{0,2}$ reaches a maximum before 5 min of reaction have elapsed. A higher production of dead polymer of type 2 ($Z_{0,2}$) than that for type 1 ($Z_{0,1}$) was predicted by the model. Additionally, M_n for $Z_{0,2}$ was five-fold higher than that for $Z_{0,1}$ (Figure 8a), which reached a plateau after 5 min. As illustrated, the value of M_n for the final polymer product is very similar to that obtained to that predicted for $Z_{0,2}$. Notably, the experimental values for M_n and M_w for the final product (symbols) show an excellent agreement with the predicted values (lines) with dispersity of 2.03. Such difference in the values of M_n between $Z_{0,1}$ and $Z_{0,2}$ resulted in two populations, which are observed in the MWD of the final product measured by SEC (Figure 8b), being the long tail of the distribution attributable to $Z_{0,1}$. The two polymer populations have been obtained by statistical deconvolution of the SEC experimental data, and they should be taken with wariness. More advanced deconvolution procedures can be found in the literature [37], and these will be explored in further works.

At 50 °C, the R_p profile was more sustained (Figure 7a); however, its maximum value was lower than that predicted at 40 °C, which was probably due to a fast transfer chain to polymer rate. This produces similar concentrations for $Z_{0,1}$ and $Z_{0,2}$ species throughout the polymerization, as shown in Figure 7b. At this temperature, the gap of M_n values between the $Z_{0,1}$ and $Z_{0,2}$ species was smaller than at lower temperatures (Figure 8a), and the populations of both species similarly contributed to the MWD (Figure 8a) with a dispersity of 2.65.

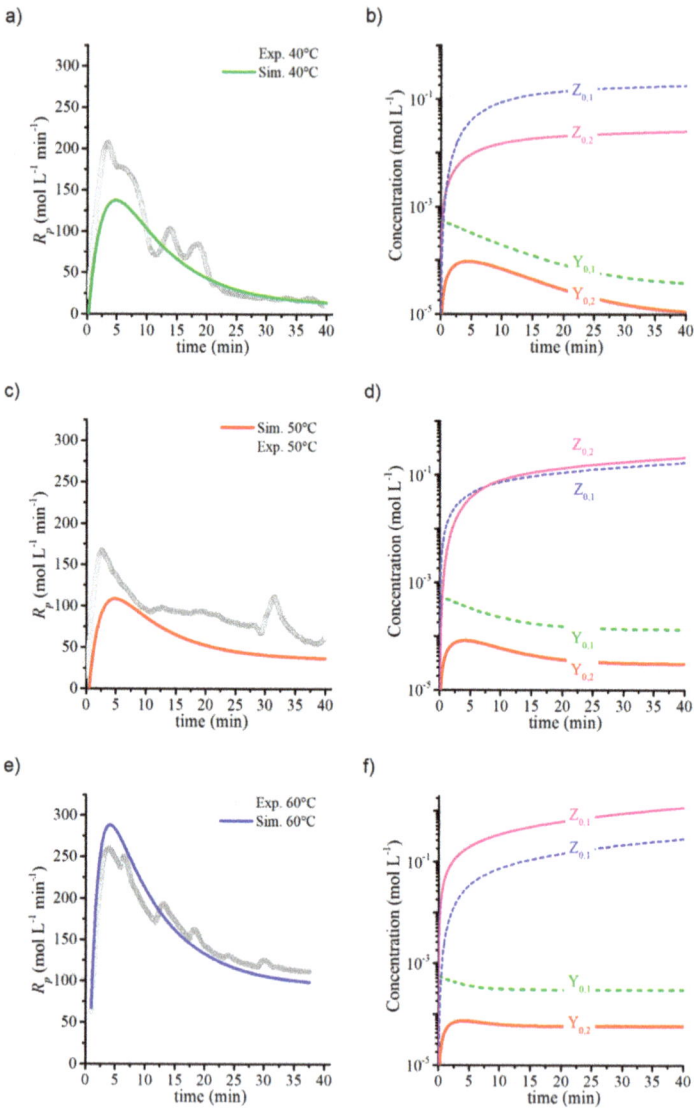

Figure 7. Effect of the temperature variation at 40 °C (**a,b**), 50 °C (**c,d**), and 60 °C (**e,f**) on: R_p profiles (**a,c,e**) with the experimental data (gray symbols) and the simulation (continues lines), and total concentration of polymer species (zeroth moment) (**b,d,f**). The ratio of $[B]_0/[Zr]_0 = 2.5$ and $P = 150$ bar. Other operating conditions are shown in Table 6.

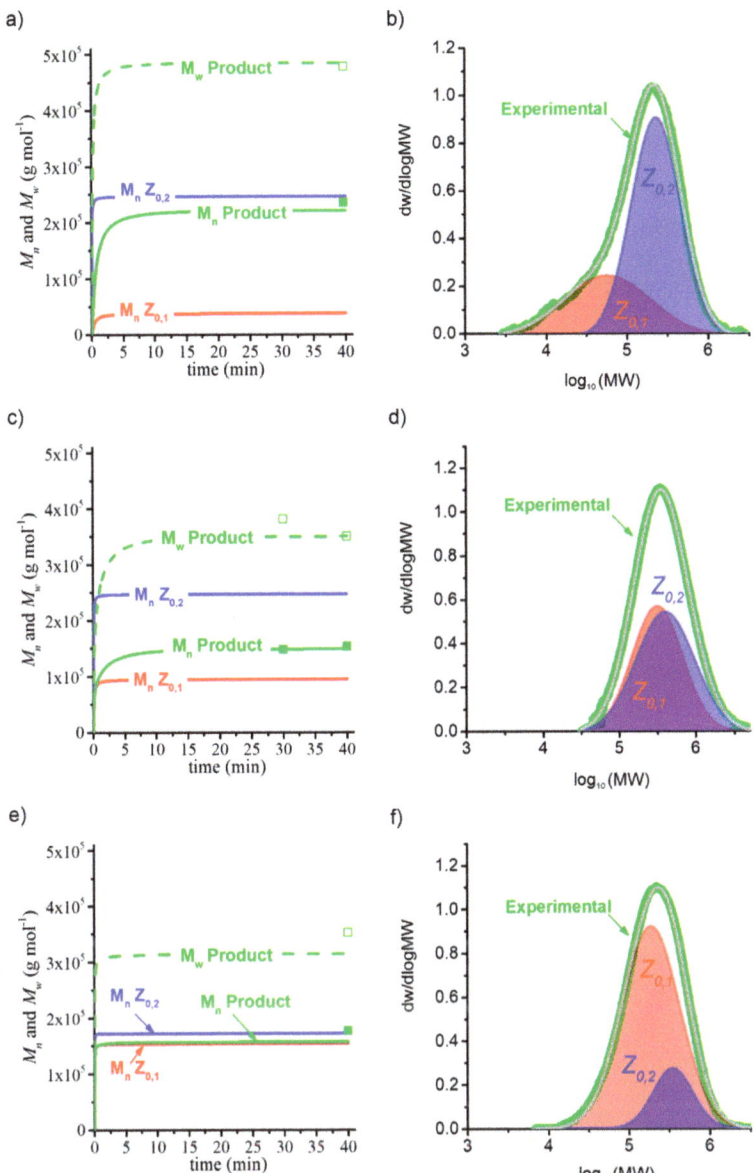

Figure 8. Effect of the temperature variation at 40 °C (**a,b**), 50 °C (**c,d**), and 60 °C (**e,f**) on M_n and M_w for the final product and both types of polymer species (**a,c,e**) with the experimental data (green symbols) and the simulations (lines), and molecular weight distributions (**b,d,f**) for SEC data (green line) and partial populations (blue and red lines). The ratio of $[B]_0/[Zr]_0 = 2.5$ and $P = 150$ bar. Other operating conditions are shown in Table 6.

At 60 °C, the model accurately predicted the R_p experimental profile, presenting a maximum point at approximately 5 min: after that time, the curve suddenly falls until 100 mol L^{-1} min^{-1} at 40 min of reaction (Figure 7e). The concentration of $Y_{0,2}$ reached a maximum at 5 min, followed by a plateau where the concentration remained almost constant throughout the whole reaction (Figure 7f). Here, the concentration of $Z_{0,1}$ was higher than $Z_{0,2}$, but both species presented very similar M_n—such values overlap with the experimental values of M_n (Figure 8e). Additionally, Figure 8f presents the MWD and its deconvolution in two populations, corresponding to $Z_{0,1}$ and $Z_{0,2}$, with a higher concentration of $Z_{1,0}$, as determined previously. The dispersity value was 2.00; therefore, so it is possible to assume that a single population, such as $Z_{0,2}$, is only generated in the system, and $Y_{0,1}$ fulfills the role of intermediary species or transition state of $Y_{0,2}$.

5. Conclusions

A systematic study about the influence of different parameters over the catalytic activity and polymer final properties was performed, using organoboron compounds as activators of zirconocenes with different ligand substituents. Carbocation organoboron compounds ($[(C_6H_5)_3C]^+[B(C_6F_5)_4]^-$) promoted the ethylene polymerization to the greatest extent, while zirconocenes with cyclopentadienyl rings having non-substituted-ligands were found to yield the highest activity, suggesting that, in this catalytic system, the steric hindrance from the substituents plays a greater role than their electron-releasing effect over the polymerization. Toluene was found to be the best solvent among those compared for this kind of system.

A comprehensive kinetic model for ethylene polymerization using a $CP_2Zr(CH_3)_2$/$[CPh_3][B(C_6F_5)_4]$ catalyst system and toluene as solvent was developed, based on a proposed reaction mechanism considering two active sites. A parameter estimation, involving the fit of the polymerization rate profiles using a nonlinear least square optimization, was carried out. A comparison between the two models revealed that the reactivation of the deactivated sites of type 2 is a very important step, obtaining coefficients of determination around the unit. Additionally, the chain transfer rate constants were fitted by using the method of moments, resulting in an excellent agreement with the molecular weight experimental data. Moreover, E_A and A_0 parameters were estimated in two series of experiments at ratios of B3/Zr1 of 1 and 2.5, and they presented high energetic barriers for rate constants corresponding to the active site 2 with the higher ratio. The model and the kinetic parameters were validated with the experimental data, confirming that a multiple (two) active site consideration is a possible explanation for the multimodal MWD, observed in SEC measurements. The predictions of the total concentrations expose a higher generation of the dead polymer of type 1 than type 2 when the temperature is higher.

Supplementary Materials: The following are available online at https://www.mdpi.com/2227-9717/9/1/162/s1.

Author Contributions: Conceptualization, R.D.d.L. and I.Z.-G.; methodology, J.D.-E. and F.E.-M.; software, I.Z.-G. and E.S.-G.; validation, I.Z.-G. and P.Q.-Á.; formal analysis, R.D.d.L. and I.Z.-G.; investigation, L.V. and R.L.-G.; resources, R.D.d.L.; data curation, I.Z.-G.; writing—original draft preparation, L.V., F.E.-M. and I.Z.-G.; writing—review and editing, L.V., R.D.d.L. and I.Z.-G.; visualization, L.V., P.Q.-Á. and I.Z.-G.; supervision, R.L.-G. and E.S.-G.; project administration, R.D.d.L. and I.Z.-G.; funding acquisition, R.D.d.L. All authors have read and agreed to the published version of the manuscript.

Funding: The authors acknowledge the financial support of the Mexican National Council of Science and Technology (CONACyT) through the Basic Science project 258278.

Institutional Review Board Statement: Not applicable.

Informed Consent Statement: Not applicable.

Data Availability Statement: Data will be made available on request.

Acknowledgments: The authors thank Ricardo Mendoza for their technical support. I.Z.-G. thanks the financial support of CONACyT, though Cátedras CONACYT, project 707.

Conflicts of Interest: The authors declare that they have no known competing financial interests or personal relationships that could have appeared to influence the work reported in this paper.

References

1. Sauter, D.W.; Taoufik, M.; Boisson, C. Polyolefins, a success story. *Polymers* **2017**, *9*, 185. [CrossRef]
2. Kaminsky, W. The discovery and evolution of metallocene-based olefin polymerization catalysts. *Rend. Lincei* **2017**, *28*, 87–95. [CrossRef]
3. Linnolahti, M.; Collins, S. Formation, Structure, and Composition of Methylaluminoxane. *ChemPhysChem* **2017**, *18*, 3369–3374. [CrossRef]
4. Kaminsky, W. Discovery of methylaluminoxane as cocatalyst for olefin polymerization. *Macromolecules* **2012**, *45*, 3289–3297. [CrossRef]
5. Velthoen, M.E.Z.; Muñoz-Murillo, A.; Bouhmadi, A.; Cecius, M.; Diefenbach, S.; Weckhuysen, B.M. The Multifaceted Role of Methylaluminoxane in Metallocene-Based Olefin Polymerization Catalysis. *Macromolecules* **2018**, *51*, 343–355. [CrossRef]
6. Yang, X.; Stern, C.L.; Marks, T.J. "Cation-like" Homogeneous Olefin Polymerization Catalysts Based upon Zirconocene Alkyls and Tris(pentafluorophenyl)borane. *J. Am. Chem. Soc.* **1991**, *113*, 3623–3625. [CrossRef]
7. Yang, X.; Stern, C.L.; Marks, T.J. Cationic Zirconocene Olefin Polymerization Catalysts Based on the Organo-Lewis Acid Tris(pentafluorophenyl)borane. A Synthetic, Structural, Solution Dynamic, and Polymerization Catalytic Study. *J. Am. Chem. Soc.* **1994**, *116*, 10015–10031. [CrossRef]
8. Piers, W.E.; Chivers, T. ChemInform Abstract: Pentafluorophenylboranes: From Obscurity to Applications. *Chem. Soc. Rev.* **1997**, *26*, 345–354. [CrossRef]
9. Chen, E.Y.X.; Marks, T.J. Cocatalysts for metal-catalyzed olefin polymerization: Activators, activation processes, and structure-activity relationships. *Chem. Rev.* **2000**, *100*, 1391–1434. [CrossRef]
10. Mehdiabadi, S.; Soares, J.B.P.; Bilbao, D.; Brinen, J.L. A Polymerization Kinetics Comparison between a Metallocene Catalyst Activated by Tetrakis(pentafluorophenyl) Borate and MAO for the Polymerization of Ethylene in a Semi-batch Solution Reactor. *Macromol. React. Eng.* **2011**, *5*, 418–430. [CrossRef]
11. Yang, X.; Stern, C.L.; Marks, T.J. Models for Organometallic Molecule-Support Complexes. Very Large Counterion Modulation of Cationic Actinide Alkyl Reactivity. *Organometallics* **1991**, *10*, 840–842. [CrossRef]
12. Chien, J.C.W.; Tsai, W.M.; Rausch, M.D.; Rausch, M.D. Isospecific Polymerization of Propylene Catalyzed by rac-Ethylenebis(indenyl)methylzirconium "Cation". *J. Am. Chem. Soc.* **1991**, *113*, 8570–8571. [CrossRef]
13. Chien, J.C.W.; Wang, B.-P. Metallocene-methylaluminoxane catalysts for olefin polymerization. V. Comparison of Cp_2ZrCl_2 and $CpZrCl_3$. *J. Polym. Sci. Part A Polym. Chem.* **1990**, *28*, 15–38. [CrossRef]
14. Valencia López, L.A.; Enríquez-Medrano, F.J.; Mendoza Carrizales, R.; Soriano Corral, F.; Castañeda Facio, A.; Diaz De Leon Gomez, R.E. Influence of organoboron compounds on ethylene polymerization using cp_2zrcl_2/mao as catalyst system. *Int. J. Polym. Sci.* **2014**, *2014*. [CrossRef]
15. Correa, S.A.; Daniliuc, C.G.; Stark, H.S.; Rojas, R.S. Nickel Catalysts Activated by rGO Modified with a Boron Lewis Acid to Produce rGO-Hyperbranched PE Nanocomposites. *Organometallics* **2019**, *28*, 3327–3337. [CrossRef]
16. Ewen, A.J.A. *Catalytic Polymerization of Olefins*; Keii, T., Soga, K., Eds.; Elsevier: Tokyo, Japan, 1986; p. 271.
17. Charles, R.; González-Hernández, R.; Morales, E.; Revilla, J.; Elizalde, L.E.; Cadenas, G.; Pérez-Camacho, O.; Collins, S. Novel supported catalysts for ethylene polymerization based on aluminohydride-zirconocene complexes. *J. Mol. Catal. A Chem.* **2009**, *307*, 98–104. [CrossRef]
18. Favero, C.; Mignoni, M.L.; De Souza, R.F.; Bernardo-Gusmão, K. Polymerization of ethylene with zirconocene heterogenized on spherical ZSM-5. *J. Braz. Chem. Soc.* **2015**, *26*, 1405–1410. [CrossRef]
19. González-Hernández, R.; Chai, J.; Charles, R.; Pérez-Camacho, O.; Kniajanski, S.; Collins, S. Catalytic system fot homogeneous ethylene polymerization based on aluminohydride-zirconocene complexes. *Organometallics* **2006**, *25*, 5366–5373. [CrossRef]
20. Charoenchaidet, S.; Chavadej, S.; Gulari, E. Borane-functionalized silica supports. *J. Mol. Catal. A Chem.* **2002**, *185*, 167–177. [CrossRef]
21. Estrada, J.M.V.; Hamielec, A.E. Modelling of ethylene polymerization with Cp_2ZrCl_2 MAO catalyst. *Polymer (Guildf)* **1994**, *35*, 808–818. [CrossRef]
22. Jiang, S.; Wang, L.; Zhang, P.; Feng, L. New kinetic model of ethene polymerization with Cp_2ZrCl_2/MAO catalyst. *Macromol. Theory Simul.* **2002**, *11*, 77–83. [CrossRef]
23. Ahmadi, M.; Nekoomanesh, M.; Jamjah, R.; Zohuri, G.; Arabi, H. Modeling of slurry polymerization of ethylene using a soluble Cp_2ZrCl_2/MAO catalytic system. *Macromol. Theory Simul.* **2007**, *16*, 557–565. [CrossRef]
24. Mehdiabadi, S.; Soares, J.B.P. Ethylene homopolymerization kinetics with a constrained geometry catalyst in a solution reactor. *Macromolecules* **2012**, *45*, 1777–1791. [CrossRef]
25. Soares, J.B.P.; McKenna, T.F.L.; Cheng, C.P. Coordination Polymerization. In *Polymer Reaction Engineering*; Asua, J., Ed.; Blackwell Publishing Ltd.: Kuala Lumpur, Malaysia, 2007; pp. 29–117, ISBN 978-1-4051-4442-1.
26. Gaw, W.J.; Swinton, F.L. Thermodynamic Properties of Binary Systems containing. *Trans. Faraday Soc.* **1967**, *64*, 637–647. [CrossRef]
27. National Institute of Standars and Technology. Toluene. Available online: https://webbook.nist.gov/cgi/cbook.cgi?ID=C108883&Mask=4&Type=ANTOINE&Plot=on (accessed on 11 November 2020).
28. Michels, A.; Wassenaar, T. The vapour pressure of ethylene. *Physica* **1950**, *16*, 221–224. [CrossRef]
29. National Institute of Standards and Technology. Ethylene. Available online: https://webbook.nist.gov/cgi/cbook.cgi?ID=C74851&Mask=4&Type=ANTOINE&Plot=on (accessed on 11 November 2020).

30. MathWorks®. Fmincon. Available online: https://la.mathworks.com/help/optim/ug/fmincon.html?lang=en (accessed on 11 November 2020).
31. Bates, D.M.; Watts, D.G. *Nonlinear Regression Analysis and Its Applications*; John Wiley & Sons, Inc.: Hoboken, NJ, USA, 1998.
32. Bochmann, M.; Lancaster, S.J. Cationic group IV metal alkyl complexes and their role as olefin polymerization catalysts: The formation of ethyl-bridged dinuclear and heterodinuclear zirconium and hafnium complexes. *J. Organomet. Chem.* **1995**, *497*, 55–59. [CrossRef]
33. Crossley, N.S.; Green, J.C.; Nagy, A.; Stringer, G. Electronic structure of transition-metal indenyl compounds: A He I and He II photoelectron spectroscopic study of [Mn(η^5-C_9H_7)(CO)$_3$], [Fe(η^5-C_9H_7)$_2$], [Ru(η^5-C_9H_7)$_2$], and [Ru(η^5-C_9H_7)η-C_5Me_5)]. *J. Chem. Soc. Dalton Trans.* **1989**, *11*, 2139–2147. [CrossRef]
34. Frankcom, T.M.; Green, J.C.; Nagy, A.; Kakkar, A.K.; Marder, T.B. Electronic structure and photoelectron spectroscopy of d8 rhodium indenyl complexes. *Organometallics* **1993**, *12*, 3688–3697. [CrossRef]
35. Nguyen, K.T.; Lane, E.E.; McMillen, C.D.; Pienkos, J.A.; Wagenknecht, P.S. Is Indenyl a Stronger or Weaker Electron Donor Ligand than Cyclopentadienyl? Opposing Effects of Indenyl Electron Density and Ring Slipping on Electrochemical Potentials. *Organometallics* **2020**, *39*, 670–678. [CrossRef]
36. Lee, L.S.; Ou, H.J.; Hsu, H.L. The experiments and correlations of the solubility of ethylene in toluene solvent. *Fluid Phase Equilib.* **2005**, *231*, 221–230. [CrossRef]
37. Soares, J.B.P.; McKenna, T.F.L. *Polyolefin Reaction Engineering*; Wiley-VCH: Weinheim, Germany, 2012; pp. 187–269, ISBN 9783527317103.

Article

Mathematical Modeling of the Production of Elastomers by Emulsion Polymerization in Trains of Continuous Reactors

Enrique Saldívar-Guerra [1,*], Ramiro Infante-Martínez [1] and José María Islas-Manzur [2]

1. Centro de Investigación en Química Aplicada, Blvd. Enrique Reyna 140, Saltillo Coahuila CP 25294, Mexico; ramiro.infante@ciqa.edu.mx
2. Dynasol Group, Planta Emulsión, Km 13.5 Carretera Tampico-Mante Int. A, Altamira, Tamaulipas CP 89608, Mexico; jose.islas@hotmail.com
* Correspondence: enrique.saldivar@ciqa.edu.mx; Tel.: +52-844-419-1854

Received: 12 October 2020; Accepted: 13 November 2020; Published: 20 November 2020

Abstract: A mechanistic model is proposed to describe the emulsion polymerization processes for the production of styrene–butadiene rubber (SBR) and acrylonitrile–butadiene rubber (NBR) elastomers in trains of continuous stirred tank reactors (CSTRs). A single model was used to describe both processes by choosing the proper physicochemical parameters of each system. Most of these parameters were taken from literature sources or estimated a priori; only one parameter (the entry rate coefficient) was used as an adjustable value to reproduce the kinetics (mainly conversion), and another parameter (the transfer to polymer rate coefficient) was used to fit the molecular weight distribution (MWD) experimental values from plant data. A 0-1-2 model for the number of particles and for the moments of the MWD was used to represent with more fidelity the compartmentalization effects. The model was based on approaches used in previous emulsion polymerization models published in the literature, with the premise of reaching a compromise between the level of detail, complexity, and practical value. The model outputs along the reactor train included conversion, remaining monomer composition, instantaneous and accumulated copolymer composition, the number of latex particles and particle diameter, polymerization rate, the average number of radicals per particle, average molecular weights, and the number of branches per chain.

Keywords: emulsion polymerization; styrene–butadiene rubber; nitrile rubber; mathematical modeling

1. Introduction

Two of the most important rubber products (copolymers) from the economical point of view—styrene–butadiene rubber (SBR) and acrylonitrile–butadiene rubber or nitrile rubber (NBR)—are produced via emulsion polymerization, mainly via the continuous cold process (5–15 °C) in trains of continuous stirred tank reactors (CSTRs). Some types of styrene–butadiene copolymer rubbers can also be produced via living anionic polymerization in solution, with a better control of the composition microstructure along the polymer chain, but the subject of this paper is the random copolymer produced via radical polymerization in emulsion. The world production of SBR was estimated to be around 8.3×10^6 ton/year in 2020 (based on [1]), and that of NBR was estimated to be 1.56×10^6 ton/year by 2023 [2], together representing roughly 3% of the total world polymer production (assuming a total world production of 300×10^6 ton/year).

The main use of SBR is in tires and automotive applications, while NBR is used in seals, gaskets, hoses, etc., where chemical resistance, higher strength, and lower gas permeability are needed.

As mentioned above, a large portion of these copolymers is produced worldwide via continuous cold processes that, due to the lower temperatures used, avoid undesired crosslinking during polymerization. The processes for the production of SBR and NBR have many similarities, basically differing in the formulation of the emulsion components fed to the process. Both processes are performed by free-radical emulsion polymerization in a train of CSTRs at low temperatures, and they use a redox system of initiators to maintain a flux of radicals at these low temperatures. Additionally, they both are designed to reach maximum monomer conversions between 70 and 75% to avoid crosslinking due to branching reactions and to limit the composition drift. Side streams of the monomer that are consumed faster can also be included in the process to correct the composition drift in both the SBR and the NBR processes.

Concerning the cold SBR process, early models were published by the Hamielec and MacGregor group in the 1980s. Both steady-state (Kanetakis et al. [3]) and dynamic models (Broadhead et al. [4] and Penlidis et al. [5]) were presented, capable of calculating the monomer conversion, the copolymer composition, the molecular weight averages, and the long-chain branching frequency, as well as the particle size distribution (PSD), in each reactor. In the steady-state model [3], the PSD was calculated based on the known residence time distribution of a CSTR using a convolution integral to estimate the PSD in reactors downstream of the train. Only micellar nucleation was included, considering radical entry competition between micelles and already formed particles (Smith–Ewart theory); Smith–Ewart case II kinetics were assumed with 0.5 as the average number of radicals per particle ($\bar{n}$) [6]. The molecular weight averages were calculated by using steady-state expressions derived by applying the method of moments. The dynamic model [4,5] had similar bases to the steady-state one, although it contains some differences in certain respects. Naturally, ordinary differential equations (ODEs) for the components of the redox initiation system, monomers, and converted monomers, surfactant, number of particles, and molecular weight distribution (MWD) moments per reactor were explicitly included. The treatment of the MWD equations was made by using the pseudo-homopolymer approach [7] (also known as the pseudo-kinetic [8] or apparent rate constants [9] methods). The evolution of the PSD was calculated in an approximated way via the discretization of the population by particle age (generations) and the use of several simplifying assumptions. In this way, only one differential equation was solved for the volume growth of a particle of the first generation, and this information was used to estimate the growth of the particles of any other generation. Interestingly, this model included the energy balances of the reactor and the cooling jacket, as well as an equation for a temperature controller. Smith–Ewart case II kinetics were also used in this model during the nucleation period and at low conversions, while at higher conversions, the extended fraction solution of the Smith–Ewart recurrence equation, proposed by Ugelstad et al. [10], was utilized to estimate $\bar{n}$. These models were used in several applications: to determine operation policies aimed at improving the productivity of the reactor train or the quality of the SBR produced, to manipulate the PSD, to enhance the operation stability (avoiding sustained oscillations for example), and to design start-up procedures to minimize the quantity of the off-spec product [4,5]; the dynamic model adapted for semibatch operation was later used to design midcourse correction policies for the control of molecular weight and cross-linking density in semibatch SBR reactors [11].

Years later, Gugliotta et al. [12] published a model based on the Broadhead et al. model with added detail in some respects. The main differences in Guggliota et al. model with respect to [4] were:

- The effect of impurities in the process was included.
- The PSD was modeled by using a population balance approach for the continuous distribution $N(v,t)dv$, which represents the number of particles having a volume between v and $v + dv$ at time t, resulting in a partial differential equation (PDE). This equation was approximately solved by discretizing the v domain into small subdomains Δv and converting the PDE into a set of ODEs.
- The monomer partitioning among phases was calculated by using the Flory–Huggins theory [13] extended to emulsion systems by Morton et al. [14] to include the effect of the interfacial energy between particles and the continuous aqueous phase.

- The possibility of considering side streams fed to reactors along the CSTR train was included.

Similarly to the Broadhead et al. model [4], the Gugliotta et al. model [12] was applied to design policies for increasing production and transient optimization [15–17]. More recently, it was also used for the calibration of a soft sensor for the monitoring and control of an SBR production process using seven CSTRs in series. The model replaced actual plant data by simulating the operation of the process in an extended variable space [18].

Zubov et al. presented a detailed dynamic model for the SBR process in a train of CSTRs based on population balances and the method of moments for the description of the MWD, besides mass balances for other species [19]. Several non-conventional features are used in the model. Interestingly, they use an empirical relation for the average number of radicals in particles ($\bar{n}$) as a function of the initiator concentration; as a consequence, no radical entry or radical desorption terms are used. Additionally, the transport of the monomer from the droplets to the particles is modeled by using mass transfer terms, where the concentration gradients are calculated using partition coefficients for the estimation of the equilibrium concentrations. The model is validated with industrial data for conversion, molecular weight, and Mooney viscosity for the steady-state and transient case, and its usefulness is illustrated by its application to generating a grade transition procedure to minimize the production of off-spec product. More recently, Mustafina et al. developed a rather simple model to describe the copolymer composition and other characteristics of SBR, and they used it to fit the corresponding kinetic parameters to represent intrinsic viscosity, conversion, and molecular weight data [20]. Apparently, the used data came from a train of CSTRs, but a batch model was implemented to approximately represent the process. The used MWD model was too simple and did not take branching reactions into account.

Regarding the NBR case, the apparently first mathematical models for batch and semibatch industrial processes were published, almost simultaneously, by Dubé et al. [21] and by Vega et al. [22]. The Dubé et al. model was built to represent batch and semibatch reactors while considering redox initiation, the thermodynamic partitioning of monomers among phases using the partition coefficients model, and the pseudo-kinetic rate constants method for the copolymerization kinetics; $\bar{n}$ was calculated by using the Ugelstad [10] expression. The model was capable of predicting conversion, the number of particles per L of water N_p, the rate of polymerization R_p, the copolymer composition, the number and weight average molecular weights M_n and M_w (by the method of moments), the branching averages, and the average particle diameter D_p.

The model of Vega et al. [22] was also written for batch and semibatch reactors, and it was an extension of the previous model built by this group for the SBR process [12] (which was itself based on the Broadhead et al. model), with some modifications to capture the mechanistic features generated by the presence of the more water-soluble acrylonitrile (AN) monomer. The modifications of the model included the polymerization of AN in the aqueous phase, radical desorption, and homogeneous nucleation. However, some of these modifications were only applied to the polymerization rate but not to the MWD calculations, making these two parts of the model somewhat inconsistent. The latter model was further extended in [23] to describe in more detail the bivariate chain length and branching distribution in the batch case. It was also used to simulate plant data for testing an open-loop estimator based on calorimetric measurements to monitor NBR properties in a semibatch process [24]. Notice that none of the previous cited works addressed continuous NBR production processes; it was only in the works of Minari et al. [25,26] that the Vega et al. model was extended to the case of a train of CSTRs for NBR production to study the effect of intermediate feed streams of chain transfer agents (CTAs) and AN on the polymer quality [25] and to reduce transients during grade transitions [26]. More recently, the model of Vega et al. was used to evaluate a closed-loop control strategy to produce the NBR of uniform copolymer composition in a semibatch reactor [27].

Another model for NBR production in a train of CSTRs (also used for batch and semibatch operations) was proposed by Washington et al. [28,29]. This model was based on that of Dubé et al. [21] with many similarities: redox initiation, the partitioning of monomers described by partition coefficients, the use of pseudo-kinetic rate constants for the reaction kinetics, and the calculation of $\widetilde{n}$ by using the extended fraction expression [10]. It also used an average particle size and consisted of differential equations (dynamic model). As in the case of other models for continuous reactors, it was used to study start-up procedures and the control of polymer properties (copolymer composition, molecular weight, and particle size).

More recently, the Washington et al. model was used to study, in more detail, the effect of the monomer partitioning model, the radical desorption, and the monomer soluble impurities of several polymer properties in the batch and the CSTR train case. [30] The model was also applied to study dynamic aspects of a train of eight CSTRs and the effect of different operation policies on the off-spec material produced during transient operation. [31] Scott et al. also used this model to demonstrate the application of the Bayesian experimental design technique to maximize the information that can be experimentally obtained for an NBR process carried out in a train of CSTRs [32].

In summary, there are few models for trains of CSTRs for the production of SBR ([3,4,12,19,20]) and NBR ([25,26,29]). Many of the models ([12,25,26]) stemmed from the same model of Broadhead et al. [4] with some adaptations, and others ([3,29]) are not very different either in terms of assumptions and structure.

Given the similarities of the SBR and NBR processes, it was the goal of this work to develop and present a single model that is flexible enough to represent both processes with relatively small variations and to demonstrate that the same mathematical structure could be applied with minor adjustments to both processes. Some previous models have addressed only one of the two polymerization systems at a time (not necessarily for the continuous process but for batch or semibatch modes), although, in some cases, they used essentially the same model with adaptations. We believe that it is instructive to treat the two processes with the same model simultaneously in order to emphasize their similar underlying structures. Other goals of this work were to introduce improvements in some key aspects of the models (to be specified below) and to reach a good balance between the level of detail, the ease of implementation, and the predictive power of the model. To do this, a different model is proposed that is intended to study the effect of the configuration of the CSTR's train in the production (rate and quality) of both SBR and NBR elastomers. Configuration refers to the number of reactors in the train (in the range of 7–16), as well as the possible presence and position (reactor number) of side feed streams of one of the monomers or CTAs in the reactor train.

The goal of the study was to compare steady-state operations; nonetheless, a dynamic model was written in the form of ordinary differential equations (ODEs) since, in our experience, the numerical solution of this kind of system is more robust than that of the corresponding steady-state non-linear algebraic equation system. This selection was based on the experience of many years of our group working on the mathematical modeling of industrial emulsion polymerization processes.

2. Mathematical Model

In this section, the different components of the used mathematical model are presented and briefly discussed. Since some of these components were taken or adapted from existing models—only the main features of them are provided, and their details can be found in the given references and in the Supplementary Materials (SM).

The philosophy of the present work during the definition of the different components of the model was to use only the level of complexity deemed necessary to have a reliable description of each of the involved phenomena. In some cases, as in the calculation of $\widetilde{n}$, which was based on a 0-1-2 particle model adapted from one by Vale and McKenna [33], this choice resulted in an increased level of detail compared with previous models. Additionally, the present model uses a more detailed calculation of the leading moments (0–2) of the MWD which, for consistency with the particle model, were also calculated using a 0-1-2 model based on a subset of the population balance model of Saldívar et al. [9] The model is not just a collection of existing pieces from other models, but the equations were re-derived and/or adapted as needed. In other cases, our choice was a simpler model than those in other works; this was the case, for example, of the monomer partitioning model, which is quite simple but still effective. A more detailed discussion of these differences is provided in the specific sections of the paper.

The used kinetic scheme is given in Table 1. Notice that the medium part of the table shows, in detail, the terminal model on which the pseudo-homopolymer approach was based (for propagation and chain transfer to monomers and CTAs). Diffusion-controlled termination using a single coefficient was used. The lower part of the table shows the kinetics in the pseudo-homopolymer form, which is particularly useful for the MWD (moments) calculations. Notice that the internal and terminal double bond reactions only occur on the butadiene units of the dead polymer, and this is reflected in the multiplication of the corresponding rate constant by $F^c_{1,mo}$, which is the accumulated molar fraction of butadiene in the polymer.

Table 1. Kinetics in the aqueous and particle phase.

Reaction	Kinetics
Aqueous phase	
Redox initiation	$I_w + Y^r_1 \xrightarrow{k_{r1}} R_w + Y^o_1 + products$
	$Y^o_1 + Y_2 \xrightarrow{k_{r2}} Y^r_1 + products$
	$R_w + M_w \xrightarrow{\overline{k_p}} P_w$
	$P_w + M_w \xrightarrow{\overline{k_p}} P_w$
	$P_w P_w \xrightarrow{\overline{k_t}} D_w$
Particle phase, detailed terminal model [1]	
Propagation and cross-propagation (i,j = 1,2)	$P_i + M_j \xrightarrow{k_{p,ij}} P_j$
Chain transfer to monomer (i,j = 1,2)	$P_i + M_j \xrightarrow{k_{trMij}} P_j + D$
Chain transfer to chain transfer agent T (i,j = 1,2)	$P_i + T \xrightarrow{k_{trTi}} P_i + D$
Particle phase, molecular weight distribution (MWD) pseudo-homopolymer form [2]	
Propagation (r = 1, ..., ∞)	$P^r + M \xrightarrow{\overline{k_p}} P^{r+1}$
Chain transfer to monomer (r = 1, ..., ∞)	$P^r + M \xrightarrow{\overline{k_{trM}}} P^1 + D^r$
Chain transfer to chain transfer agent (r = 1, ..., ∞)	$P^r + T \xrightarrow{\overline{k_{trT}}} P^1 + D^r$
Termination by combination and disproportionation (r,q = 1, ..., ∞) [2]	$P^r + P^q \xrightarrow{\overline{k_{tc}}} D^{r+q}$
	$P^r + P^q \xrightarrow{\overline{k_{td}}} D^r + D^q$
Chain transfer to polymer (q,r = 1,...,∞) [3]	$P^q + D^r \xrightarrow{r\overline{k_{trp}}} P^r + D^q$
Internal or terminal double bond polymerization (q,r = 1, ..., ∞) [3]	$P^q + D^r \xrightarrow{F^c_{1,mo} r k_{db}} P^{r+q}$

[1] A pseudo-homopolymer approach with apparent rate constants is used to represent the copolymerization kinetics based on the terminal model. [2] A single value $\overline{k_t} = \overline{k_{tc}} + \overline{k_{td}}$ was used for the termination step, calculated as $\overline{k_t} = \sqrt{k_{t11}k_{t22}}$ while assuming diffusion-controlled termination, where k_{tjj}, $j = 1, 2$, are the homotermination constants for monomers 1 and 2. [3] The transfer-to-polymer and double bond polymerization reactions are only considered for the MWD calculations.

The symbols used in Table 1 for the aqueous phase are: I_w, initiator; Y^r_1 and Y^o_1, reducing agents in the reduced and oxidized states, respectively; Y_2, secondary reducing agent; R_w, primary radicals in the aqueous phase; P_w, polymeric radicals in the aqueous phase; and D_w, the dead polymer in the aqueous phase. For the particle phase, the nomenclature is: P_i, polymeric radicals of type i terminal unit; M_j, type j monomer; D, the dead polymer; P^r and D^r, length (r) of live and dead polymers, respectively; M, total monomer ($M = M_1 + M_2$); and T, chain-transfer agent. Notice that $P^r = P^r_1 + P^r_2$, where P^r_i are r-length polymeric radicals of type i ($i = 1,2$).

The following equations are valid for any reactor in the train, except where noted.

2.1. Monomer Mass Balances

The remaining mass of monomer M_i^m ($i = 1,2$)

$$\frac{dM_i^m}{dt} = F_{in}w_{Mi,\,in} - \frac{M_i^m F_{out}}{M_T} - F_{i,ma}R_p V_w M_{wi} + F_{ri,Mi} \tag{1}$$

Monomer bound in the polymer chain H_i^m ($i = 1,2$)

$$\frac{dH_i^m}{dt} = F_{in}w_{Hi,\,in} - \frac{H_i^m F_{out}}{M_T} + F_{i,ma}R_p V_w M_{wi} \tag{2}$$

where M_{wi} is the molecular weight of monomer i; $w_{s,\,in}$ is the inlet mass fraction of species s entering the reactor; $F_{i,ma}$ is the mass fraction of monomer i instantaneously formed in the reactor according to the Mayo–Lewis equation in mass form; F_{in} and F_{out} are the inlet and outlet total mass flows entering and exiting the reactor, respectively; M_T is the total mass in the reactor; R_p is the reaction rate, mol/(L s); and V_w is the water volume in the reactor. Notice that the reaction rate is defined per liter of the aqueous phase. $F_{ri,Mi}$ are side feed streams (mass flow) of monomer i to reactor ri.

The Mayo–Lewis equation in molar units is:

$$F_{i,mo} = \frac{r_1 f_1^2 + f_1 f_2}{r_1 f_1^2 + 2 f_1 f_2 + r_2 f_2^2} \tag{3}$$

where $F_{i,mo}$ ($i = 1,2$) is the mole fraction of i-monomer units instantaneously incorporated in the copolymer; f_1 ($i = 1,2$) is the mole fraction of monomer i remaining in the reaction site; and r_i ($i = 1,2$) comprises the reactivity ratios of the copolymer system. The two forms (molar and mass) of the equation are related by a simple function f_M that converts the molar fraction into the mass fraction:

$$F_{i,ma} = f_M(F_{i,mo},\, f_1,\, M_{wi}) \tag{4}$$

where M_{wi} is the molecular weight of monomer i ($i = 1,2$).

Notice that for practical reasons, mass, instead of molar units, is used for some of the balances, since the recipes are handled on a mass basis in industrial practice.

2.2. Rate of Polymerization

The polymerization rate can be calculated with the classical expression for emulsion polymerization:

$$R_p = \frac{\bar{k}_p \bar{n} [M_p] N_p}{N_a} \tag{5}$$

where $[M_p]$ is the monomer concentration in particles, mol/L; $\bar{n}$ is the average number of radicals per particle; N_A is the Avogadro number, mol^{-1}; N_p is the number of particles per L of water, L^{-1}; and $\bar{k}_p$ is the apparent rate constant for propagation, L (mol s)$^{-1}$, calculated as:

$$\bar{k}_p = \sum_{i=1}^{2} \sum_{j=1}^{2} p_i \varnothing_j k_{p_{ij}} \tag{6}$$

where p_i is the probability of type i radical in particle ($i = 1,2$) and $\varnothing_i$ is the mole fraction of monomer i in particle ($i = 1,2$). The probabilities p_i can be calculated as shown in [34].

2.3. Population Balance Equations for Particles and Calculation of R_p and $\bar{n}$

Defining $F_{n,ri}$ as the number of particles with n radicals ($n = 0,1,2$) per liter of water, L^{-1}, in the reactor r_i in the train, population balance equations (PBEs) based on those postulated by Vale and McKenna [33] for a 0-1-2 system can be obtained. These equations are different from those of Vale and McKenna in two aspects: they do not consider the full-size distribution (partial derivatives with respect to size) but an average particle size instead that can increase with time and/or reactor. Another difference is that these equations include the inflow and outflow terms not present in the batch model of Vale and McKenna:

$$\frac{dV_w F_{0,ri}}{dt} = -\rho F_{0,ri} + k_{des} F_{1,ri} + \frac{2k'_t}{v} F_{2,ri} + Q_{in,ri} F_{0,ri-1} - Q_{out,ri} F_{0,ri} \quad (7)$$

$$\frac{dV_w F_{1,ri}}{dt} = \rho F_{0,ri} - (\rho + k_{des}) F_{1,ri} + (\rho + 2k_{des}) F_{2,ri} + \rho_{mic} M_{ic} + Q_{in,ri} F_{1,ri-1} - Q_{out,ri} F_{1,ri} \quad (8)$$

$$\frac{dV_w F_{2,ri}}{dt} = \rho F_{1,ri} - (\rho + 2k_{des} + \frac{2k'_t}{v}) F_{2,ri} + Q_{in,ri} F_{2,ri-1} - Q_{out,r} F_{2,ri} \quad (9)$$

where ρ, k_{des}, and k'_t are the radical entry, radical exit, and termination rate coefficients, respectively (see details and units below); v is the volume of an average particle, L; M_{ic} is the micelle concentration L^{-1}; $Q_{in,ri}$ and $Q_{out,ri}$ are the total in and out volumetric flows to and from reactor ri, respectively; and $k'_t = \frac{k_t}{N_A}$. Notice that the radical entry and exit coefficients are particle size-dependent as detailed below. The micellar nucleation mechanism is included in Equation (8); in general, the radicals in the aqueous phase may enter micelles or particles in a competitive way.

A dimensionless version of these equations, used for the numerical solution of the model, can be found in the Supplementary Materials.

From the solution of Equations (7)–(9) it is possible to calculate $\bar{n}$ and N_p from the following expressions:

$$\bar{n} = \frac{\sum_{n=0}^{2} n F_n}{\sum_{n=0}^{2} F_n} \quad (10)$$

$$N_p = \sum_{n=0}^{2} F_n \quad (11)$$

2.4. Radical Entry and Exit Coefficients

The used specific expressions were taken from [9], but they were based on concepts previously proposed and generally accepted [35–37]. See the Supplementary Materials for specific details; Supplementary Material Equations (S1a)–(S1o).

2.5. Mass Balances of Species

In all the balance equations, $w_{S_p,in}$ represents the mass fraction of species S_p in the reactor feed and M_{S_p} is the molecular weight of S_p (except for M_T that is total mass in the reactor). Differential equations representing the balances are written in molar units, except where noted. They are written for the initiator I_w, reducing agents Y_1 and Y_2, surfactant S, and chain transfer agent T_r, and they are summarized below. More details on the assumptions and derivation of theses equations are included in the Supplementary Materials.

Initiator I_w:

$$\frac{dI_w}{dt} = -k_{d1} Y_1^r I_w / V_w + F_{in} \frac{w_{I,\,in}}{M_I} - \frac{I_w F_{out}}{M_T} \quad (12)$$

Reducing agents Y_1 and Y_2:

$$\frac{dY_1}{dt} = \frac{F_{in} w_{y1,\,in}}{M_{y1}} - \frac{Y_1 F_{out}}{M_T} \quad (13)$$

$$\frac{dY_2}{dt} = \frac{F_{in}W_{y2,in}}{M_{y2}} - \frac{Y_2 F_{out}}{M_T} - k_{d2} Y_1^{\circ} Y_2 / V_w \tag{14}$$

For further details on the use of these equations, see Supplementary Material Equations (S1a–S1r) in the Supplementary Materials.

Surfactant S:

$$\frac{dS}{dt} = \frac{F_{in}w_{s,in}}{M_s} - F_{out}\frac{S}{M_T} \tag{15}$$

The surfactant balance is connected and solved simultaneously with the corresponding adsorption equilibrium (Langmuir isotherm) described by the Supplementary Material Equations (S1u)–(S1x) in the Supplementary Materials.

Given the total amount of surfactants by the mass balance, the partitioning of the surfactant adsorbed in particles (S_a), and free in water (S_f) can be defined; S_f is given by a quadratic equation and includes the surfactant in solution and the surfactant in micelles. The micelle concentration is obtained by Equation (16) or (17):

$$M_{ic} = \frac{(S_f/V_w - [S]^{cmc}) N_A a_{em}}{4\pi r_m^2} \quad \text{if } S_f/V_w \geq [S]^{cmc} \tag{16}$$

$$M_{ic} = 0 \text{ if } S_f/V_w < [S]^{cmc} \tag{17}$$

where $[S]^{cmc}$ is the critical micelle concentration of the surfactant (mol L^{-1}), a_{em} is the surface area per surfactant molecule, and r_m is the radius of a micelle.

Chain transfer agent (Tr):

$$\frac{dT_r}{dt} = \frac{F_{in}w_{Tr,in}}{M_{Tr}} - \frac{F_{out}T_r}{M_T} - k_{trT}[T_r]_p \frac{\sum_{n=1}^{2} nF_n}{N_A} V_w + \frac{F_{ri,Tr}}{M_{Tr}} \tag{18}$$

where $F_{r,Tr}$ is a side feed stream (mass flow) of CTAs to reactor ri. The partitioning equilibrium of the CTAs defines their concentration in the particles. This is based on the assumption that the component partitions between the particles and the aqueous phase according to the partition coefficient are defined as:

$$K_T = \frac{[Tr]_p}{[Tr]_w} \tag{19}$$

where $[Tr]_p$ and $[Tr]_w$ are the concentrations of CTAs in the particle and aqueous phase, respectively.

The simultaneous solution of the mass balance and the CTA partitioning is described in detail by Supplementary Material Equations (S1y)–(S1aa) in the Supplementary Materials.

Water (in mass units):

$$\frac{dW}{dt} = F_{in}w_{w1\ in} - \frac{F_{out}W}{M_T} \tag{20}$$

Differential mass balance equations were also derived for primary radicals R_w and polymeric radicals P_w (aqueous phase), but the quasi-steady state approximation (QSSA) was assumed for these species, resulting in a simple quadratic equation for P_w:

$$- k_t P_w^2 / V_w - k_{mp} 4\pi r^2 P_w \sum_{n=0}^{2} F_n - k_{mm} a_m M_{ic} P_w + \frac{K_{des}}{N_A} \sum_{n=0}^{2} F_n + k_{d1} f Y_1^r I_w / V_w = 0 \tag{21}$$

where k_{mp} is an entry coefficient for radicals in particles (m/s^{-1}); r is the radius of an average particle; k_{mm} is the entry coefficient for radicals in micelles; a_m is the total micellar surface area; and M_{ic} (L^{-1}) is the micelle concentration.

2.6. Monomer Partitioning

The modeling approach used here was relatively simple but still effective. Madhuranthakam and Penlidis [30] concluded that the use of different models for monomer partitioning (e.g., partition coefficients and models based on equations of state) does not make a significant difference in the case of the more complex NBR system in which the water solubility of AN is important. For the simpler SBR case, an equipartition approach (similar proportions of the monomers in the droplets and particle phases) equivalent to that proposed by Dougherty [38], is used. Additionally, the approach used here allows for the a priori estimation of key parameters of the partitioning model based on the known physicochemical properties of the system components. The used modeling approach was based on the simple limit conversion for saturation, x_{sat}, used by Gardon [39] as the conversion at which the monomer droplets disappear (end of interval 2). Up to this point, in the presence of excess monomers, the polymer particles are saturated with monomers and maintain a nearly constant concentration ($[M]_p$).

$[M]_p$ is calculated differently depending on the stage (interval) of the polymerization. The decision is made based on the equivalent conversion, x, defined as:

$$x = \frac{P_1^m + P_2^m}{M_1^m + M_2^m - M_{2w} + P_1^m + P_2^m} \tag{22}$$

where M_{2w} stands for the mass of monomer 2 in water. This latter quantity is nearly zero for SBR, but it is significant for NBR due to the partial solubility of AN in water.

If $x \leq x_{sat}$ (intervals 1 and 2), $[M]_p$ is calculated by the Equation (23). The details of how to arrive to this equation and how to calculate each of the involved quantities can be found in the Supplementary Materials Section 2, Supplementary Material Equations (S2a)–(S2c) of the Supplementary Materials.

$$[M]_p = [M]_{psat} = \frac{(M_1/M_{w1}) + (M_2/M_{w2})}{V_p'} \tag{23}$$

where V_p' is the volume of this hypothetical particle, which can be calculated assuming volume additivity:

$$V_p' = \frac{M_1}{\rho_1} + \frac{M_2}{\rho_2} + \frac{P_1}{\rho_{p1}} + \frac{P_2}{\rho_{p2}} \tag{24}$$

where M_{wi} = molecular weight of monomer i and ρ_i and ρ_{pi} ($i = 1,2$) are the densities of monomer i and homopolymer i, respectively.

If $x > x_{sat}$ (mass conversion x, interval 3), $[M]_p$ is calculated assuming that all the remaining monomers, except for the possible presence of AN in the water phase (NBR case), are in the particles:

$$[M]_p = \frac{(M_1^m/M_{w1}) + ((M_2^m - M_{2w}^m)/M_{w2})}{V_p} \tag{25}$$

where:

$$V_p = \frac{M_1^m}{\rho_1} + \frac{M_2^m - M_{2w}^m}{\rho_2} + \frac{H_1^m}{\rho_{p1}} + \frac{H_2^m}{\rho_{p2}} \tag{26}$$

and the superindex m indicates mass units.

An approximate way of estimating the amount of AN in the water phase (M_{2w}) in the NBR case is based on a partition coefficient, defined as the ratio of mass concentrations of monomer 2 in the particles and the water phase:

$$\chi_{AN} = \frac{[(M_2^m)]_p}{[(M_2^m)]_w} \tag{27}$$

As shown in Supplementary Materials Section 2 (see Supplementary Material Equation (S2d)), M_{2w} can be obtained by solving the following quadratic equation:

$$\frac{X_{AN}}{\rho_2} M_{2w}^2 - \left[X_{AN} \left(\frac{M_1^m}{\rho_1} + \frac{M_2^m}{\rho_2} + \frac{P_1^m}{\rho_{p1}} + \frac{P_2^m}{\rho_{p2}} \right) + V_w \right] M_{2w} + M_2^m V_w = 0 \tag{28}$$

2.7. Total Mass Balance

Since there might be side-feed streams of monomers and/or CTAs along the train, a total mass balance per reactor (in mass units) is necessary:

$$\frac{dM_{T,ri}}{dt} = F_{in,ri} - F_{out,ri} + F_{ri,tr} + F_{ri,M1} + F_{ri,M2} \tag{29}$$

where $M_{T,ri}$ is the total mass present in reactor ri (reactors of different volumes can be considered along the train). Notice that in previous equations, the second subindex ri was omitted when it was clear that the balances were written for any reactor in the train. The same convention is used for the total mass flows entering and leaving the reactor ($F_{in,ri}$ and $F_{out,ri}$, respectively). As an approximation, it is assumed that the time derivative is zero; that is, a constant mass is maintained in each reactor. In reality, a constant volume operation is used instead (level control); nonetheless, since the density variations are relatively small, this approximation introduces little error during the transient calculations and no error when the steady-state is achieved. In this way, the total output mass flow can be explicitly calculated from Equation (29) because the other terms are known (naturally, $F_{in,ri} = F_{out,ri-1}$ for $ri > 1$).

2.8. Molecular Weight Distribution

The quantities of two different distributions, one for the live polymer and the other for the dead polymer, are defined as follows:

N_n^r = number of length-r radicals in particles having n radicals, per L of water.

D_n^r = number of length-r inactive polymer chains radicals in particles having n radicals, per L of water.

For the assumptions and details of the derivation of the corresponding PBEs, the reader is referred to [9]. Though the derivations used here follow the general assumptions in [9], the model was adapted to comply with the 0-1-2 scheme. Some of the terms in the balances represent a shift in the classification of radicals in the presence of a given kinetic or physicochemical event.

A general balance for the live polymer $r = 1, 2, 3\ldots$ and $n = 1, 2, 3\ldots$, in any reactor, where the identification of the reactor number (ri) is kept only for the inlet and outlet mass flows, is:

$$\begin{aligned}
\frac{d V_w N_n^r}{dt} =\ & k_p N_n^{r-1}[M]_p V_w - k_p N_n^r[M]_p V_w - k_{trM} N_n^r[M]_p V_w + k_{des}\, n\, N_{n+1}^r \\
& - k_{des} N_n^r(n-1) - K_{trT} N_n^r[Tr]_p V_w + \rho N_o\, \delta(n=1)\, \delta(r=1) \\
& + \rho_{mic}\, Mic\, \delta(n=1)\delta(r=1) - k_{des}\, N_n \delta(r=1)\delta(n=1) + \rho\, N_{n-1}^r \\
& - \rho N_n^r/n + k_{trT} N_n[Tr]_p\, V_w \delta(r=1) + k_{trM} N_n[M]_p V_w \delta(r=1) \\
& + \frac{k_t V_w}{2 N_A v}\left[n(n+1)\, N_{n+2}^r - n(n-1)N_n^r \right] + \left(\frac{\rho N_2^r}{2}\right)\delta(n=1) \\
& + k_{db} F_1 \left[-\left(\sum_{j=1}^{\infty} j D_n^j\right)N_n^r + \sum_{j=1}^{r} j D_n^j N_n^{r-j} \right]\frac{n V_w}{N_n N_A v} + \left[-k_{trp}\left(\sum_{j=1}^{\infty} j D_n^j\right)N_n^r \right. \\
& \left. + k_{trp} r D_n^r \sum_{j=1}^{\infty} N_n^j \right]\frac{n V_w}{N_n N_A v} + F_{in,ri} w_w^f N_n^{r,f}/\rho_w - \frac{N_n^r V_w F_{out,ri}}{M_T}
\end{aligned} \tag{30}$$

where $\delta(j = j_0)$ represents the Kronecker delta function, which is 1 when a given integer variable $j = j_0$ and is 0 elsewhere. For a dead polymer, the balance is, in principle, valid for $r = 1, 2, 3\ldots$ and $= 0, 1, 2, 3\ldots$; the resulting PBE is:

$$\begin{aligned}\frac{dV_w D_n^r}{dt} &= \rho\left(D_{n-1}^r - D_n^r\right) + k_{des}(n+1)D_{n+1}^r - k_{des}n D_n^r + V_w k_{trM}[M]_p N_n^r \\ &+ V_w k_{trT}[Tr]_p N_n^r + \frac{V_w}{2N_A v}k_t(n+2)(n+1)D_{n+2}^r - \frac{V_w}{2N_A v}k_t\, n(n-1)D_n^r + \\ &\quad \frac{k_{td}V_w}{2N_A v}(n+1) \\ &\quad N_{n+2}^r + \frac{k_{tc}V_w}{2N_A v}\frac{(n+1)}{N_n+2}\sum_{j=1}^{r-1} N_{n+2}^j N_{n+2}^{r-j} \\ &\quad + \left[-k_{trp}r D_n^r \sum_{j=1}^{\infty} N_n^j + k_{trp}N_n^r \sum_{j=1}^{\infty} j D_n^j\right]\frac{nV_w}{N_n N_A v} \\ &\quad - k_{db}F_1 r D_n^r \left(\sum_{j=1}^{\infty} N_n^j\right)\frac{nV_w}{N_n N_A v} + F_{in,ri}w_w^f D_n^{r,f}/\rho_w - \frac{D_n^r V_w F_{out,ri}}{M_T}\end{aligned}\quad (31)$$

where w_w^f is the mass fraction of water at the inlet and $P_n^{r,f}$ and $D_n^{r,f}$ are the values of the distribution in the inlet stream for the live and dead polymers, respectively.

Since a 0-1-2 model was to be used, Equation (30) was applied only to $n = 1,2$, while Equation (31) was applied to $n = 0,1,2$. Then, the method of moments was applied to derive equations for the following moments:

$$\mu_{K,n} = \sum_{r=1}^{\infty} r^K\, n = 1, 2,\ K = 0, 1, 2 \quad (32)$$

$$\lambda_{K,n} = \sum_{r=1}^{\infty} r^K D_n^r\, n = 0, 1, 2,\ K = 0, 1, 2 \quad (33)$$

where $\mu_{K,n}$ and $\lambda_{K,n}$ are the K-th moments of the live and dead polymers, respectively, in particles having n radicals. Only the first three moments (0, 1, and 2) of both populations are needed to calculate the number- and weight-average molecular weights. The final equations for these moments, six for the live polymer (Supplementary Material Equations (S3d)–(S3i)) and nine for the dead polymer (Supplementary Material Equations (S3j)–(S3r)), are provided in the Supplementary Materials, both in their original and dimensionless versions. The modeling approach followed here involved much more detail than previous modeling efforts since a full set of moments (0–2) had to be calculated for particles with 0, 1, and 2 radicals and then added together, taking compartmentalization effects into account more faithfully.

2.9. Number of Branches

The number of branches, B, per L of water in any reactor, can be estimated simply by the following ODE:

$$\frac{dBV_w}{dt} = \frac{F_{in}w_w^f B^f}{\rho_w} - \frac{F_{out}V_w B}{M_T} + \frac{k_{trp}+k_{db}}{N_A V_p}\sum_{n=1}^{2} n\lambda_{1,n}\, V_w \quad (34)$$

In steady-state, this can be written as:

$$0 = \frac{F_{in}B^f}{M_T} - \frac{F_{out}B}{M_T} + \frac{k_{trp}+k_{db}}{N_A V_p}\sum_{n=1}^{2} n\lambda_{1,n} \quad (35)$$

or, equivalently:

$$B = \frac{F_{in}B^f}{F_{out}} + \frac{M_T}{F_{out}}\frac{k_{trp}+k_{db}}{N_A V_p}\sum_{n=1}^{2} n\lambda_{1,n} \quad (36)$$

2.10. Strategy of Solution

The model described in the previous section resulted in a first set of 13 ODEs per reactor (*ri*) associated to the states M_1, M_2, H_1, H_2, S, W, Tr, I_w, Y_1, Y_2, $F_{0,ri}$, $F_{1,ri}$, and $F_{2,ri}$, as well as 15 additional ODEs per reactor for the moments of live (6) and dead polymers (9) (Supplementary Material Equations (S2d)–(S2r)). Note, however, that the equations for the live polymer moments $\mu_{0,1}$ and $\mu_{0,2}$ did not need to be solved, since equivalent information could be obtained from the solution of Equations (8) and (9) by realizing that $\mu_{0,1} \equiv F_{1,ri}$ and $\mu_{0,2} \equiv 2F_{2,ri}$. Though the number of branches B could be dynamically calculated using Equation (34), it was decided to calculate its value only at the final steady-state by using Equation (36). In summary, the model resulted in a set of 26 ODEs per reactor. A summary of the ODEs and the most important auxiliary equations, which are the set of working equations implemented in the solution code, is included in Table 2.

To reduce the stiffness of the ODE system, the QSSA was applied to the remaining 4 live polymer moments. Simple explicit algebraic expressions were obtained for these variables, reducing the effective number of ODEs per reactor to 22.

Given the relative complexity of the equations (especially those related to the MWD) and to facilitate the programming and debugging steps of the implementation of the numerical solution of the ODEs, this task was implemented in two stages. In the first stage (computer program), only the first 13 ODEs (for each reactor) were integrated by using a given set of initial conditions in all the reactors in the chain until the steady-state was reached. The set of initial conditions was designed to reach the steady-state fast; linear profiles of the species along the reactor train, according to the approximate profiles expected at the steady-state, were used as the initial conditions. The evolution of the states was recorded at regular time intervals in a file, as illustrated in Figure 1. The variables in the first set of ODEs are independent of the MWD, and, therefore, the corresponding equations could be solved first. Once the solution of this set was obtained, the set of ODEs corresponding to the MWD was numerically solved in a dynamic way in a second computer program by using the dynamic information of the states obtained from the previous integration stage. To obtain more accuracy in the second stage, linear interpolation was used as needed to obtain values of the variables in the first set at integration times between the recorded time intervals (every ~20 s). The numerical routine DDASSL (Differential-Algebraic System Solver) [40] was used in the two stages (programs) for the integration of the equations, which were coded in Fortran. DDASSL is capable of solving either a system of differential-algebraic equations or a system of pure ordinary differential equations, and, since the present system is numerically a system of ODEs (all the coupled algebraic equations are explicit), any ODE integrator for stiff equations could have been used. In this case, all the variables were declared as double-precision in Fortran, and the used numerical relative tolerance was 10^{-7}–10^{-6}. These numerical parameters allowed for safe and robust convergence in all cases. The execution of each stage took ~1–2 s from time zero to the achievement of the steady-state in a standard laptop, even in the case of simulations for the maximum number of implemented reactors (16 in a train). The simulated time needed to achieve the steady-state was around 2–3 times the overall residence time of the reaction in the reactor train, which is usually close to 5 h. The implemented solution strategy turned out to be efficient, robust, and relatively easy to debug.

Table 2. Working equations for implementation in the solution code.

Variable	Equation(s)	Number of Equations
Ordinary differential equations (ODEs) (mass or population balance equations)		
Monomers, M_i	1	2
Monomer bound in polymer, H_i	2	2
Initiator, I_w	12	1
Reducing agent 1, Y_1	13	1
Reducing agent 2, Y_2	14	1
Surfactant, S	15	1
Chain transfer agent, Tr	18	1
Water, W	20	1
Number of particles con n radicals, F_n	7–9	3
Live polymer moments	Supplementary Material Equations (S3f)–(S3i) with extensions Supplementary Material Equation (S3u), Supplementary Material Equation (S3v), Supplementary Material Equation (S3x), and Supplementary Material Equation (S3y)	4
Dead polymer moments	Supplementary Material Equations (S3j)–(S3r) with extensions Supplementary Material Equations (S3aa)–(S3af)	9
TOTAL ODEs		26
Main auxiliary (algebraic) equations		
Radicals in the aqueous phase, P_w	21	1
Number of micelles per L of water, Mic	16 or 17	1
Entry coefficient of radicals to particles, ρ	Supplementary Material Equation (S1a)	1
Entry coefficient of radicals to micelles, ρ_{mic}	Supplementary Material Equation (S1b)	1
Desorption coefficient, k_{des}	Supplementary Material Equation (S1c)	1
Chain transfer agent (CTA) concentration in particles, $[Tr]_p$	Supplementary Material Equation (S1aa)	1
Monomer concentration in particles, $[M]_p$	23 or 25	1
Total mass, M_T	29	1
Number of branches per chain, B	36	1
Closure expression for 3rd moments	Supplementary Material Equation (S3ag)	1
Total number of algebraic equations		10

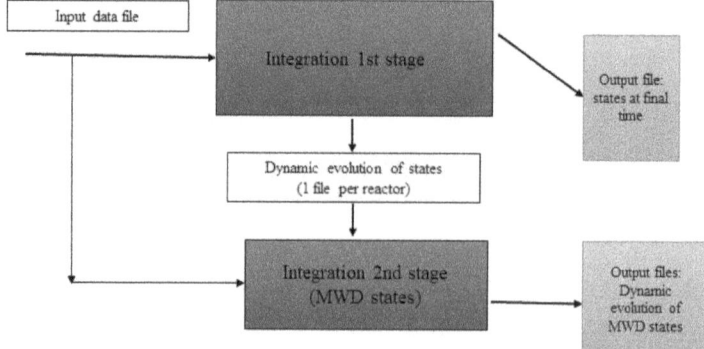

Figure 1. Data flow and computer programs for the implementation of the numerical solution of the mathematical model. Each stage (darker blocks) represents a computer program that is run separately and sequentially. See the text for more details.

3. Results

3.1. Parameter Values

The first problem to solve for the practical use of the developed computer programs was the selection of the physicochemical parameters of the simulated systems (SBR and NBR). Many of them were found in literature sources, but in several cases, the information was scarce or nonexistent. In the latter cases, most of the parameters were estimated a priori based on reasonable assumptions or indirect information; for example, the kinetic constants of the redox initiation system were estimated a priori based on data on the concentration of initiator at the last reactor exit. Notice though that these should not be considered adjustable parameters, since they were not fitted after comparison with the experimental data corresponding to the model outputs but estimated before the simulations were performed. A possible concern with this approach is how significant is the effect of these parameters on the final calculations and, therefore, if the error introduced with these estimations is significant. To answer this, the discussion can be better organized by classifying the prior-estimated parameters into three categories that are discussed next:

- Kinetic parameters (k_{d1}, k_{d1}, f, C_{M_1}, C_{M_2}, C_{T_1}, C_{T_2}): These were based on independent kinetic data (e.g., the kinetic constants of the redox initiation system, as discussed above) or assumed similar to values published for chemically similar systems; therefore, it is expected that little error could be introduced through these parameters.
- Surfactant-related parameters (r_m, a_{em}): These were also estimated by assuming similar values to those of chemically similar systems, which vary within relatively narrow ranges, so no significant errors are expected associated with these estimations.
- Desorption-related coefficients (D_{wT}, D_{pT}, m_d, m_{d_T}): These parameters enter into the expressions of the desorption coefficient (see Supplementary Material Equations (S1h)–(S1i)) and do not have strong influence on the final responses, according to our experience.
- Partition coefficients (K_T, χ_{AN}): These parameters were inferred from known values of the solubilities of the respective components in water and organic media, so it is expected that the estimations used were not far from the real values. Indirect evidence of the adequacy of these estimations came from the behavior observed in the model for responses affected by these parameters, such as the copolymer composition (NBR case) and the molecular weights, which were close to experimental values.

Essentially, only one parameter was left as an adjustable value (entry rate coefficients to micelles and particles which were assumed equal, $k_{mm} = k_{mp}$) to fit the kinetic data (mainly conversion), and a second parameter (transfer to polymer rate coefficient k_{trp}) was used to fit the MWD dispersity. This was justified since the value of the last parameter is usually model-dependent, as it is quite difficult to estimate it from independent experiments; see, for example, [28]. Since both the k_{trp} and the k_{db} affect the MWD dispersity and the number of branches B but it is practically impossible to discern their individual values based solely on the experimental values of the dispersity and B, it was decided to arbitrarily set $k_{db} = 0$ and to fit only the value of k_{trp}, which should then be considered as an effective value. In principle, it should be possible to measure by ^{13}C NMR techniques the number of trifunctional and tetrafunctional branches, mainly associated with transfer to polymer (one tertiary carbon) and internal double bond (two tertiary carbons) reactions, respectively, and from these values to independently obtain a rough estimate of both kinetic coefficients; however, this fell out of the scope of the present work. Regarding propagation and termination coefficients, it is well-known that the most reliable values for these parameters are those obtained by pulse laser polymerization (PLP) (when available), and there have been some reported for these coefficients for styrene and AN (see for example [41,42] for propagation coefficients); however, it was also interesting to compare the model results with kinetic parameters that had been used before by other groups working with emulsion systems, in particular for styrene polymerization, [43] or specifically for NBR polymerization, [24,28], even though they were not determined by PLP. Both types of parameters were tested in the present model; however, for comparison with previous work, the values shown in the simulations were not necessarily obtained with PLP-determined parameters (see details in Table 3). Nonetheless, the sensitivity of the model outputs to the used parameter type (determination method) was quite low. For example, for the SBR system, by changing the value of the propagation coefficient $k_{p,22}$ for styrene from the non-PLP parameter value (151 L mol^{-1} s^{-1} at 10 °C) [43] to the PLP parameter value (43 L mol^{-1} s^{-1} at 10 °C) [41] for simulated conditions very similar to those in Table 4, there were only minor changes in the final conversion in a 10-reactor train (from 0.740 to 0.732, ~1%) and in the number of particles (from 2.74×10^{18} to 2.76×10^{18} L^{-1}, less than 0.5%), as well as even smaller changes in all the other outputs. This low sensitivity can be explained in part due to the relatively low amount of styrene used in the SBR formulation (~17% on a molar basis), but other mechanistic features of this kind of system could also play a role. For the NBR system, a similar situation occurs. A change of the propagation coefficient value $k_{p,22}$ for AN from the non-PLP parameter (6630 L mol^{-1} s^{-1} at 10 °C) [24] to the PLP parameter value (2604 L mol^{-1} s^{-1} at 10 °C) [42] for an initial monomer feed of $f_1 = 0.68$ (wt. basis (without additional side stream of AN)), results in changes smaller than 0.2% in all the model outputs. Unfortunately, for butadiene, the available PLP-determined expression for $k_{p,11}$ [44] is not applicable in the temperature range of interest for these processes; the same is true for the termination coefficient for butadiene, $k_{t,11}$, but in this case, it was decided to extrapolate the Arrhenius expression in [45] out of its valid temperature range because the activation energy and, therefore, the temperature effect is much smaller for the termination than for the propagation step. Table 3 contains the values or range of values used in the simulations. For proprietary reasons, some of the details of the formulation and specific parameters used are not disclosed.

Table 3. Parameter values tested or used in the simulations for the styrene–butadiene rubber (SBR) and acrylonitrile–butadiene rubber (NBR) systems.

Parameter	Value	Units	Source/Comments
\multicolumn{4}{SBR System: monomer 1 butadiene, monomer 2 styrene}			
r_1	1.4–1.58		1.4 from [46] (@ 50 C)
r_2	0.54–0.58		0.58 from [46] (@ 50 C)
$^1E, k_{p,11}$	8540	cal mol^{-1}	[47]
$^1A, k_{p,11}$	1.12×10^8	L mol^{-1} s^{-1}	[47]
$^2E, k_{p,22}$	6220	cal mol^{-1}	[46]
$^2A, k_{p,22}$	4.50×10^6	L mol^{-1} s^{-1}	[46]
$E, k_{p,22}$	7770	cal mol^{-1}	[41] Pulse laser polymerization (PLP)-determined
$A, k_{p,22}$	4.27×10^7	L mol^{-1} s^{-1}	[41] PLP-determined
k_{d1}	100–250	L mol^{-1} s^{-1}	Estimated
k_{d2}	50–200	L mol^{-1} s^{-1}	Estimated
$k_{t,11}$ @ 10 C	9.16×10^8	L mol^{-1} s^{-1}	[45] PLP-determined
$k_{t,22}$ @ 10 C	9.2×10^6	L mol^{-1} s^{-1}	[46]
C_{M_1}	9×10^{-5}	-	Estimated from [46]
C_{M_2}	9×10^{-5}	-	Assumed equal to C_{M_1}
C_{T_1}	2	-	Estimated
C_{T_2}	2	-	Estimated
r_m	5×10^{-9}	m	Estimated
$[S]^{cmc}$	9.8×10^{-4}	mol L^{-1}	[48]
a_{em}	2.2×10^{-19}	m^2	Estimated
b	2000	L mol^{-1}	[9]
Γ_∞	3.5×10^{-6}	mol m^{-2}	[9]
x_{sat}	0.55	-	Estimated
f	0.5	-	Estimated
k_{mm}			This work
k_{mp}			This work
D_w	3.55×10^{-15}	m^2 s^{-1}	[49]
D_p	3.55×10^{-15}	m^2 s^{-1}	[49]
D_{wT}	2×10^{-15}	m^2 s^{-1}	Estimated
D_{pT}	2×10^{-15}	m^2 s^{-1}	Estimated
m_d	20		Estimated
m_{d_T}	100		Estimated
K_T	200		Estimated
k_{trp}	0.005–0.05	L mol^{-1} s^{-1}	This work
ρ_1	618	g L^{-1}	Estimated from [50]
ρ_2	906	g L^{-1}	[49]
ρ_{p1}	910	g L^{-1}	Estimated from MSDS Sigma Aldrich St. Louis MO polybutadiene Mw = 200,000
ρ_{p2}	1040	g L^{-1}	[46]

Table 3. Cont.

Parameter	Value	Units	Source/Comments
Density surfactant	918	$g\ L^{-1}$	Estimated
NBR System: monomer 1 butadiene, monomer 2 acrylonitrile			
[3]NBR System: monomer 1 butadiene, monomer 2 acrylonitrile [3]			
r_1	0.30		[51]
r_2	0.04		[51]
[2]$k_{p,22}$	6633	$L\ mol^{-1}\ s^{-1}$	[24]
$E, k_{p,22}$	3690	$cal\ mol^{-1}$	[42] PLP-determined
$A, k_{p,22}$	1.83×10^6	$L\ mol^{-1}\ s^{-1}$	[42] PLP-determined
$k_{t,22}$ @ 10 C	3.4×10^7	$L\ mol^{-1}\ s^{-1}$	[28]
$[S]^{cmc}$	6.5×10^{-4}	$mol\ L^{-1}$	[48]
a_{em}	3.8×10^{-19}	m^2	Estimated
m_d	10		Estimated
ρ_2	822	$g\ L^{-1}$	[28] @ 10 C
ρ_{p2}	1167	$g\ L^{-1}$	[28] @ 10 C
χ_{AN}	3.5		Estimated

[1] In all kinetic constants k_x where an Arrhenius expression was available, the used notation is: $k_x = A \exp\left(-\frac{E}{RT}\right)$ with R = 1.987 cal mol^{-1} K^{-1} and the temperature T in K. [2] Values used in all the simulations in figures. [3] Only the used values that are different from those of the SBR system are listed.

Table 4. Typical recipe in parts per hundred monomer (pphm). Approximately the medium value of the range for each component was used in the simulations.

Component	Parts per Hundred Monomer (pphm)				
Butadiene (M_1)	72				
Styrene (M_2)	28				
Initiator	0.1–0.2				
Reducing agent 1 [1]	0.02–0.05				
Reducing agent 2 [2]	0.04–0.10				
Chain transfer agent	0.10–0.17				
Surfactant	3.0–5.0				
Caustic potash	0.5–0.7				
Electrolyte (sodium carbonate, potassium chloride)	0.2–0.4				
Reactor No	R1–R5	R6	R7	R8	R9–R10
Temperature	baseT (T_b)	T_b + 10 C	T_b + 7 C	T_b + 6 C	T_b + 2 C

[1] Based on ferrous sulfate. [2] Based on sodium formaldehyde sulfoxylate.

3.2. Model Validation

Figure 2 shows a comparison of the model predictions with plant data for conversion (18 experimental points) in the SBR case for a 10-reactor train. The steady-state operation of each reactor (except for reactor 2 (R2) in the train) was sampled at two different times (data labeled as experiments 1 and 2); slight temperature variations (± 1 C) were recorded for the temperatures of the reactors at these different times, except for reactor 10, in which the temperature difference was +3 C for experiment 2. Details of the used temperature profile, as well as the formulation data used in the plant, are provided in Table 4. A similar formulation has been published before for the NBR case [29,32]. The agreement between model and experiment was quite reasonable using a single value of $k_{mm} = k_{mp}$.

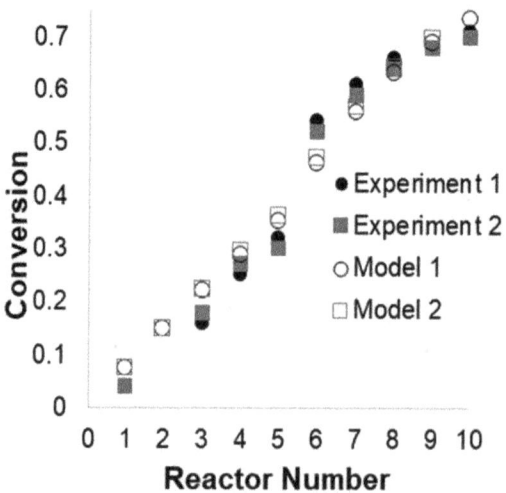

Figure 2. Comparison of plant conversion data with model predictions for two operating conditions. The two simulated operating conditions are indicated as Model 1 and Model 2, See the conditions in Table 4.

Figure 3 shows the model predictions with a few experimental data from the plant available for the number average molecular weight M_n, the MWD dispersity, and the accumulated copolymer composition F_1. The copolymer composition was measured by refractive index (ASTM D5775-95(2019)), while M_n and the MWD dispersity were measured by size exclusion chromatography (estimating absolute values from those relative to a polystyrene standard). The reactivity ratios were slightly adjusted (~10%) with respect to those reported in the open literature to get a better agreement with the copolymer composition data (Table 3), but this was justified since the literature data were obtained at a different temperature. The agreement of the model with the experimental data was reasonable but unfortunately limited to the few available experimental data points.

Figures 4–6 show the effect of variation of surfactant on the most important outputs and process parameters with respect to the base case, which corresponded to the conditions of Experiment 1 and included a side stream of CTAs in R5. The predicted effects were all expected and easy to explain. As the surfactant increased with respect to the base value, more particles were produced (Figure 5A) and the reaction rate was higher (Figure 5C), as was the conversion (Figure 4A). At a higher conversion, the composition drift increased, as shown in Figure 4B–D. A higher number of particles resulted in smaller particle diameters (Figure 5D), although the effect is barely perceptible in the plot. Finally, higher reaction rates resulted in higher molecular weights and dispersities. The jump observed in R_p in R6 was due to the abrupt increase in temperature from R5 to R6 (+10 C). Figure 5B shows the value of the average number of radicals per particle ($\widetilde{n}$) according to the model. It must be highlighted that some of the previous models ([3] and partially in [4,5]) assumed a constant value of $\widetilde{n} = 0.5$ (Smith–Ewart case II kinetics), but, as seen in the figure, this may have deviated from the presumably true value by 20–30%, thus having a direct impact on the estimation of the polymerization rate (see Equation (5)).

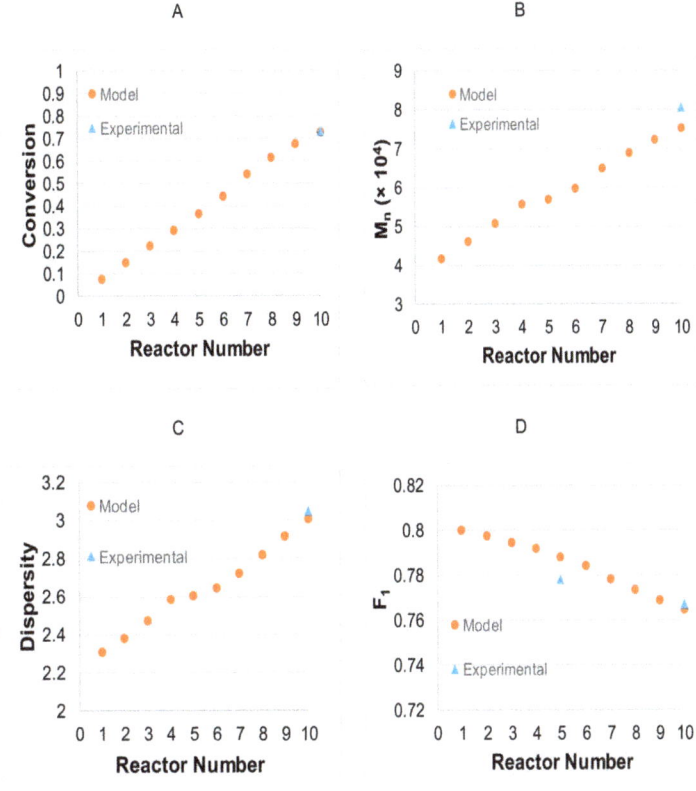

Figure 3. Comparison of model prediction with plant experimental data for conversion (**A**), number average molecular weight instantaneous (M_n) (**B**), MWD dispersity (**C**), and accumulated copolymer composition F_1 (**D**). The parameter values were those in Table 3, and the reaction conditions were similar to those in Table 4.

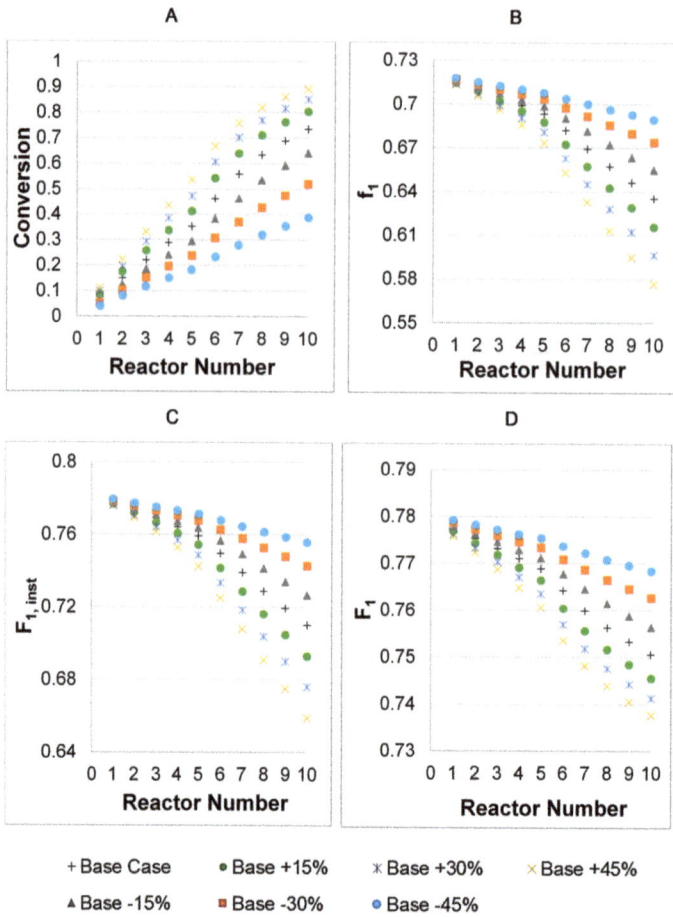

Figure 4. Effect of variation of surfactant (from −45% to +45%) on conversion (**A**), remaining monomer feed composition f_1 (**B**), instantaneous copolymer composition ($F_{1,inst}$) (**C**), and accumulated copolymer composition F_1 (**D**) along the reactor train with respect to the base case in SBR production. All compositions are weight fractions.

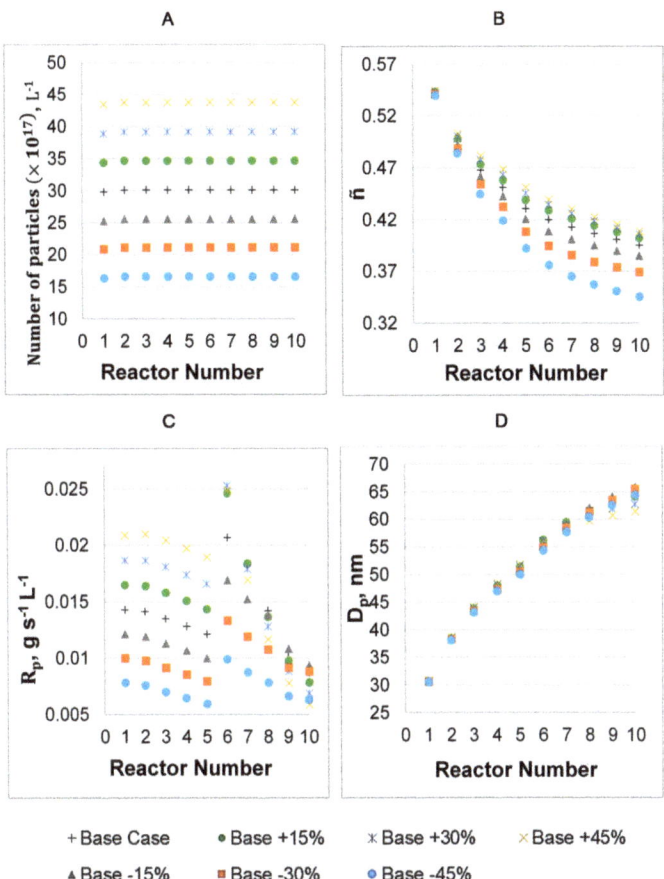

Figure 5. Effect of variation of surfactant (from −45% to +45%) on number of particles N_p (**A**), average number of radicals $\tilde{n}$ (**B**), reaction rate R_p (**C**), and particle diameter D_p (**D**) along the reactor train with respect to the base case in SBR production.

Figure 6 shows how an increase in surfactant concentration also caused an increase in the average molecular weight (Figure 6A,B). This was due to an increased number of particles that delayed the entry of a second radical to the particle (more competition for radicals), extending its life and, therefore, its molecular weight. The increased competition for radicals also resulted in an increased dispersion in the probability of capturing a second radical, leading to increased MWD dispersity (Figure 6C). Notice that this explanation has also been put forward by other researchers, albeit in the frame of miniemulsion polymerization. [52,53]

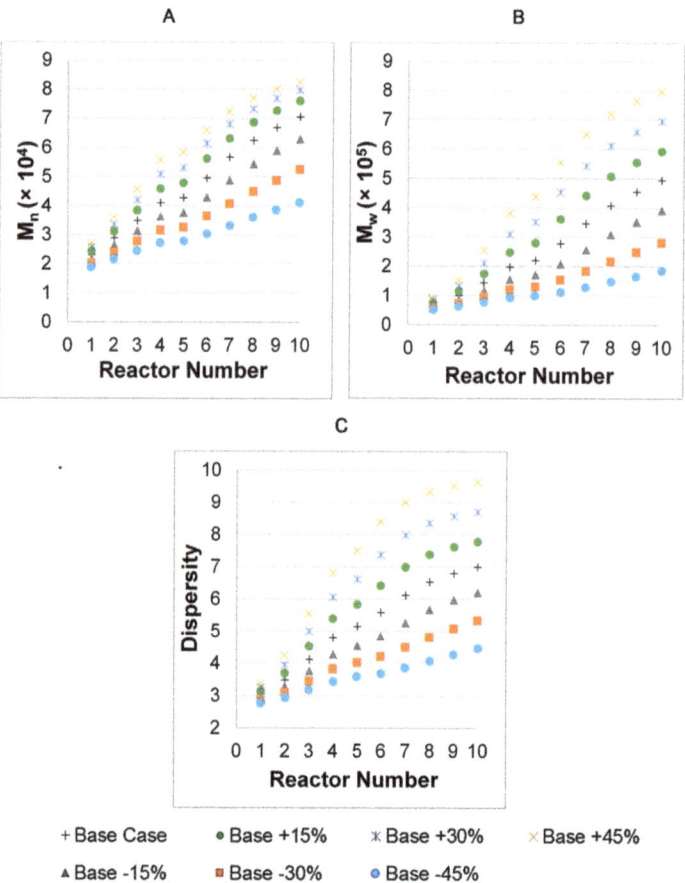

Figure 6. Effect of variation of surfactant (from −45% to +45%) on the number-average molecular weight M_n (**A**), weight-average molecular weight M_w (**B**), and MWD dispersity (**C**) along the reactor train with respect to the base case in SBR production.

Figure 7 shows the evolution of the number of particles with zero, one, and two radicals for the simulated SBR base case. At the first reactor, due to the relative abundance of initiator radicals, the number of particles with one radical (F_1) was higher than that with zero radicals (F_0); however, down the train, this situation was reversed as the initiator was depleted. From the plots, it is also clear that the number of particles with two radicals (F_2) was negligible in all the reactors, pointing to the conclusion that a 0-1 model should suffice to represent these systems. Notice that this behavior does not mean that an average value of $\tilde{n} = 0.5$ would provide a good representation of the system as assumed in early models; it rather means that $\tilde{n} < 0.5$. However, the whole situation could change if a gel effect was included in the model.

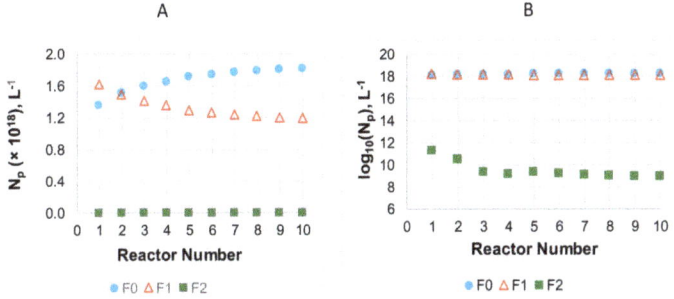

Figure 7. Number of particles with zero (F_0), one (F_1), and two (F_2) radicals per L of water at each reactor for the SBR base case. (**A**) Linear scale; (**B**) log scale.

Figure 8 shows the effect of the variation of the amount of CTAs on conversion, average molecular weights, and MWD dispersity. The effect was almost negligible on the conversion since the desorption of CTAs was not significant; however, the effect on the molecular weight averages and dispersity was as expected, with large increases in these quantities as the amount of CTAs decreased. The plot for a 45% decrease of CTAs is not shown since the calculation diverged for the last reactor as the system reached the gelation point (the second moment of the MWD tended to infinity).

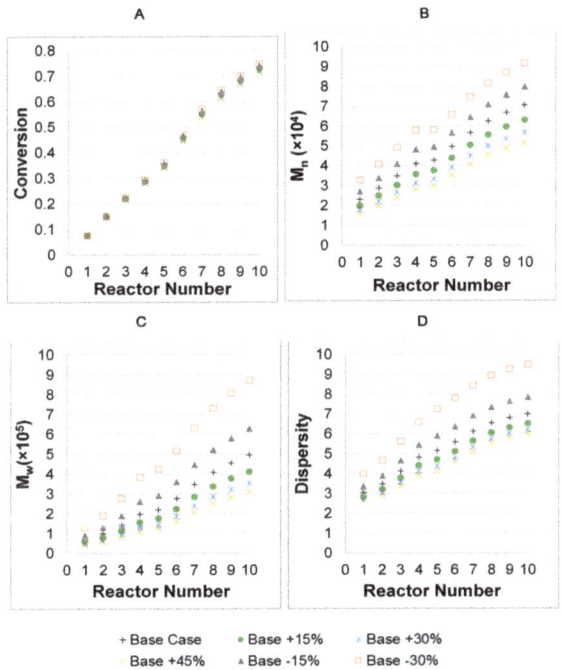

Figure 8. Effect of variation of chain transfer agents (CTAs) (from −30% to +45%) on the conversion (**A**), number-average molecular weight M_n (**B**), weight-average molecular weight M_w (**C**), and MWD dispersity (**D**) along the reactor train with respect to the base case in SBR production.

Finally, Figure 9 shows the effect of the side stream of CTAs on the MWD. As mentioned above, in the base case, a (correction) side stream of CTAs was fed to R5. Figure 9 compares the behavior of the train with and without the intermediate CTA feed. As previously discussed, the effect was negligible for the progress of the conversion along the train (Figure 9A), but the final M_n and the dispersity increased from ~70,000 to ~90,000 and from ~7 to ~9, respectively, when the CTA side stream was suppressed. Similarly, the final number of branches increased from ~1 to ~1.2 in the absence of the side stream. Clearly, the intermediate CTA stream avoided an excessive increase in the molecular weight and the eventual formation of gel in the last reactors of the train that operate at high conversions.

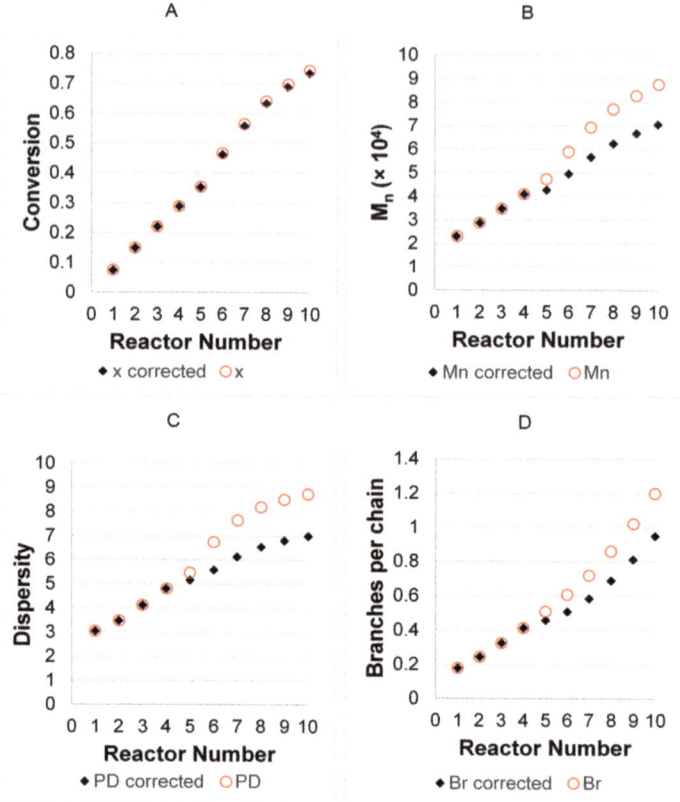

Figure 9. Effect of suppressing the side stream of CTAs of the base case on the conversion (**A**), number-average molecular weight M_n (**B**), MWD dispersity (**C**), and number of branches per chain (**D**) along the reactor train with respect to the base case (corrected) in SBR production.

The effects of the changes in the redox initiator system were milder than those found for the surfactant and CTAs and are shown in the Supplementary Materials (Figure S1).

Similar behavior was found for the NBR system upon changes in the surfactant, CTA, and initiator. The effects of changes in the surfactant for an NBR system are shown in Figures 10–12 for the same important outputs and process parameters shown for the SBR system in Figures 4–6. The variations of the surfactant were made with respect to a base case, which corresponded to a composition feed of $f_1 = 0.68$ (wt. basis) and weight feed ratios of 1.25/0.082/0.35/100 for S/I$_w$/CTAs/monomers. The base case also included a side stream of an additional 20% flow rate of CTAs at R6. The used temperature was 10 °C in all reactors except in R1 (12 °C) and R10 (15 °C). The effects were very similar to those found for the SBR system, with a few exceptions. In Figure 10, the butadiene composition of the remaining monomer f_1, as well as those of the copolymer (accumulated and instantaneous (F_1 and

$F_{1,\text{inst}}$, respectively)), increased instead of decreasing like in the SBR case. This was expected given the differences in the reactivity ratios between the two systems. In this region of the monomer feed composition (above the azeotropic point for NBR) in the NBR case, acrylonitrile was consumed faster than butadiene (see the Mayo–Lewis plot for this system in Figure S2), and the composition of the remaining monomer therefore increased in butadiene with conversion, also leading to a richer butadiene composition in the copolymer.

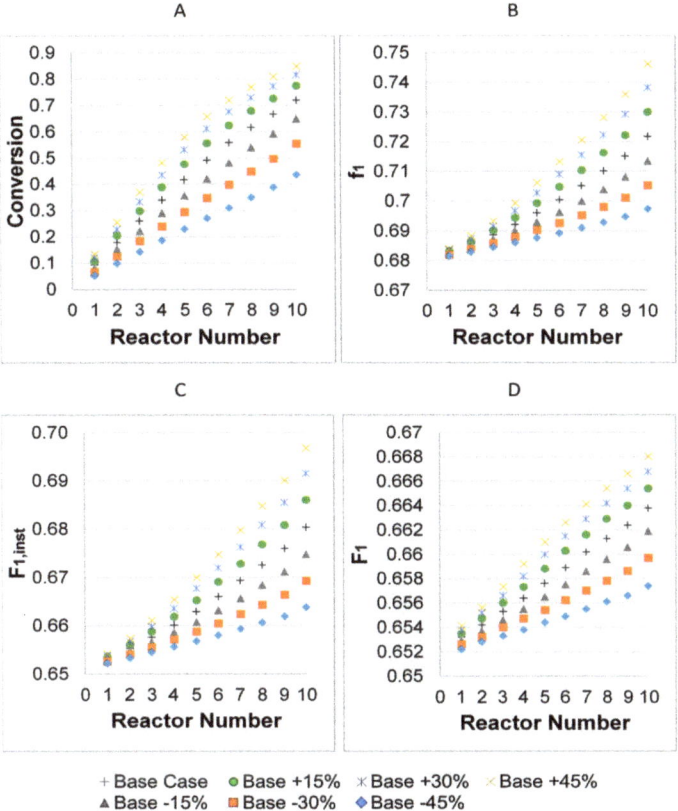

Figure 10. Effect of variation of surfactant (from −45% to +45%) on conversion (**A**), remaining monomer feed composition f_1 (**B**), instantaneous copolymer composition ($F_{1,\text{inst}}$) (**C**), and accumulated copolymer composition F_1 (**D**) along the reactor train with respect to the base case in NBR production. All compositions are weight fractions.

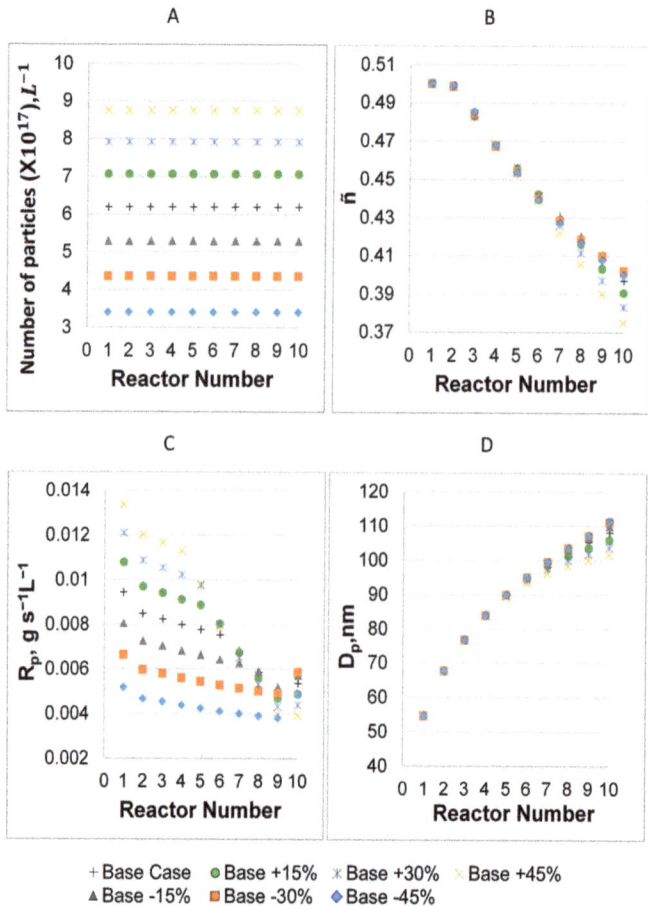

Figure 11. Effect of variation of surfactant (from −45% to +45%) on number of particles N_p (**A**), average number of radicals $\tilde{n}$ (**B**), reaction rate R_p (**C**), and particle diameter D_p (**D**) along the reactor train with respect to the base case in NBR production.

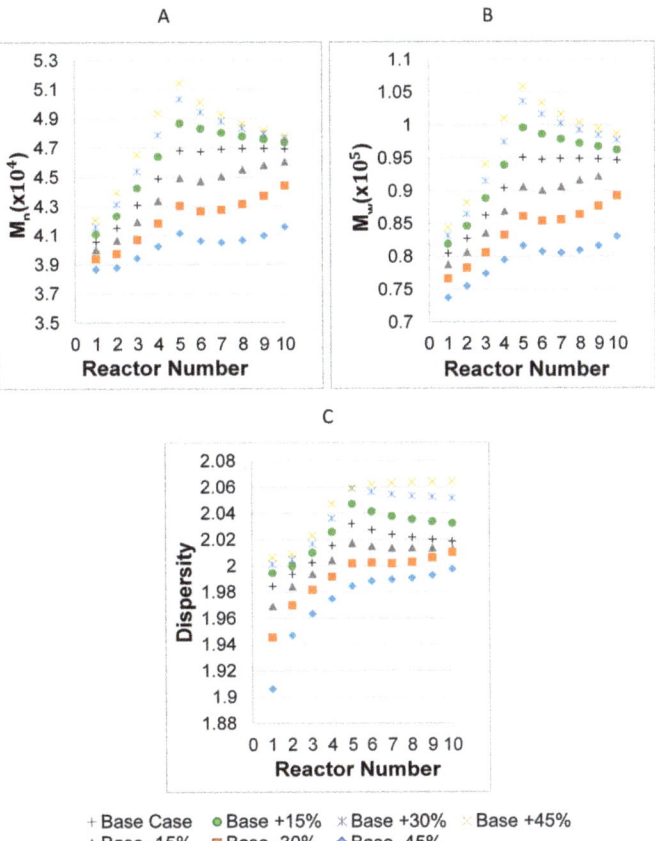

Figure 12. Effect of variation of surfactant (from −45% to +45%) on the number-average molecular weight M_n (**A**), weight-average molecular weight M_w (**B**), and MWD dispersity (**C**) along the reactor train with respect to the base case in NBR production.

The effects of the surfactant changes on N_p, R_p, $\tilde{n}$, and D_p shown in Figure 11 for the NBR system were qualitatively very similar to those observed and already discussed for the SBR case (Figure 5). The only minor difference was that the effects were smaller for $\tilde{n}$ in the NBR than in the SBR case. Another difference is that in the NBR case, there was not a jump in the reaction rate as in the SBR case; however, this was simply because in the former case, there was not a jump in an intermediate reactor temperature as in the latter. Figure 12 shows similar effects of the surfactant concentration for the NBR system compared to those in the SBR system (Figure 6). A difference was that in the NBR case, the molecular weights reached a peak in R5 to decrease in R6 due to the side stream of CTAs. This effect was less important in the SBR base case, but in that system, the higher temperatures in the last reactors also significantly contributed to higher molecular weights. Additionally, the molecular weights and dispersities were lower for NBR compared to those exhibited by the SBR system.

The effects of changes in the CTA feed with respect to the NBR base case are shown in Figure 13, which is comparable to Figure 8 for the SBR system. Again, the behavior of both systems was qualitatively similar with the only exception of the peak exhibited by the molecular weights in R5 for the NBR case due to the side stream of CTAs in R6, as in Figure 12, and the different temperature profiles used for the SBR and the NBR cases.

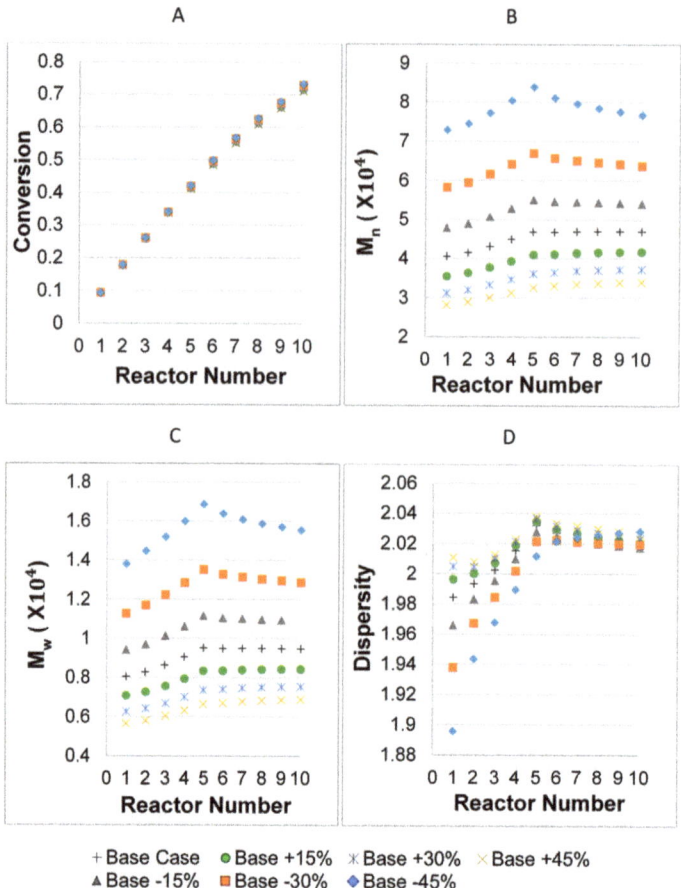

+ Base Case • Base +15% ✳ Base +30% ✕ Base +45%
▲ Base -15% ■ Base -30% ♦ Base -45%

Figure 13. Effect of variation of CTAs (from −45% to +45%) on the conversion (**A**), number-average molecular weight M_n (**B**), weight-average molecular weight M_w (**C**), and MWD dispersity (**D**) along the reactor train with respect to the base case in NBR production.

More interesting behavior for the NBR system was related to the drift in copolymer composition along the train. Given the reactivity ratios of this system (0.30 and 0.04), the system exhibited an azeotropic composition at $f_{1,az} = F_{1,az} = 0.578$ (molar basis, 0.582 weight basis) (see the F_1 vs. f_1 Mayo–Lewis curve in the Supplementary Materials, Figure S2) Therefore, composition drift depends on what side of the azeotrope the composition of the feed is located. If the feed composition is above $f_{1,az}$, as in most commercial grades of NBR, without correction, the instantaneous product composition will initially deviate towards copolymers with a lower Bd composition than the feed (R1); however, as the reaction mixture proceeds along the reactor train, the Bd-enriched environment will tend to increase the Bd content in the copolymer. This can be seen in Figure 14 for a product grade in which f_1 in the feed was 0.68 (wt. basis). Since composition drift may lead to a product with heterogeneous composition or a product deviated from the target average composition, it is common to introduce a side feed stream of AN in an intermediate reactor of the train to "correct" or attenuate the composition drift. Figure 10 also shows the composition profile when an additional AN side stream (20% of the initial feed) was fed to reactor R7. Without the correction, the final F_1 was above 0.66; on the other hand, with the side feed stream at R7, the final F_1 was slightly below 0.65. A second example is shown in Figure 15; in this case, the composition of the feed was $f_1 = 0.5$ on a wt. basis, a composition which was located on the other side of the azeotrope (see the F_1 vs. f_1 curve in the Supplementary Materials). The behavior of this system was the opposite of that in the previous example, exhibiting a composition drift that initially produced copolymers with a higher Bd composition than the feed (R1, F_1 ~0.58) and a decreasing Bd content in the copolymer along the reactor train to end up with an F_1 slightly above 0.56, although the instantaneous F_1 fell below 0.54 at R10. Figure 15 also shows how this drift could be attenuated by feeding a side stream of Bd (20% of the initial feed). This correction worked very well, resulting in a final average composition near the target (F_1 ~0.58) while avoiding large deviations in the instantaneous composition produced in each reactor (better composition homogeneity).

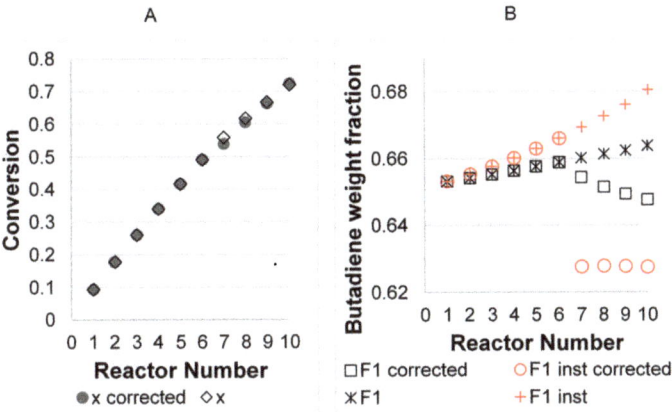

Figure 14. Effect of the operation of the reactor train for NBR production with (corrected) and without a side stream of additional AN in R7 (20% with respect to the AN feed in R1) on conversion (**A**) and instantaneous butadiene copolymer composition ($F_{1,inst}$) and accumulated butadiene copolymer composition (F_1) (**B**). The initial monomer feed was $f_1 = 0.68$ (wt. basis).

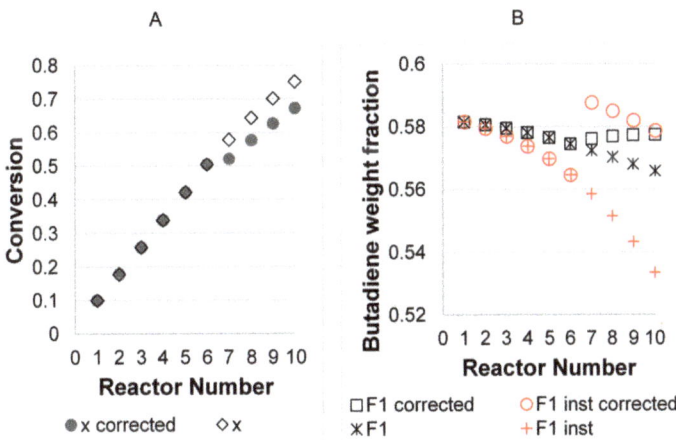

Figure 15. Effect of the operation of the reactor train for NBR production with (corrected) and without a side stream of additional Bd in R7 (20% with respect to the Bd feed in R1) on conversion (**A**) and instantaneous butadiene copolymer composition ($F_{1,inst}$) and accumulated butadiene copolymer composition (F_1) (**B**). The initial monomer feed is $f_1 = 0.5$ (wt. basis).

4. Conclusions

A single mechanistic model was built and used to describe the production of both SBR and NBR elastomers in trains of emulsion polymerization CSTRs. The aim was to balance the level of complexity of the model with its capability of describing the effects observed at the plant on the process parameters in a quantitative way and the product quality upon variable changes. Some novel aspects of this model compared to previous models are summarized below, along with additional comments:

- The radical compartmentalization was considered in more detail, both in the particle kinetics description and in the calculation of the moments of the MWD, and for the first time, a 0-1-2 model was used to describe these two important aspects in these processes. It was found that at least a 0-1 model should be used for an accurate description of the compartmentalization effects, especially for the first reactors in the train. A similar conclusion was recently put forward by Marien et al. [52] using stochastic simulation, although for a general miniemulsion system.
- The monomer partitioning model was as simple as possible, and its parameters could be estimated a priori based on published physicochemical data. As mentioned by Madhuranthakam and Penlidis, [30] the use of more sophisticated thermodynamic models for monomer partitioning seems to be unnecessary for this kind of system.
- A single modeling framework was presented in a unified form for the SBR and NBR systems. Though previous models have been applied with adaptations to both systems in some series of papers, it is believed that by presenting them together, it is easier to appreciate the similarities and differences in both systems.

The model is presently used in the plant to design changes of configurations of the reactor train, such as the number of reactors in the train (up to 16 reactors in the existing facility), the reactor volume, the location and flows of side streams of CTAs and monomers, and grade recipes, among other variables, to maximize the plant operation versatility maintaining high productivity and product quality. The model is quite flexible, and it has been implemented with the help of a friendly graphical interphase. Current work involves the generation of product quality correlations that link model outputs, e.g., M_n and M_w, and number of branches, with product performance parameters, such as Mooney viscosity.

Supplementary Materials: The following are available online at http://www.mdpi.com/2227-9717/8/11/1508/s1, File with figures and additional equations.

Author Contributions: Conceptualization, E.S.-G. and J.M.I.-M.; methodology, E.S.-G., R.I.-M., and J.M.I.-M; software, E.S.-G. and R.I.-M.; validation, E.S.-G., R.I.-M., and J.M.I.-M.; writing—original draft preparation, E.S.-G. and J.M.I.-M; writing—review and editing, E.S.-G. and J.M.I.-M.; project administration, E.S.-G. and J.M.I.-M; funding acquisition, E.S.-G. and J.M.I.-M. All authors have read and agreed to the published version of the manuscript.

Funding: This research was funded by Dynasol Elastomers and the publication fee by CIQA.

Acknowledgments: The authors thank Dynasol Elastomers for industrial data and CIQA for administrative support.

Conflicts of Interest: The authors declare no conflict of interest.

References

1. Production Capacity of Styrene-Butadiene-Rubber Worldwide in 2018 and 2023. Available online: https://www.statista.com/statistics/1063647/styrene-butadiene-rubber-production-capacity-globally/ (accessed on 8 January 2020).
2. The Global Nitrile Butadiene Rubber (NBR) Market to 2023: Global Production of NBR is Likely to Reach ~1,562 Thousand Tons. Available online: https://www.businesswire.com/news/home/20181126005386/en/Global-Nitrile-Butadiene-Rubber-NBR-Market-2023 (accessed on 8 January 2020).
3. Kanetakis, J.; Wong, F.Y.C.; Hamielec, A.E.; MacGregor, J.F. Steady State Modeling of a Latex Reactor Train for the Production of Styrene-Butadiene Rubber. *Chem. Eng. Commun.* **1985**, *35*, 123–140. [CrossRef]

4. Broadhead, T.O.; Hamielec, A.E. Dynamic modelling of the batch, semi-batch and continuous production of styrene/butadiene copolymers by emulsion polymerization. *Die Makromol. Chem.* **1985**, *10*, 105–128. [CrossRef]
5. Penlidis, A.; MacGregor, J.F.; Hamielec, A.E. Dynamic modelling of emulsion polymerization reactors. *AIChE J.* **1985**, *31*, 881–889. [CrossRef]
6. Smith, W.V.; Ewart, R.H. Kinetics of Emulsion Polymerization. *J. Chem. Phys.* **1948**, *16*, 592–599. [CrossRef]
7. Storti, G.; Carrá, S.; Morbidelli, M.; Vita, G. Kinetics of multimonomer emulsion polymerization. The pseudo-homopolymerization approach. *J. Appl. Polym. Sci.* **1989**, *37*, 2443–2467. [CrossRef]
8. Tobita, H.; Hamielec, A.E. Kinetics of free-radical copolymerization: The pseudo-kinetic rate constant method. *Polymer* **1991**, *32*, 2641–2647. [CrossRef]
9. Saldívar, E.; Dafniotis, P.; Ray, W.H. Mathematical Modeling of Emulsion Copolymerization Reactors. I. Model Formulation and Application to Reactors Operating with Micellar Nucleation. *J. Macromol. Sci. Part C Polym. Rev.* **1998**, *38*, 207–325. [CrossRef]
10. Ugelstad, J.; Mork, P.C.; Aasen, J.E. Kinetics of emulsion polymerization. *J. Polym. Sci.* **1967**, *A5*, 2281–2288. [CrossRef]
11. Yabuki, Y.; MacGregor, J.F. Product Quality Control in Semibatch Reactors Using Midcourse Correction Policies. *Ind. Eng. Chem. Res.* **1997**, *36*, 1268–1275. [CrossRef]
12. Gugliotta, L.M.; Brandolini, M.C.; Vega, J.R.; Iturralde, E.O.; Azum, J.L.; Meira, G.R. Dynamic Model of a Continuous Emulsion Copolymerization of Styrene and Butadiene. *Polym. React. Eng.* **1995**, *3*, 201–233.
13. Flory, P.J. *Principles of Polymer Chemistry*; Cornell University Press: Ithaca, NY, USA, 1953.
14. Morton, M.; Kaizerman, S.; Altier, M.W. Swelling of latex particles. *J. Colloid Sci.* **1954**, *9*, 300–312. [CrossRef]
15. Vega, J.R.; Gugliotta, L.M.; Brandolini, M.C.; Meira, G.R. Steady-State Optimization in a Continuous Emulsion Copolymerization of Styrene and Butadiene. *Lat. Am. Appl. Res.* **1995**, *25*, 207–214.
16. Vega, J.R.; Gugliotta, L.M.; Meira, G.R. Continuous Emulsion Polymerization of Styrene and Butadiene. Reduction of Off-Spec Product between Steady-State. *Lat. Am. Appl. Res.* **1995**, *25*, 77–82.
17. Minari, R.J.; Vega, J.R.; Gugliotta, L.M.; Meira, G.R. Continuous Emulsion Styrene-Butadiene Rubber (SBR) Process: Computer Simulation Study for Increasing Production and for Reducing Transients between Steady States. *Ind. Eng. Chem. Res.* **2006**, *45*, 245–257. [CrossRef]
18. Godoy, J.L.; Minari, R.J.; Vega, J.R.; Marchetii, J.L. Multivariate statistical monitoring of an industrial SBR process. Soft-sensor for production and rubber quality. *Chemom. Intell. Lab. Syst.* **2011**, *107*, 258–268. [CrossRef]
19. Zubov, A.; Pokorny, J.; Kosek, J. Styrene–butadiene rubber (SBR) production by emulsion polymerization: Dynamic modeling and intensification of the process. *Chem. Eng. J.* **2012**, *207*, 414–420. [CrossRef]
20. Mustafina, S.; Miftakhov, E.; Mikhailova, T. Mathematical Study of Copolymer Composition and Compositional Heterogeneity during the Synthesis of Emulsion-Type Butadiene-Styrene Rubber. *Int. J. Chem. Sci.* **2014**, *12*, 1135–1144.
21. Dubé, M.A.; Penlidis, A.; Mutha, R.K.; Cluett, W.R. Mathematical Modeling of Emulsion Copolymerization of Acrylonitrile/Butadiene. *Ind. Eng. Chem. Res.* **1996**, *46*, 4434–4448. [CrossRef]
22. Vega, J.R.; Gugliotta, L.M.; Bielsa, R.O.; Brandolini, M.C.; Meira, G.R. Emulsion Copolymerization of Acrylonitrile and Butadiene. Mathematical Model of an Industrial Reactor. *Ind. Eng. Chem. Res.* **1997**, *36*, 1238–1246. [CrossRef]
23. Rodríguez, V.I.; Estenoz, D.A.; Gugliotta, L.M.; Meira, G.R. Emulsion Copolymerization of Acrylonitrile and Butadiene. Calculation of the Detailed Macromolecular Structure. *Int. J. Polym. Mater.* **2002**, *51*, 511–529. [CrossRef]
24. Vega, J.R.; Gugliotta, L.M.; Meira, G.R. Emulsion Copolymerization of Acrylonitrile and Butadiene. Semibatch Strategies for Controlling Molecular Structure on the Basis of Calorimetric Measurements. *Polym. React. Eng.* **2002**, *10*, 59–82. [CrossRef]
25. Minari, R.J.; Gugliotta, L.M.; Vega, J.R.; Meira, G.R. Continuous emulsion copolymerization of acrylonitrile and butadiene: Computer simulation study for improving the rubber quality and increasing production. *Comput. Chem. Eng.* **2007**, *31*, 1073–1080. [CrossRef]
26. Minari, R.J.; Gugliotta, L.M.; Vega, J.R.; Meira, G.R. Continuous emulsion copolymerization of acrylonitrile and butadiene: Simulation study for reducing transients during changes of grade. *Ind. Eng. Chem. Res.* **2007**, *46*, 7677–7683. [CrossRef]

27. Clementi, L.A.; Suvire, R.B.; Rossomando, F.G.; Vega, J.R. A Closed-Loop Control Strategy for Producing Nitrile Rubber of Uniform Chemical Composition in a Semibatch Reactor: A Simulation Study. *Macromol. React. Eng.* **2018**, *12*, 1700054. [CrossRef]
28. Washington, I.D. Dynamic Modelling of Emulsion Polymerization for the Continuous Production of Nitrile Rubber. Master's Thesis, University of Waterloo, Waterloo, ON, Canada, 2008.
29. Washington, I.D.; Duever, T.A.; Penlidis, A. Mathematical Modeling of Acrylonitrile-Butadiene Emulsion Copolymerization: Model Development and Validation. *J. Macromol. Sci. Part A Pure Appl. Chem.* **2010**, *47*, 747–769. [CrossRef]
30. Madhuranthakam, C.M.R.; Penlidis, A. Modeling Uses and Analysis of Production for Acrylonitrile-Butadiene (NBR) Emulsions. *Polym. Eng. Sci.* **2011**, *51*, 1909–1918. [CrossRef]
31. Madhuranthakam, C.M.R.; Penlidis, A. Improved Operating Scenarios for the Production of Acrylonitrile-Butadiene Emulsions. *Polym. Eng. Sci.* **2012**, *53*, 9–20. [CrossRef]
32. Scott, A.J.; Nabifar, A.; Madhuranthakam, C.M.R.; Penlidis, A. Bayesian Design of Experiments Applied to a Complex Polymerization System: Nitrile Butadiene Rubber Production in a Train of CSTRs. *Macromol. Theory Simul.* **2015**, *24*, 13–27. [CrossRef]
33. Vale, H.M.; McKenna, T.F. Synthesis of bimodal PVC latexes by emulsion polymerization: An experimental and modeling study. *Macromol. Symp.* **2006**, *243*, 261–267. [CrossRef]
34. Dubé, M.A.; Saldívar-Guerra, E.; Zapata-González, I. Copolymerization. In *Handbook of Polymer Synthesis, Characterization and Processing*; Saldívar-Guerra, E., Vivaldo-Lima, E., Eds.; John Wiley and Sons: Hoboken, NJ, USA, 2013.
35. Blackley, D.C. *Emulsion Polymerization*; Applied Science Publishers: London, UK, 1975.
36. Friis, N.; Nyhagen, L. A Kinetic Study of the Emulsion Polymerization of Vinyl Acetate. *J. Appl. Polym. Sci.* **1973**, *17*, 2311–2327. [CrossRef]
37. Nomura, M. Desorption and Reabsorption of Free Radicals in Emulsion Polymerization. In *Emulsion Polymerization*; Piirma, I., Ed.; Academic Press: New York, NY, USA, 1982; pp. 191–219.
38. Dougherty, E. The SCOPE Dynamic Model for Emulsion Polymerization. I. Theory. *J. Appl. Polym. Sci.* **1986**, *32*, 3051–3078. [CrossRef]
39. Gardon, J.L. Emulsion polymerization. VI. Concentration of monomers in latex particles. *J. Polym. Sci. Part A-1 Polym. Chem.* **1968**, *6*, 2859–2879. [CrossRef]
40. Ascher, U.M.; Petzold, L.R. *Computer Methods for Ordinary Differential Equations and Differential-Algebraic Equations*; Society for Industrial and Applied Mathematics: Philadelphia, PA, USA, 1998.
41. Buback, M.; Gilbert, R.G.; Hutchinson, R.A.; Klumperman, B.; Kuchta, F.-D.; Manders, B.G.; O'Driscoll, K.F.; Russell, G.T.; Schweer, J. Critically evaluated rate coefficients for free-radical polymerization, 1. Propagation rate coefficients for styrene. *Macromol. Chem. Phys.* **1995**, *196*, 3267–3280. [CrossRef]
42. Junkers, T.; Koo, S.P.S.; Barner-Kowollik, C. Determination of the propagation rate coefficient of acrylonitrile. *Polym. Chem.* **2010**, *1*, 438–441. [CrossRef]
43. Saldívar, E.; Araujo, O.; Giudici, R.; López-Barrón, C. Modeling and Experimental Studies of Emulsion Copolymerization Systems. II. Styrenics, J. Appl. Polym. Sci. 2001, 79, 2380–2397. **2001**, *79*, 2380–2397.
44. Deibert, S.; Bandermann, F.; Schweer, J.; Sarnecki, J. Propagation rate coefficient of free-radical polymerization of 1,3-butadiene. *Makromol. Chem. Rapid Commun.* **1992**, *13*, 351–355. [CrossRef]
45. Bachmann, R.; Pallaske, U.; Schmidt, A.; Muller, H.G. *DECHEMA Monographs*; VCH: Weinham, Germany, 1995; Volume 131, pp. 65–74.
46. Deibert, S.; Bandermann, F. Rate coefficients of the free-radical polymerization of 1,3-butadiene. *Die Makromol. Chem.* **1993**, *194*, 3287–3299. [CrossRef]
47. Brandrup, J.; Immergut, E.H.; Grulke, E.A. (Eds.) *Polymer Handbook*, 4th ed.; John Wiley and Sons: New York, NY, USA, 1994.
48. Mukerjee, P.; Mysels, K.J. *Critical Micelle Concentrations of Aqueous Surfactant Systems*; NSRDS-NBS 36; National Bureau of Standards: Washington, DC, USA, 1971.
49. Rawlings, J.B.; Ray, W.H. The modeling of batch and continuous emulsion polymerization reactors. Part II: Comparison with experimental data from continuous stirred tank reactors. *Polym. Eng. Sci.* **1988**, *28*, 257–274. [CrossRef]
50. Arve, D.S.C.; Manley, D.B.; Poling, B.E. Relative volatilities from mixture liquid densities for the *n*-butane/1,3-butadiene system. *Fluid Phase Equilib.* **1987**, *35*, 117–126. [CrossRef]

51. Guyot, A.; Guillot, J.; Graillat, C.; Llauro, M.F. Controlled composition in emulsion copolymerization application to butadiene-acrylonitrile copolymers. *J. Macromol. Sci.-Chem.* **1984**, *A21*, 683–699. [CrossRef]
52. Marien, Y.W.; Van Steenberge, P.H.M.; D'hooge, D.; Marin, G.B. Particle by Particle Kinetic Monte Carlo Tracking of Reaction and Mass Transfer Events in Miniemulsion Free Radical Polymerization. *Macromolecules* **2019**, *52*, 1408–1423. [CrossRef]
53. Marien, Y.W.; Van Steenberge, P.H.M.; Pich, A.; D'hooge, D. Coupled stochastic simulation of the chain length and particle size distribution in miniemulsion radical copolymerization of styrene and *N*-vinylcaprolactam. *React. Chem. Eng.* **2019**, *4*, 1935–1947. [CrossRef]

Publisher's Note: MDPI stays neutral with regard to jurisdictional claims in published maps and institutional affiliations.

© 2020 by the authors. Licensee MDPI, Basel, Switzerland. This article is an open access article distributed under the terms and conditions of the Creative Commons Attribution (CC BY) license (http://creativecommons.org/licenses/by/4.0/).

Article

Kinetics and Modeling of Aqueous Phase Radical Homopolymerization of 3-(Methacryloylaminopropyl)trimethylammonium Chloride and its Copolymerization with Acrylic Acid

Ikenna H. Ezenwajiaku, Emmanuel Samuel and Robin A. Hutchinson *

Department of Chemical Engineering, Queen's University, Kingston, ON K7L3N6, Canada;
ikenna.ezenwajiaku@queensu.ca (I.H.E.); emmanuel.samuel@queensu.ca (E.S.)
* Correspondence: robin.hutchinson@queensu.ca; Tel.: +1-613-533-3097

Received: 30 September 2020; Accepted: 21 October 2020; Published: 26 October 2020

Abstract: The radical homopolymerization kinetics of 3-(methacryloylaminopropyl) trimethylammonium chloride (MAPTAC) and its batch copolymerization with nonionized acrylic acid (AA) in aqueous solution are investigated and modeled. The drift in monomer composition is measured during copolymerization by in situ NMR over a range of initial AA molar fractions and monomer weight fractions up to 0.35 at 50 °C. The copolymer becomes enriched in MAPTAC for monomer mixtures containing up to 60 mol% MAPTAC, but is enriched in AA for MAPTAC-rich mixtures; this azeotropic behavior is dependent on initial monomer content, as electrostatic interactions from the cationic charges influence the system reactivity ratios. Models for MAPTAC homopolymerization and AA-MAPTAC copolymerization are developed to represent the rates of monomer conversion and comonomer composition drifts over the complete range of experimental conditions.

Keywords: aqueous phase polymerization; polyelectrolytes; radical polymerization; modeling and simulation

1. Introduction

Recent investigations have led to an improved understanding of the radical aqueous-phase polymerization kinetics of nonionized and ionizable water-soluble monomers. These water-soluble monomers are utilized to synthesize polyelectrolytes with tailor-made properties for a wide range of consumer and industrial applications such as flocculants in wastewater treatments, and personal care and pharmaceutical products [1–5]. However, monomer-solvent interactions during polymerization in aqueous phase result in complexities not encountered in organic phase, with the rate coefficients dependent on monomer concentration, ionic strength, and pH. Application of specialized techniques such as pulsed laser polymerization (PLP) used in combination with size exclusion chromatography (SEC), near infrared (NIR) spectroscopy, and electron paramagnetic resonance (EPR) spectroscopy has resulted in tremendous progress in understanding the polymerization kinetics of these water-soluble monomers [6,7].

PLP investigations have revealed that both the propagation (k_p) and termination (k_t) rate coefficients of water-soluble monomers such as methacrylic acid (MAA) [8–11] and acrylic acid (AA) [12–15] decrease with increasing monomer content in aqueous solution when the monomer is in its nonionized form, effects that have been captured in mechanistic models formulated to represent their batch and semi-batch radical homopolymerization under natural pH conditions [16–18]. Both rate coefficients decrease by orders of magnitude as pH is increased to ionize the monomer, due to the influence of electrostatic repulsion on propagation between a charged monomer and charged radical species and on the termination between two charged radical species [10,13,19,20]. The net effect on AA polymerization is a significant decrease in the rate of monomer conversion with increasing pH; once the monomer is fully ionized, however, addition of excess salt leads to a recovery in polymerization rate [21]. This complex relationship between degree of ionization and ionic strength (through addition of salt) also greatly influences the copolymerization behavior of AA with a nonionized monomer such as acrylamide [22–25].

Similar studies of the kinetic behavior of fully ionized monomers are limited. A recent PLP-SEC investigation revealed that k_p of the permanently-charged cationic monomer 2-(methacryloyloxyethyl) trimethylammonium chloride (TMAEMC) [26] increases with increasing monomer content in aqueous solution, the opposite trend to what was observed for nonionized water-soluble monomers. This behavior was attributed to increased screening of charges on the monomer and radical structures as the concentration of counterions in solution increases. The relative influence of the charge screening on propagation and termination differs, and is influenced by various factors including the distance between the reacting radical and the location of the cationic charge. Thus, the rate of monomer conversion for TMAEMC homopolymerization was found to decrease with increased monomer content despite the increased k_p, indicating that counterion concentration has a stronger influence on the termination of charged macroradicals than propagation [27]. Polymerization rate of the permanently-charged TMAEMC remained unchanged with solution pH, demonstrating that the kinetics do not have the pH dependence found for ionizable monomers (AA and MAA). Using these observations to develop correlations for k_p and k_t as a function of solution composition, a mechanistic model for TMAEMC radical homopolymerization was developed to represent polymerization rates as well as polymer molar mass distributions (MMD) produced under a range of batch and semi-batch operating conditions [27].

Polyelectrolyte charge density and hence application properties are controlled by radical copolymerization of cationic monomers with neutral monomers such as acrylamide or ionizable monomers such as AA. While the number of studies is limited, it has been shown that the relative consumption rates of the two monomers are dependent on the initial amount of cationic monomer in the system [28–30]. Capturing these influences in mechanistic models developed to represent the polymerization system can aid selection of appropriate synthesis conditions to control the copolymer composition. Thus, the comonomer composition drifts for TMAEMC copolymerized with nonionized AA were measured using an in situ NMR technique over a range of initial compositions and monomer loadings to develop a description of how the system reactivity ratios varied as a function of aqueous solution conditions [29]. The analysis indicated that copolymer composition could be well represented using the standard model of terminal copolymerization kinetics, as long as the influence of charge-screening on TMAEMC k_p and k_t values was properly accounted for. This insight was used to develop an AA-TMAEMC copolymerization model that also included the influence of AA based midchain radicals formed through intramolecular chain transfer on reaction rates [29].

In this work, we examine whether the insights gained from the study of AA-TMAEMC copolymerization extend to similar systems. MAPTAC is an amide-based cationic monomer with a methacrylate structure similar to TMAEMC; in addition to the different functionality, the spacing between the monomer double bond and the cationic charge is increased by one CH_2 unit, as shown in Scheme 1. Given their similar structure, one would expect similarities in the influence of electrostatic interactions on the rate coefficients for these two cationic monomers and subsequently their (co)polymerization kinetics. Indeed, the recent PLP-SEC investigation revealed that increased counterion concentration (C_{Cl^-}), achieved by either increasing the initial monomer concentration or adding NaCl, increases the value of k_p for both TMAEMC and MAPTAC [26]. As shown in Figure 1, the k_p values for MAPTAC are lower than those of TMAEMC, in agreement with other comparisons of k_p values for ester vs amide-based monomers polymerized in aqueous phase [31,32]. In addition, the dependence of MAPTAC k_p on C_{Cl^-} is not as strong as that observed for TMAEMC across the entire concentration range studied, equivalent to varying the initial weight fraction of monomer in aqueous solution ($w_{mon,0}$) between 0.05 and 0.40. This difference may be explained by reduced electrostatic effects resulting from the increased distance of the charged moiety from the reactive double bond in MAPTAC compared to TMAEMC (7 bond length vs 6 bond length) [33].

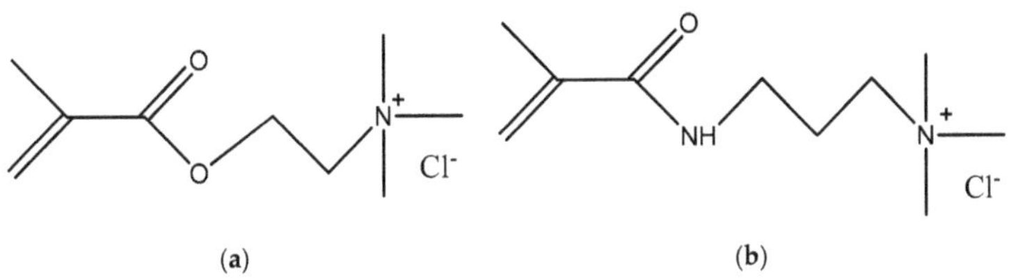

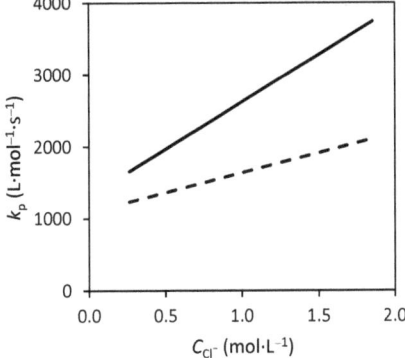

Scheme 1. Structures of the cationic monomers: (**a**) 2-(methacryloyloxyethyl)trimethylammonium chloride (TMAEMC); (**b**) 3-(methacryloylaminopropyl)trimethylammonium chloride (MAPTAC).

Figure 1. The dependence of k_p on counterion concentration (C_{Cl^-}) in aqueous solution for TMAEMC (—) and MAPTAC (- - -) at 50 °C, as generated using representations developed for TMAEMC [27] and MAPTAC (this work) based on PLP-SEC experimentation [26].

Under continuous initiation, a decreased rate of monomer conversion was found for both monomers when the initial monomer content was increased, with MAPTAC batch homopolymerization occurring at a lowered rate compared to TMAEMC [26]. In the present study we extend the comparison to copolymerization, applying the in situ NMR technique to investigate the influence of initial monomer composition, total monomer content ($w_{mon,0}$ between 0.05 and 0.40) and added salt (NaCl) on the copolymerization of AA-MAPTAC. The results are compared and contrasted to the recent AA-TMAEMC study conducted under a similar range of conditions [29] to provide insights on the influence of the cationic monomer structure on reactivity, and to determine whether the modeling strategy developed previously for TMAEMC homo- and copolymerization can also be used represent MAPTAC homopolymerization and its copolymerization with AA.

2. Materials and Methods

All in situ NMR experiments were carried out in deuterium oxide (D$_2$O, 99.9%, Cambridge Isotope Laboratories Inc., Montreal, QC, Canada) as solvent, while samples for physical measurements were prepared in deionized water. Other materials used as received from Sigma Aldrich (Oakville, ON, Canada) include: 3-(methacryloylaminopropyl)trimethylammonium chloride (MAPTAC, 50 wt% solution in H$_2$O), acrylic acid (AA, 99%), propionic acid (PA, ≥99.5%), sodium chloride BioXtra (NaCl, ≥99.5%, AT), sodium hydroxide reagent grade (NaOH, ≥98%, anhydrous pellets), 2,2-azobis(2-methylpropionamidine) dihydrochloride (V-50, 97%). Hydrochloric acid (HCl, reagent, ACS-PUR, Fisherbrand, Mississauga, ON, Canada) was used for adjusting solution pH. Solution preparation for bench-scale and in situ batch polymerizations as well as the subsequent data analyses were carried out using the methods developed for the TMAEMC and AA-TMAEMC studies [27,29].

Figure 2 shows the NMR spectrum at 50 °C for an equimolar mixture of AA and MAPTAC ($f_{MAPTAC,0}$ = 0.5) with a monomer weight fraction ($w_{mon,0}$) of 0.05, and 0.80 wt% V-50 initiator in solution. The decrease in intensities of peaks from AA and MAPTAC terminal double bond protons relative to the HOD peak (4.71 ppm) was followed over time and used to calculate monomer conversion profiles and the change in the MAPTAC molar fraction, as detailed in Supporting Information. Repeat experiments carried out for AA-MAPTAC copolymerization under selected conditions showed good reproducibility, as shown in Figure S1.

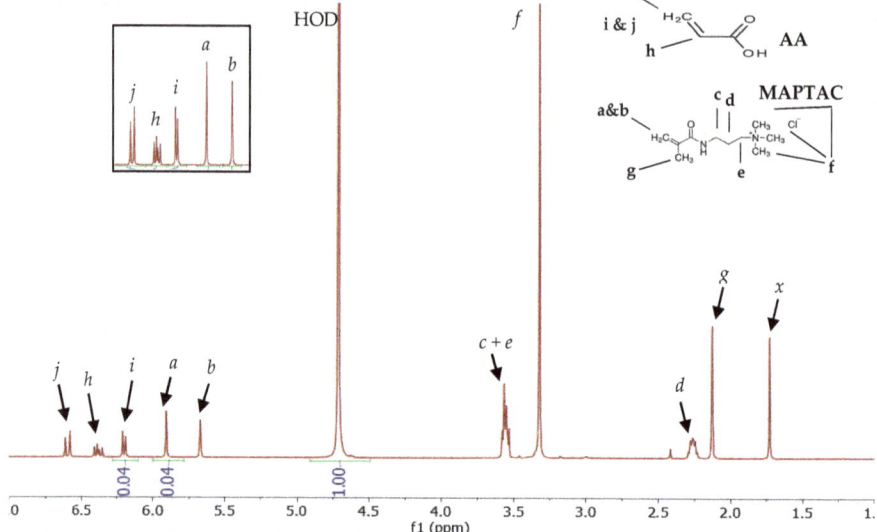

Figure 2. ^{1}H NMR peaks assignments for AA-MAPTAC comonomer mixture with $f_{MAPTAC,0}$ = 0.5, $w_{mon,0}$ = 0.05 and 0.80 wt% initiator in D$_2$O at 50 °C (1st scan). Peak x was identified as arising from V-50 initiator. Inset shows the separation of the double bond peaks used to track the relative consumption of the two monomers during reaction.

Polymer samples used for physical characterization were synthesized under the same reaction conditions with the in situ NMR method using a mixed 20 mL lab reactor, as the previous TMAEMC homopolymerization study revealed good agreement in polymer properties and reaction rates between the two techniques [27]. An Orion™ Versa Star Pro™ pH/ISE/Conductivity/Dissolved Oxygen Multiparameter Benchtop Meter was used to measure pH, and a calibrated Cannon–Fenske viscometer was utilized to obtain the dynamic viscosity of the samples.

3. Experimental Results

While there are a few investigations of the copolymerization of acrylamide with MAPTAC [5,34], no study on the copolymerization of nonionized AA with MAPTAC could be found. Thus, this section provides first results for AA-MAPTAC copolymerization, comparing them to the recent AA-TMAEMC copolymerization study [29]. Addition of AA to the MAPTAC system greatly lowers system pH to ~2 such that AA remains in its nonionized form, as it also was for copolymerization with TMAEMC [29]. Batch copolymerization experiments with $w_{mon,0}$ of 0.10 and initial MAPTAC molar fractions ($f_{MAPTAC,0}$) varied between 0.3 and 0.9 were investigated at 50 °C and 0.80 wt% V-50 in D$_2$O using in situ NMR to measure polymerization rates and copolymer composition. As shown by the conversion profiles in Figure 3a, the rate of polymerization systematically decreases with an increase in initial MAPTAC fraction up to 90% mol. The polymerization rate at $f_{MAPTAC,0} = 0.9$ matches that of MAPTAC homopolymerization, despite the significant difference in solution pH, 2.8 for $f_{MAPTAC,0} = 0.9$ and 6.7 for MAPTAC homopolymerization, respectively. Figure 3b plots the change in f_{MAPTAC}, the molar fraction of MAPTAC remaining in the comonomer solution, as a function of conversion. MAPTAC is preferentially incorporated into the copolymer (as indicated by the decrease in f_{MAPTAC}) for the experiments conducted with $f_{MAPTAC,0} = 0.3$ and 0.5, but there is a preferential incorporation of AA into the copolymer with $f_{MAPTAC,0} = 0.7$ and 0.9. Initial rates could not be captured by the in situ NMR technique at the higher V-50 level used for AA-MAPTAC copolymerization with $f_{MAPTAC,0} = 0.10$.

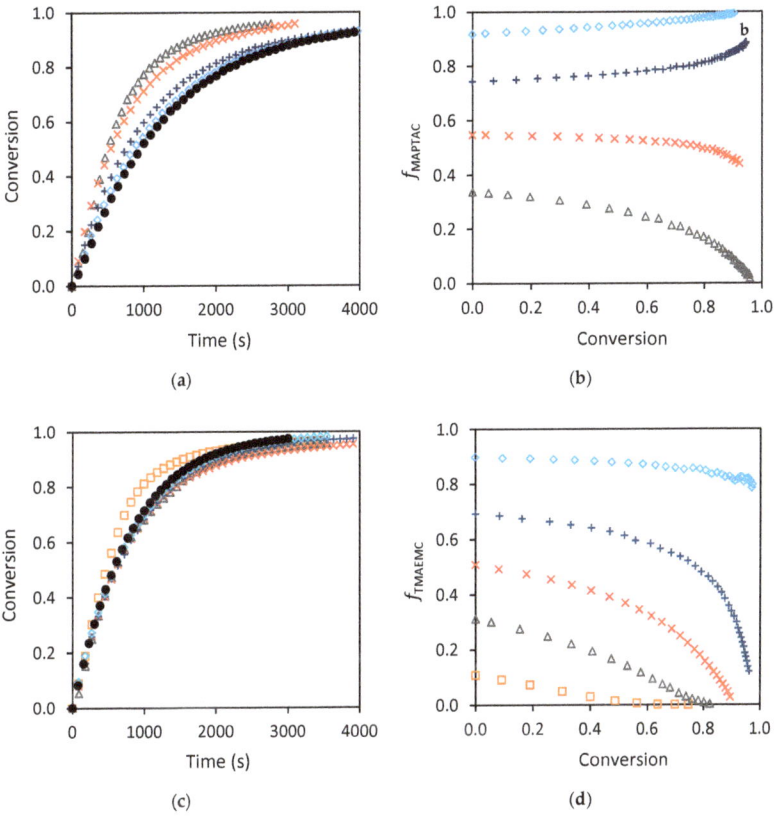

Figure 3. Overall monomer conversion profiles for (**a**) AA-MAPTAC and (**c**) AA-TMAEMC, and composition drift of (**b**) MAPTAC and (**d**) TMAEMC mole fraction when copolymerized with AA at 50 °C with 0.80 (for MAPTAC) and 0.40 wt% (for TMAEMC) V-50, $w_{mon,0} = 0.10$ and different initial monomer compositions ($f_{MAPTAC,0}/f_{TMAEMC,0} = 0.1(\square), 0.3(\triangle), 0.5(\times), 0.7(+), 0.9(\lozenge), 1(\bullet)$).

It is informative to compare these results to those obtained for TMAEMC copolymerized with AA at identical conditions, although with a lowered V-50 level; the corresponding conversion profiles are shown in Figure 3c, and TMAEMC composition drift in Figure 3d. The influence of TMAEMC content on the conversion rates is minimal for AA-TMAEMC copolymerization, except for the slightly higher rate observed with $f_{TMAEMC,0} = 0.10$. In contrast, addition of AA at any level significantly increases the rate of monomer conversion for AA-MAPTAC compared to MAPTAC homopolymerization. The azeotropic behavior (i.e., minimal drift in comonomer composition) occurs at a lowered MAPTAC molar fraction (between 0.5 and 0.7) compared to that observed for TMAEMC at $f_{TMAEMC,0} = 0.90$ as initial monomer concentration was decreased from 10 to 5 wt% [29]. The relative reactivity of

MAPTAC was also reduced from that of TMAEMC when copolymerized with acrylamide [5,34,35], a result consistent with MAPTAC's lowered homopolymerization rate and k_p values (Figure 1), and its copolymerization with AA in this study relative to AA-TMAEMC.

Both the counterion concentration (C_{Cl^-}) and the charge density of the cationic polyelectrolyte formed during polymerization depends on initial MAPTAC molar fraction and total monomer concentration [36]. Thus, AA-MAPTAC copolymerizations were also carried out at $w_{mon,0}$ levels of 0.05, 0.20 and 0.40 to study the combined influence of initial monomer composition and total monomer concentration (both affecting C_{Cl^-}) on rates of polymerization and comonomer composition drifts. The polymerization rates are grouped according to $w_{mon,0}$ in Figure S2 and $f_{MAPTAC,0}$ in Figure 4, which also includes a plot of the comonomer composition drifts for the complete set of experiments. Figure S2 demonstrates that there is a decrease in the polymerization rate with increased MAPTAC fraction at all monomer contents, as seen for $w_{mon,0} = 0.10$ in Figure 3a. The decrease in polymerization rate with increasing $w_{mon,0}$ from 0.05 to 0.40 is more pronounced as $f_{MAPTAC,0}$ increases from 0.3 (Figure 4a) to 0.9 (Figure 4c), with the relative decrease smaller for all comonomer compositions than observed for MAPTAC homopolymerization [26].

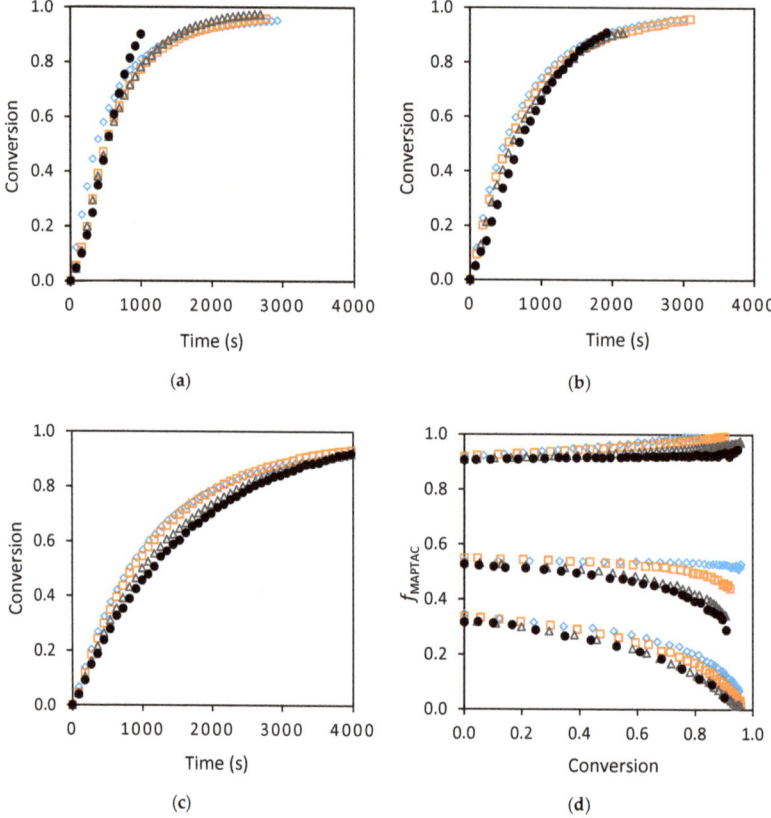

Figure 4. Overall monomer conversion profiles for AA-MAPTAC copolymerization with $f_{MAPTAC,0}$ of (**a**) 0.3 (**b**) 0.5 and (**c**) 0.9 at 50 °C and 0.80 wt% V-50 with varying $w_{mon,0}$ of 0.05(◊), 0.10(□), 0.20(∆), 0.40(●). The corresponding changes in comonomer composition are shown in (**d**) as a function of overall monomer conversion.

The drift in f_{TMAEMC} for AA-TMAEMC copolymerization was observed to be more pronounced as $w_{mon,0}$ increased from 0.05 to 0.40 [29], and the same is found for the AA-MAPTAC system, as seen for f_{MAPTAC} in Figure 4d. However, the influence of $w_{mon,0}$ is weaker for the MAPTAC copolymerization system, with little difference seen between the curves for $w_{mon,0}$ at 0.20 and 0.40. There is no observable composition drift with $f_{MAPTAC,0} = 0.5$ and $w_{mon,0} = 0.05$, indicating that these conditions lead to an azeotropic copolymerization. The difference in azeotropic behavior of MAPTAC (Figure 3b) and TMAEMC (Figure 3d and [29]) is likely related to the lowered k_p of MAPTAC relative to TMAEMC, as well as its lowered sensitivity to C_{Cl^-} (Figure 1), as will be further illustrated in the modeling section of this study.

AA-MAPTAC copolymerizations were also carried out with 1.0 mol·L^{-1} added NaCl and $w_{mon,0}$ of 0.10 at $f_{MAPTAC,0} = 0.3$, 0.5 and 0.9, to explore how manipulating C_{Cl^-} by addition of NaCl affects copolymerization kinetics. Figure 5a–c demonstrates that there is a consistent slight decrease in conversion rates with added NaCl. As shown in Figure 5d, addition of NaCl increases the relative rate of incorporation of MAPTAC into the copolymer (i.e. lowers f_{MAPTAC} compared to the cases without salt), although to a lesser extent compared to AA-TMAEMC copolymerization [29].

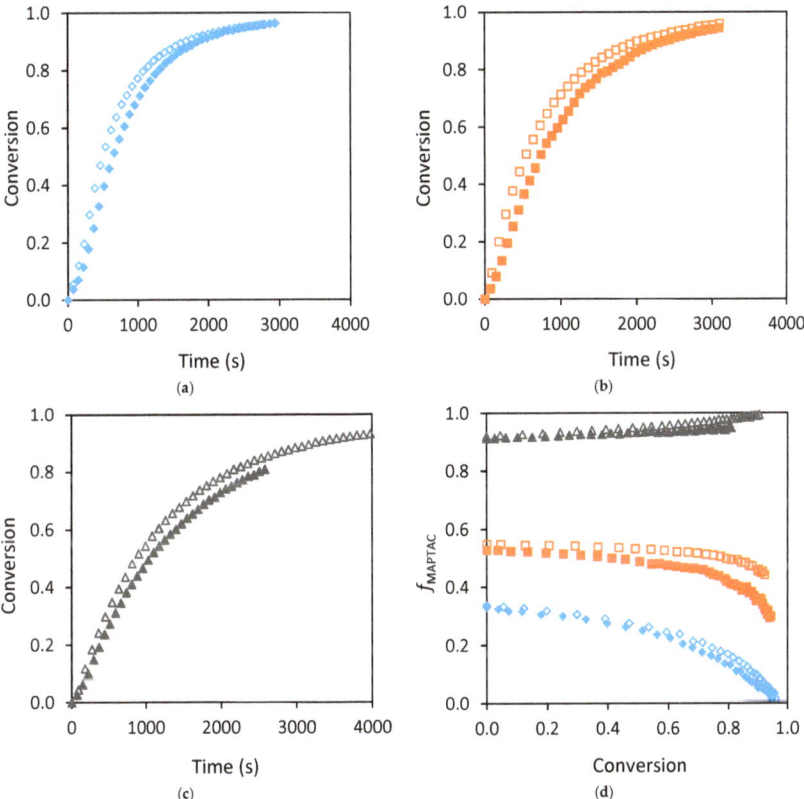

Figure 5. Overall monomer conversion profiles for AA-MAPTAC copolymerization with $f_{MAPTAC,0}$ of (**a**) 0.3 (**b**) 0.5 and (**c**) 0.9 at 50 °C, 0.80 wt% V-50, and $w_{mon,0} = 0.10$ without NaCl (open symbols) and with 1.0 mol·L^{-1} NaCl (closed symbols). The corresponding comonomer composition drifts are shown in (**d**) as a function of overall monomer conversion.

As shown in Figure S3a, the pH of MAPTAC monomer in aqueous solution is close to 7 and relatively independent of w_{mon}, while the values are below 3 for AA-MAPTAC comonomer mixtures and decrease with increasing AA content; the pH of the comonomer mixture also decreases with an increase in $w_{mon,0}$, as shown in Figure S3b. This suggests that polymerization rate is controlled by the influence of C_{Cl^-} on MAPTAC reactivity, as the AA remains in nonionized form under all reaction conditions; the same conclusion was reached for the AA-TMAEMC system [29]. To further study the specific influence of AA on kinetics, conversion profiles were compared for experiments that substituted propionic acid (PA) as a non-reactive version of AA while maintaining the total MAPTAC and acid weight fraction at 0.10; the copolymerization of MAPTAC with AA at $f_{AA,0}$ = 0.7 and 0.5 was compared to MAPTAC homopolymerization with PA molar fraction of 0.7 and 0.5 (replacing AA monomer), and also MAPTAC homopolymerization at the natural pH. The faster reaction rate of the copolymerization reaction relative to the homopolymerization cases (with and without PA) shown in Figure S4 highlights that it is the relative reactivity of the comonomer that influences copolymerization rate under the nonionized conditions of AA, not the decreased pH of the solution.

4. Model Development

With a better understanding of the influence of the reaction environment on kinetics, the approach to represent the copolymerization behavior of cationic monomers can be generalized. To this end, the modeling framework previously implemented in the mechanistic modeling package Predici® (version 11) for TMAEMC homopolymerization [27,37] and AA-TMAEMC copolymerization [29] is used here for the MAPTAC systems, with adjustments to account for weaker influence of counterion concentration on the kinetic rate coefficients. The mechanisms included are initiation, propagation (terminal model for copolymerization), termination (geometric mean for copolymer cross-termination), transfer to monomer and reactions associated with AA backbiting, as summarized in Table 1. The influence of C_{Cl^-} on MAPTAC propagation and termination rate coefficients are established using a combination of the current experimental results and k_p values obtained from PLP-SEC studies [26]. Other treatments of rate coefficients are the same as implemented for MAPTAC and AA [16] homopolymerization models, as summarized in Table 2. The value of the AA homotermination rate coefficient is expressed as a function of the initial monomer/solvent viscosity. Viscosity measurements of AA-MAPTAC mixtures carried out at the same overall monomer contents ($w_{mon,0}$ = 0.025, 0.05 and 0.10) were constant over the range of comonomer compositions (Figure S5). Thus, there was no need to adjust the treatment of the AA k_t value to account for the effect of MAPTAC on solution viscosity of the comonomer mixture in water. As also seen in the studies on TMAEMC homopolymerization and AA-TMAEMC copolymerization, the viscosity of pMAPTAC solution is significantly higher than that of the monomer solutions (Figure S6); although the viscosity of the systems increases substantially during polymerization, the increase does not influence termination kinetics which are dominated by electrostatic interactions [27,29].

Table 1. Reaction steps included in the AA-MAPTAC copolymerization model.

Initiation
$I \xrightarrow{k_i} 2fI^*$
$I^* + MAPTAC \xrightarrow{k_p^{MAPTAC}} P_1^{MAPTAC}; \quad I^* + AA \xrightarrow{k_p^{AA}} P_1^{AA}$
Chain Propagation
$P_n^{MAPTAC} + MAPTAC \xrightarrow{k_p^{MAPTAC}} P_{n+1}^{MAPTAC}$
$P_n^{AA} + AA \xrightarrow{k_p^{AA}} P_{n+1}^{AA}$
$P_n^{MAPTAC} + AA \xrightarrow{k_p^{MAPTAC \cdot AA}} P_{n+1}^{AA}$
$P_n^{AA} + MAPTAC \xrightarrow{k_p^{AA \cdot MAPTAC}} P_{n+1}^{MAPTAC}$

Table 1. Cont.

Transfer to Monomer

$$P_n^{MAPTAC} + MAPTAC \xrightarrow{k_{tr}^{MAPTAC}} D_n + P_1^{MAPTAC}$$

$$P_n^{AA} + AA \xrightarrow{k_{tr}^{AA}} D_n + P_1^{AA}$$

$$P_n^{AA} + MAPTAC \xrightarrow{k_{tr}^{AA \cdot MAPTAC}} D_n + P_1^{MAPTAC}$$

$$P_n^{MAPTAC} + AA \xrightarrow{k_{tr}^{MAPTAC \cdot AA}} D_n + P_1^{AA}$$

Termination SPR-SPR

$$P_n^{MAPTAC} + P_m^{MAPTAC} \xrightarrow{(1-\alpha_{ss}^{MAPTAC})k_{t,ss}^{MAPTAC}} D_{n+m} / \xrightarrow{\alpha_{ss}^{MAPTAC} k_{t,ss}^{MAPTAC}} D_n + D_m$$

$$P_n^{AA} + P_m^{AA} \xrightarrow{(1-\alpha_{ss}^{AA})k_{t,ss}^{AA}} D_{n+m} / \xrightarrow{\alpha_{ss}^{AA} k_{t,ss}^{AA}} D_n + D_m$$

$$P_n^{AA} + P_m^{MAPTAC} \xrightarrow{(1-\alpha_{ss}^{AA \cdot MAPTAC})k_{t,ss}^{AA \cdot MAPTAC}} D_{n+m} / \xrightarrow{\alpha_{ss}^{AA \cdot MAPTAC} k_{t,ss}^{AA \cdot MAPTAC}} D_n + D_m$$

Reactions Related to Backbiting

Backbiting

$$P_n^{AA} \xrightarrow{F_{AA} k_{bb}} Q_n^{AA}$$

Addition to MCR

$$Q_n^{AA} + AA \xrightarrow{k_{p,tert}^{AA \cdot AA}} P_{n+1}^{AA}$$

$$Q_n^{AA} + MAPTAC \xrightarrow{k_{p,tert}^{AA \cdot MAPTAC}} P_{n+1}^{MAPTAC}$$

Cross Termination MCR-SPR

$$P_n^{MAPTAC} + Q_m^{AA} \xrightarrow{(1-\alpha_{st}^{MAPTAC \cdot AA})k_{t,st}^{MAPTAC \cdot AA}} D_{n+m} / \xrightarrow{\alpha_{st}^{MAPTAC \cdot AA} k_{t,st}^{MAPTAC \cdot AA}} D_n + D_m$$

$$P_n^{AA} + Q_m^{AA} \xrightarrow{(1-\alpha_{st}^{AA \cdot AA})k_{t,st}^{AA \cdot AA}} D_{n+m} / \xrightarrow{\alpha_{st}^{AA \cdot AA} k_{t,st}^{AA \cdot AA}} D_n + D_m$$

Termination MCR-MCR

$$Q_n^{AA} + Q_m^{AA} \xrightarrow{(1-\alpha_{tt}^{AA})k_{t,tt}^{AA}} D_{n+m} / \xrightarrow{\alpha_{tt}^{AA} k_{t,tt}^{AA}} D_n + D_m$$

* Termination occurs by combination to yield product chains of length $n + m$, or by disproportionation. α and $(1 - \alpha)$ represent the fraction of termination by disproportionation and combination respectively, with subscripts introduced for secondary propagating radicals (s) and midchain radicals (t).

Table 2. Rate coefficients and expressions used in AA-MAPTAC copolymerization model.

Rate Expression	Reference
Initiation	
$k_d(s^{-1}) = 9.24 \times 10^{14} \exp\left(-\frac{124}{RT} \frac{kJ}{mol}\right)$ $f = 0.8$	[38]
Chain propagation	
$k_p^{MAPTAC}\left(\frac{L}{mol.s}\right) = k_{p,0}^{MAPTAC}(1.0 + 0.5 C_{Cl^-})$ $k_{p,0}^{MAPTAC}\left(\frac{L}{mol.s}\right) = 4.23 \cdot 10^5 \exp\left(-\frac{1924}{T} K\right)$	[26]
$k_p^{AA}\left(\frac{L}{mol.s}\right) = k_{p0}^{AA}(0.11 + (1-0.11) \exp(-3.0 w_m\prime))$ $k_{p0}^{AA}\left(\frac{L}{mol.s}\right) = 3.2 \cdot 10^7 \exp\left(-\frac{1564}{T}\right)$	[16]
$r_{AA} = 0.36$, $r_{MAPTAC} = a(1.0 + C_{Cl^-})$ where $a = 0.46$	this work
$k_p^{MAPTAC.AA}\left(\frac{L}{mol.s}\right) = \frac{k_{p,0}^{MAPTAC}}{0.46}$; $k_p^{AA.MAPTAC}\left(\frac{L}{mol.s}\right) = \frac{k_p^{AA}}{0.36}$	this work

Table 2. Cont.

Rate Expression	Reference
Transfer/Cross-transfer to Monomer	
$k_{tr}^{MAPTAC}\left(\frac{L}{mol \cdot s}\right) = 2.89 \cdot 10^{-6}$	This work
$k_{tr}^{AA}\left(\frac{L}{mol \cdot s}\right) = 7.5 \cdot 10^{-5} k_p^{AA}$	[16]
$k_{tr}^{AA \cdot MAPTAC}\left(\frac{L}{mol \cdot s}\right) = k_{tr}^{MAPTAC} \frac{k_p^{AA}}{r_{AA} \cdot k_p^{MAPTAC}}$	This work
$k_{tr}^{MAPTAC \cdot AA}\left(\frac{L}{mol \cdot s}\right) = k_{tr}^{AA} \frac{k_p^{MAPTAC}}{r_{MAPTAC} \cdot k_p^{AA}}$ $= C_{tr}^{AA} \frac{k_{p,0}^{MAPTAC}}{r_{MAPTAC}}$	This work
Termination SPR-SPR:	
$k_{t,ss}^{MAPTAC}(1,1)\left(\frac{L}{mol \cdot s}\right) = 4.8 \cdot 10^8 \exp\left(-\frac{998}{T}K\right)(0.19 + 1.37 C_{Cl^-})$ where $\alpha_s = 0.62; \alpha_l = 0.18; i_c = 45$	[27,39]
$k_{t,ss}^{AA}(1,1)\left(\frac{L}{mol \cdot s}\right) = 9.78 \cdot 10^{11} \exp\left(-\frac{1858}{T}K\right) viscosity_{corr}$	[39,40]
$viscosity_{corr} = 1.56 - 1.77 w_{mon,o} - 1.2 w_{mon,o}^2 + 2.43 w_{mon,o}^3$	[16]
where $\alpha_s = 0.6; \alpha_l = 0.16; i_c = 30$	[16]
$\langle k_{t,ss}^{AA \cdot MAPTAC} \rangle \left(\frac{L}{mol \cdot s}\right) = \Phi \cdot \left(\langle k_{t,ss}^{AA} \rangle \cdot \langle k_{t,ss}^{MAPTAC} \rangle\right)^{1/2}$	This work
$\alpha_{ss}^{MAPTAC} = 0.8; \alpha_{ss}^{AA} = 0.05; \alpha_{ss}^{AA \cdot MAPTAC} = 0.4$	[16,39]
Backbiting	
$k_{bb}^{AA}(s^{-1}) = F_{AA}^{inst} 9.94 \cdot 10^8 \exp\left(-\frac{4576}{T}K\right)$	[16]
Addition to MCR	
$k_{p,tert}^{AA}\left(\frac{L}{mol \cdot s}\right) = 0.755 \exp\left(-\frac{2464}{T}K\right) k_p^{AA}$	[16]
Cross Termination MCR-SPR	
$\langle k_{t,st}^{AA \cdot AA} \rangle \left(\frac{L}{mol \cdot s}\right) = 0.3 \langle k_{t,ss}^{AA} \rangle$	[16,23]
$\langle k_{t,st}^{MAPTAC \cdot AA} \rangle \left(\frac{L}{mol \cdot s}\right) = 0.3 \langle k_{t,ss}^{AA \cdot MAPTAC} \rangle$	This work
$\alpha_{st}^{MAPTAC \cdot AA} = 0.8; \alpha_{st}^{AA \cdot AA} = 0.4$	This work, [16]
Termination MCR-MCR	
$\langle k_{t,st}^{MAPTAC \cdot AA} \rangle \left(\frac{L}{mol \cdot s}\right) = 0.3 \langle k_{t,ss}^{AA \cdot MAPTAC} \rangle$ $\alpha_{tt}^{AA} = 0.8$	[16,23]
Density	
$\rho_{AA}(g \cdot mL^{-1}) = 1.0731 - 1.0826 \times 10^{-3} T(^\circ C^{-1}) - 7.2379 \times 10^{-7} T(^\circ C^{-2})$	[16]
$\rho_{H2O}(g \cdot mL^{-1}) = 0.9999 - 2.3109 \times 10^{-5} T(^\circ C^{-1}) - 5.4481 \times 10^{-6} T(^\circ C^{-2})$	[16]
$\rho_{MAPTAC}(g \cdot mL^{-1}) = 0.9806 - 4.5523 \times 10^{-4} T(^\circ C^{-1}) + 1.1040 \times 10^{-7} T(^\circ C^{-2})$	[26]

w_m' refers to AA + MAPTAC monomer content on a polymer free basis.

The systematic decrease of rate of conversion (and thus $k_p/<k_t>^{0.5}$) observed with increased MAPTAC content [26], also observed in TMAEMC homopolymerization [27], implies that $<k_t>$ increases more strongly with C_{Cl^-} in the system than k_p. As there are no independent studies of MAPTAC termination kinetics, it is assumed that k_t^{MAPTAC} follows the same functional form as used to represent k_t^{TMAEMC}, both in terms of temperature and chain-length dependencies [27,39] as well as the dependence on C_{Cl^-}. The parameters for the latter relationship were determined by fitting the conversion profiles of the MAPTAC homopolymerizations measured at 50 °C [26] to the empirical relation $k_t(1,1) = a + b C_{Cl^-} + c C_{Cl^-}^2$ using the parameter estimation capabilities of Predici. As shown in

Figure 6, and in contrast to the TMAEMC homopolymerization system, the increase in $k_t(1,1)$ values is linear with total C_{Cl^-} in solution. The best-fit values obtained for the $k_t(1,1)$ parameters are 0.19 ± 0.04, 1.37 ± 0.13, and 0.002 ± 0.08 for a, b and c, respectively; thus, the quadratic term was not used in the $k_t(1,1)$ representation. These values indicate a higher termination of charged MAPTAC macroradicals at lower monomer contents compared to TMAEMC up to $w_{mon,0}$ of 0.20 with the opposite observed at higher monomer contents. Figure 7 shows that the representation of the MAPTAC homopolymerization conversion profiles obtained for $w_{mon,0}$ between 0.05 and 0.35 at 50 °C with 0.80 wt% V-50 is excellent over the full range of polymerizations.

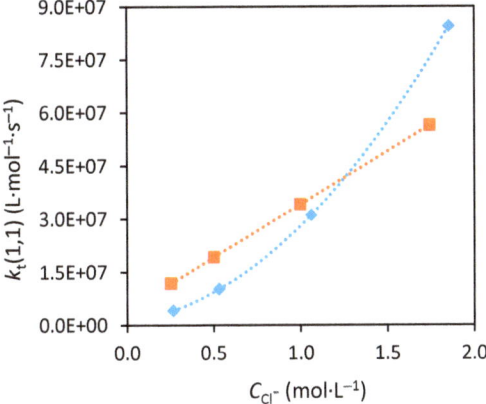

Figure 6. Estimated variation of $k_t(1,1)$ as a function of counterion concentration from fits to TMAEMC (♦) [27] and MAPTAC (■) [26] homopolymerizations carried out at 50 °C and varying initial monomer content in aqueous solution.

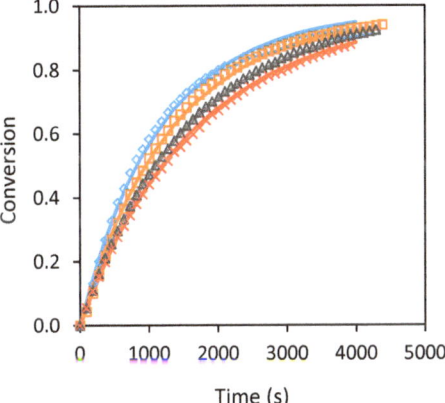

Figure 7. Comparison of experimental (symbol) and simulated (lines) batch conversion profiles collected at varying $w_{mon,0}$ of 0.05 (◊), 0.10 (□), 0.20 (Δ) and 0.35 (×) at 50 °C with 0.80 wt% V-50 in aqueous solution. Experimental results taken from [26], with simulations generated using the model summarized in Tables 1 and 2.

The rate coefficients used to represent AA [16] and MAPTAC homopolymerization are combined for the representation of AA-MAPTAC copolymerization following the approach used to develop the model of AA-TMAEMC copolymerization [29]. The model includes the backbiting mechanisms that result from presence of AA monomer and radicals in the copolymerization system. The previous study by Wittenberg et al. [16] provides the rate coefficients for AA backbiting and the subsequent reactions involving the midchain radicals. Parameter estimation in Predici is used to estimate k_t^{cross} from the copolymer conversion profiles after first fitting the comonomer composition drifts to determine the system reactivity ratios.

4.1. Estimation of Reactivity Ratios from Comonomer Composition Drifts

As the pH of the system indicates that AA does not ionize under polymerization conditions, it can be assumed that only the addition of MAPTAC to a radical with a MAPTAC terminal unit is influenced by electrostatic effects. Thus, r_{AA} is assumed to be constant, while r_{MAPTAC} varies with C_{Cl^-} following the same functional form as k_p^{MAPTAC}:

$$r_{MAPTAC} = a \cdot (1 + 0.5\, C_{Cl^-}) \tag{1}$$

The experimental comonomer composition drifts obtained over a range of $w_{mon,0}$ and $f_{MAPTAC,0}$ levels (Figures 3b and 4d) are used to estimate reactivity ratios assuming terminal model kinetics, with the instantaneous copolymer composition given by:

$$F_{MAPTAC} = \frac{r_{MAPTAC} f_{MAPTAC}^2 + f_{MAPTAC} f_{AA}}{r_{MAPTAC} f_{MAPTAC}^2 + 2 f_{MAPTAC} f_{AA} + r_{AA} f_{AA}^2} \tag{2}$$

The molar fraction of MAPTAC in the comonomer mixture is $f_{MAPTAC} = [MAPTAC]/([MAPTAC] + [AA])$, and reactivity ratios are defined by $r_{MAPTAC} = k_p^{MAPTAC}/k_p^{MAPTAC.AA}$ and $r_{AA} = k_p^{AA}/k_p^{AA.MAPTAC}$. These parameters were estimated by the model to fit the change in comonomer composition with conversion using the direct numerical integration (DNI) method [41]:

$$\frac{df_{MAPTAC}}{dx} = \frac{f_{MAPTAC} - F_{MAPTAC}}{1 - x} \tag{3}$$

with initial condition $f_{MAPTAC} = f_{MAPTAC,0}$ at $x = 0$. This technique has been previously used for the estimation of reactivity ratios for copolymerization of AA with both non-ionized AM [42] and cationic TMAEMC [29] monomers from experimental results obtained by the in situ NMR technique across a wide range of initial conditions. The only additional modification for charged systems is to express the reactivity ratio of the cationic component as a function of monomer content according to Equation (1). Estimation using the experimental results shown in Figures 3b and 4d results in estimated values of $r_{AA} = 0.36 \pm 0.006$ and $a = 0.46 \pm 0.006$, with the fit compared to the experimental data in Figure 8. The faster relative incorporation of MAPTAC into the copolymer at higher $w_{mon,0}$ (and thus C_{Cl^-}) is well captured by the model. Furthermore, excellent representations are obtained (without additional fitting) for the increased relative consumption of MAPTAC with added NaCl (Figure 8c) as well as the azeotropic behavior beyond $f_{MAPTAC,0} = 0.5$. Therefore, the effects of initial monomer concentration and added NaCl on the MAPTAC composition drifts are accurately captured using the same functional representation that was developed for AA-TMAEMC copolymerization.

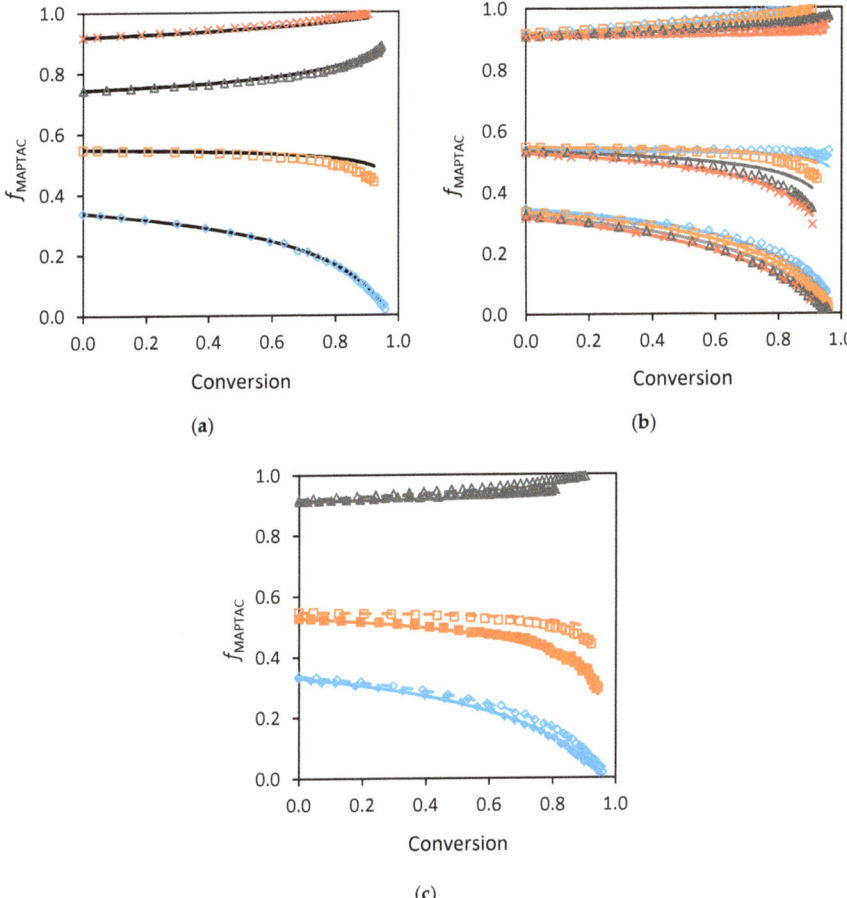

Figure 8. Comparison of batch comonomer composition drifts for AA-MAPTAC copolymerizations at 50 °C and 0.80 wt% V-50 at different initial monomer compositions with (**a**) $w_{mon,0} = 0.10$; (**b**) $w_{mon,0} = 0.05(\lozenge)$, $0.10(\square)$, $0.20(\triangle)$ and $0.40(\times)$ at different initial monomer compositions; and (**c**) for $w_{mon,0} = 0.10$ without (open symbols) and with 1.0 mol·L^{-1} NaCl (closed symbols). Lines are model representations of the composition drift developed using Equations (1)–(3), with parameters summarized in Table 2.

4.2. Model Fit of Conversion Profiles

Using the Predici software, the set of reaction mechanisms and rate coefficients listed in Tables 1 and 2 are combined with the expressions and values obtained for reactivity ratio estimations to develop a full kinetic model to simulate the conversion profiles obtained experimentally using the assumptions made to model AA-TMAEMC copolymerization [29]. It is informative to examine the terminal model propagation kinetics that leads to the following expression for the composition-averaged k_p^{cop}:

$$k_p^{cop} = \frac{r_{MAPTAC} f_{MAPTAC}^2 + 2 f_{MAPTAC} f_{AA} + r_{AA} f_{AA}^2}{r_{MAPTAC} f_{MAPTAC} / k_p^{MAPTAC} + r_{AA} f_{AA} / k_p^{AA}} \qquad (4)$$

k_p^{MAPTAC} is two orders of magnitude lower than k_p^{AA}, and thus controls the averaged rate coefficient, as shown by the plot of log k_p^{cop} against f_{MAPTAC} in Figure 9. The k_p^{cop} values for MAPTAC gradually decrease as f_{MAPTAC} increases in the system, consistent with experimental results and in contrast to the sharper decrease seen upon addition of TMAEMC to AA, with little change between $f_{TMAEMC,0}$ of 0.1 and 1.0. Also, the terminal model treatment captures the lower influence of C_{Cl^-} in MAPTAC compared to TMAEMC as seen in the smaller difference in the curves generated with w_{mon} of 0.10 and 0.40.

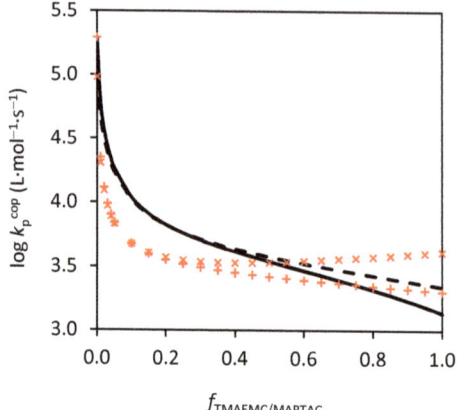

Figure 9. Comparison of $\log k_p^{cop}$ calculated at 50 °C as a function of MAPTAC/TMAEMC molar fraction at $w_{mon,0} = 0.10$ (—) and 0.40 (- - -) for MAPTAC and $w_{mon,0} = 0.10$ (+) and 0.40 (×) for TMAEMC.

Various approaches to represent cross-termination of the TMAEMC and AA copolymerization system were investigated in the previous investigation [29], an important task due to the more than two order of magnitude difference between the termination rate coefficients for homopolymerization of AA and that of the charged cationic (either TMAEMC or MAPTAC) system. The best representation of the AA-TMAEMC conversion profiles was obtained using the geometric mean treatment [29], an assumption also used here to model AA-MAPTAC copolymerization:

$$\langle k_{t,ss}^{AA,MAPTAC} \rangle = \Phi \cdot \left(\langle k_{t,ss}^{AA} \rangle \cdot \langle k_t^{MAPTAC} \rangle \right)^{1/2} \tag{5}$$

The complete set of copolymerization experiments run across different conditions (monomer concentration and composition) were used to perform parameter estimation of the single fitting parameter, with the best fit Φ value of 2.9 ± 0.2 obtained. The resulting representation of the conversion profiles is shown in Figure 10 grouped according to $w_{mon,0}$. Except for slight underpredictions of conversion rates for $w_{mon,0} = 0.05$ (Figure 10a, $f_{MAPTAC,0} = 0.3$ and 0.5), the effects of both $w_{mon,0}$ and $f_{MAPTAC,0}$ on conversion rates are well captured by the model. A similar underprediction of conversion rates at low $f_{TMAEMC,0}$ and low monomer concentrations was found in the AA-TMAEMC copolymerization study and was attributed to overprediction of the influence of AA backbiting on rate [29]. Despite this minor issue, the model, using the same assumptions and functional forms as developed for AA-TMAEMC, is able to describe the AA-MAPTAC copolymerization system over a wide range of monomer contents and the complete range of comonomer compositions, including MAPTAC homopolymerization (Figure 7).

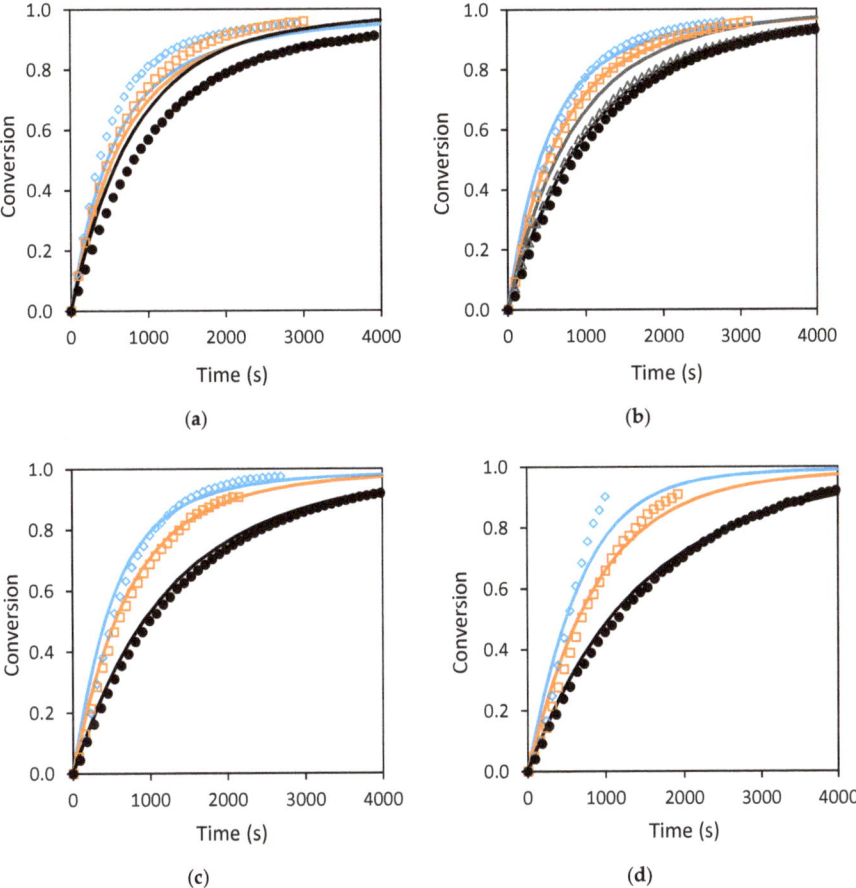

Figure 10. Simulated conversion profiles (lines) compared to experimental results (points) for $f_{MAPTAC,0}$ = 0.3(◊), 0.5(□), 0.7(Δ) and 0.9(●) at 50 °C and 0.80 wt% V-50 with $w_{mon,0}$ of (**a**) 0.05, (**b**) 0.10, (**c**) 0.20 and (**d**) 0.40. Profiles are calculated using the model summarized in Tables 1 and 2, with the copolymer termination rate coefficient calculated using the geometric mean method (best fit Φ values = 2.9).

5. Conclusions

An extensive investigation of the synthesis of polyelectrolytes from the aqueous-phase copolymerization of the cationic monomer MAPTAC with nonionized AA was conducted, to broaden and generalize the knowledge of copolymerization kinetics of cationic monomers. The rates of polymerization and relative consumption rates of AA and MAPTAC as a function of conversion were measured across varying initial monomer contents ($w_{mon,0}$ = 0.05 − 0.35) and initial comonomer compositions, with some experiments conducted with added salt to manipulate counterion concentration. Functional representations of MAPTAC propagation and termination rate coefficients were formulated to capture the effect of charge screening on the rate coefficients, extended to also represent the influence of C_{Cl^-} on comonomer composition drifts. As found for the TMAEMC system, an increase in $w_{mon,0}$ resulted in increased relative incorporation of MAPTAC into the copolymer, indicative of reduced electrostatic repulsion. The correctness of this interpretation is verified by the ability of the representation to also capture the influence of adding NaCl on the comonomer composition drift. Compared to TMAEMC, the influence of C_{Cl^-} on MAPTAC copolymer composition was not as strong, in agreement to the PLP-SEC study of homopropagation kinetics.

A comparison to the AA-TMAEMC system demonstrates that the same generalized model structure can be used to represent copolymerizations of AA with both cationic monomers, despite the amide vs ester functionality. The terminal model of propagation was combined with a treatment of AA intramolecular chain transfer to develop a complete mechanistic representation of AA-MAPTAC copolymerization. Monomer conversions profiles were well represented by combining homopolymerization models combined with the geometric mean treatment of cross-termination. However, further experiments are required to test the ability of the model to represent conversion and copolymer composition profiles, as well as copolymer molecular weights, over a range of operating temperature.

Supplementary Materials: Additional results documenting experimental reproducibility (Figure S1), solution pH (Figure S3), additional copolymerization conversion profiles (Figures S2 and S4) and solution viscosity (Figures S5 and S6) are available online at http://www.mdpi.com/2227-9717/8/11/1352/s1.

Author Contributions: Conceptualization, experimentation, analysis, draft preparation, I.H.E.; experimentation, E.S.; supervision, project administration, funding acquisition, writing review and editing, R.A.H. All authors have read and agreed to the published version of the manuscript.

Funding: This research was funded by BASF SE, Ludwigshafen, Germany.

Acknowledgments: The authors are grateful to BASF SE for financial support. Also, the technical discussions with Hugo Vale (BASF) and Igor Lacík (Polymer Institute of the Slovak Academy of Sciences) are appreciated.

Conflicts of Interest: The authors declare no conflict of interest.

References

1. Wesley, L.; Whippie, H.Z. Water-soluble free radical addition polymerizations. In *Monitoring Polymerization Reactions: From Fundamentals to Applications*; Alb, A.M., Reed, W.F., Eds.; Wiley-Interscience: Hoboken, NJ, USA, 2014.
2. Mandel, M. Polyelectrolytes. In *Encyclopedia of Polymer Science and Engineering*, 2nd ed.; Mark, H.F., Bikales, N., Overberger, C.G., Menges, G., Eds.; Wiley: New York, NY, USA, 1988; Volume 11, p. 739.
3. Heitner, N. Flocculating agents. In *Kirk-Othmer Encyclopedia of Chemical Technology*; Kroschwitz, J.I., Howe-Grant, M., Eds.; Wiley: New York, NY, USA, 2004; Volume 11.
4. BASF Home Page. Available online: http://www.basf.com/global/en.html (accessed on 12 February 2019).
5. Nicke, R. Flocculating agents help to increase the production. *Zellst. Und Pap.* **1982**, *31*, 19–23.
6. Beuermann, S.; Buback, M. Rate coefficients of free-radical polymerization deduced from pulsed laser experiments. *Prog. Polym. Sci.* **2002**, *27*, 191–254. [CrossRef]
7. Buback, M.; Schroeder, H.; Kattner, H. Detailed kinetic and mechanistic insight into radical polymerization by spectroscopic techniques. *Macromolecules* **2016**, *49*, 3193–3213. [CrossRef]
8. Beuermann, S.; Buback, M.; Hesse, P.; Lacík, I. Free-radical propagation rate coefficient of nonionized methacrylic acid in aqueous solution from low monomer concentrations to bulk polymerization. *Macromolecules* **2006**, *39*, 184–193. [CrossRef]
9. Beuermann, S.; Buback, M.; Hesse, P.; Kutcha, F.-D.; Lacík, I.; Van Herk, A.M. Critically evaluated rate coefficients for free-radical polymerization Part 6: Propagation rate coefficient of methacrylic acid in aqueous solution (IUPAC Technical Report). *Pure Appl. Chem.* **2007**, *79*, 1463–1469. [CrossRef]
10. Lacík, I.; Učňová, L.; Kukučková, S.; Buback, M.; Hesse, P.; Beuermann, S. Propagation rate coefficient of free-radical polymerization of partially and fully ionized methacrylic acid in aqueous solution. *Macromolecules* **2009**, *42*, 7753–7761. [CrossRef]
11. Beuermann, S.; Buback, M.; Hesse, P.; Hutchinson, R.A.; Kukučková, S.; Lacík, I. Termination kinetics of the free-radical polymerization of nonionized methacrylic acid in aqueous solution. *Macromolecules* **2008**, *41*, 3513–3520. [CrossRef]
12. Lacík, I.; Beuermann, S.; Buback, M. PLP–SEC study into free-radical propagation rate of nonionized acrylic acid in aqueous solution. *Macromolecules* **2003**, *25*, 9355–9363. [CrossRef]

13. Lacík, I.; Beuermann, S.; Buback, M. PLP-SEC study into the free-radical propagation rate coefficients of partially and fully ionized acrylic acid in aqueous solution. *Macromol. Chem. Phys.* **2004**, *205*, 1080–1087. [CrossRef]
14. Lacík, I.; Beuermann, S.; Buback, M. Aqueous phase size-exclusion-chromatography used for PLP–SEC studies into free-radical propagation rate of acrylic acid in aqueous solution. *Macromolecules* **2001**, *34*, 6224–6228. [CrossRef]
15. Barth, J.; Meiser, W.; Buback, M. SP-PLP-EPR study into termination and transfer kinetics of non-ionized acrylic acid polymerized in aqueous solution. *Macromolecules* **2012**, *45*, 1339–1345. [CrossRef]
16. Wittenberg, N.F.G.; Preusser, C.; Kattner, H.; Stach, M.; Lacík, I.; Hutchinson, R.A.; Buback, M. Modeling acrylic acid radical polymerization in aqueous solution. *Macromol. React. Eng.* **2016**, *10*, 95–107. [CrossRef]
17. Buback, M.; Hesse, P.; Hutchinson, R.A.; Kasák, P.; Lacík, I.; Stach, M.; Utz, I. Kinetics and modeling of free-radical batch polymerization of nonionized methacrylic acid in aqueous solution. *Ind. Eng. Chem. Res.* **2008**, *47*, 8197–8204. [CrossRef]
18. Wittenberg, N.F.G.; Buback, M.; Hutchinson, R.A. Kinetics and modeling of methacrylic acid radical polymerization in aqueous solution. *Macromol. React. Eng.* **2013**, *7*, 267–276. [CrossRef]
19. Barth, J.; Buback, M. Termination and transfer kinetics of sodium acrylate polymerization in aqueous solution. *Macromolecules* **2012**, *45*, 4152–4157. [CrossRef]
20. Kattner, H.; Drawe, P.; Buback, M. Chain-length-dependent termination of sodium methacrylate polymerization in aqueous solution studied by SP-PLP-EPR. *Macromolecules* **2017**, *50*, 1386–1393. [CrossRef]
21. Drawe, P.; Buback, M.; Lacík, I. Radical polymerization of alkali acrylates in aqueous solution. *Macromol. Chem. Phys.* **2015**, *216*, 1333–1340. [CrossRef]
22. Preusser, C.; Ezenwajiaku, I.H.; Hutchinson, R.A. The combined influence of monomer concentration and ionization on acrylamide/acrylic acid composition in aqueous solution radical batch copolymerization. *Macromolecules* **2016**, *49*, 4746–4756. [CrossRef]
23. Preusser, C. Kinetics and Modeling of Free-Radical Aqueous Phase Polymerization of Acrylamide with Acrylic Acid at Varying Degrees of Ionization. Ph.D. Thesis, Queen's University, Kingston, ON, Canada, August 2015.
24. Riahinezhad, M.; Mcmanus, N.; Penlidis, A. Effect of monomer concentration and ph on reaction kinetics and copolymer microstructure of acrylamide/acrylic acid copolymer. *Macromol. React. Eng.* **2015**, *9*, 100–113. [CrossRef]
25. Riahinezhad, M.; Kazemi, N.; McManus, N.; Penlidis, A. Effect of ionic strength on the reactivity ratios of acrylamide/acrylic acid (sodium acrylate) copolymerization. *J. Appl. Polym. Sci.* **2014**, *131*. [CrossRef]
26. Chovancová, A.; Ezenwajiaku, I.H.; Nikitin, A.; Sedlák, M.; Hutchinson, R.A.; Lacík, I. Radical polymerization of cationic methacrylate monomers TMAEMC and MAPTAC: Propagation rate coefficients obtained by PLP-SEC and batch polymerization. *Macromolecules* **2020**. manuscript in preparation.
27. Ezenwajiaku, I.H.; Chovancová, A.; Lister, K.; Lacík, I.; Hutchinson, R.A. Experimental and modeling investigation of radical homopolymerization of 2-(methacryloyloxyethyl) trimethylammonium chloride in aqueous solution. *Macromol. React. Eng.* **2020**, *14*. [CrossRef]
28. Cuccato, D.; Storti, G.; Morbidelli, M. Experimental and modeling study of acrylamide copolymerization with quaternary ammonium salt in aqueous solution. *Macromolecules* **2015**, *48*, 5076–5087. [CrossRef]
29. Ezenwajiaku, I.H.; Zigelstein, R.; Chovancová, A.; Lacík, I.; Hutchinson, R.A. Experimental and modeling investigations of aqueous-phase radical copolymerization of 2-(methacryloyloxyethyl)trimethylammonium chloride with acrylic acid. *Ind. Eng. Chem. Res.* **2020**, *59*, 3359–3374. [CrossRef]
30. Losada, R.; Wandrey, C. Copolymerization of a cationic double-charged monomer and electrochemical properties of the copolymers. *Macromolecules* **2009**, *42*, 3285–3293. [CrossRef]
31. Lacík, I.; Chovancová, A.; Uhelská, L.; Preusser, C.; Hutchinson, R.A.; Buback, M. PLP-SEC studies into the propagation rate coefficient of acrylamide radical polymerization in aqueous solution. *Macromolecules* **2016**, *49*, 3244–3253. [CrossRef]
32. Lacík, I.; Sobolčiak, P.; Stach, M.; Chorvát, D.; Kasák, P. Propagation rate coefficient for sulfobetaine monomers by PLP–SEC. *Polymer* **2016**, *87*, 38–49. [CrossRef]
33. Dobrynin, A.V.; Rubinstein, M. Theory of polyelectrolytes in solutions and at surfaces. *Prog. Polym. Sci.* **2005**, *30*, 1049–1118. [CrossRef]

34. Tanaka, H. Copolymerization of cationic monomers with acrylamide in an aqueous solution. *J. Polym. Sci. Part A Polym. Chem.* **1986**, *24*, 29–36. [CrossRef]
35. Hunkeler, D.; Hamielec, A.E.; Baade, W. The polymerization of quaternary ammonium cationic monomers with acrylamide. *Adv. Chem. Ser.* **1989**, *223*, 175–192.
36. McCormick, C.L.; Lowe, A.B.; Ayres, N. Water-soluble polymers. In *Kirk-Othmer Encyclopedia of Chemical Technology*, 5th ed.; Kroschwitz, J., Ed.; Wiley-Interscience: New York, NY, USA, 2001; Volume 20, pp. 435–504.
37. Drawe, P.; Kattner, H.; Buback, M. Kinetics and modeling of the radical polymerization of trimethylaminoethyl methacrylate chloride in aqueous solution. *Macromol. Chem. Phys.* **2016**, *217*, 2755–2764. [CrossRef]
38. Wako Speciality Chemicals Home Page. Available online: https://www.wakospecialtychemicals.com/brand/wako/product/v-50-azo-initiator/ (accessed on 15 February 2017).
39. Kattner, H. Radical Polymerization Kinetics of Nonionized and Fully Ionized Monomers Studied by Pulsed-Laser EPR. Ph.D. Thesis, Georg-August-Universität Göttingen, Göttingen, Germany, June 2016.
40. Smith, G.B.; Russell, G.T.; Heuts, J.P.A. Termination in dilute-solution free-radical polymerization: A composite model. *Macromol. Theory Sim.* **2003**, *12*, 299–314. [CrossRef]
41. Kazemi, N.; Duever, T.A.; Penlidis, A. Reactivity ratio estimation from cumulative copolymer composition data. *Macromol. React. Eng.* **2011**, *5*, 385–403. [CrossRef]
42. Preusser, C.; Hutchinson, R.A. An in situ NMR study of radical copolymerization kinetics of acrylamide and non-ionized acrylic acid in aqueous solution. *Macromol. Symp.* **2013**, *333*, 122–137. [CrossRef]

Publisher's Note: MDPI stays neutral with regard to jurisdictional claims in published maps and institutional affiliations.

© 2020 by the authors. Licensee MDPI, Basel, Switzerland. This article is an open access article distributed under the terms and conditions of the Creative Commons Attribution (CC BY) license (http://creativecommons.org/licenses/by/4.0/).

Article

Initiator Feeding Policies in Semi-Batch Free Radical Polymerization: A Monte Carlo Study

Ali Seyedi [1], Mohammad Najafi [1,*], Gregory T. Russell [2,*], Yousef Mohammadi [3,*], Eduardo Vivaldo-Lima [4] and Alexander Penlidis [5,*]

1. Department of Polymer Engineering, School of Chemical Engineering, College of Engineering, University of Tehran, P.O. Box 11155-4563, Tehran 1417466191, Iran; seyedi.a@ut.ac.ir
2. School of Physical and Chemical Sciences, University of Canterbury, Private Bag 4800, Christchurch 8140, New Zealand
3. Petrochemical Research and Technology Company (NPC-rt), National Petrochemical Company (NPC), P.O. Box 14358-84711, Tehran 1993834557, Iran
4. Facultad de Química, Departamento de Ingeniería Química, Universidad Nacional Autónoma de México, CU, Mexico City 04510, Mexico; vivaldo@unam.mx
5. Department of Chemical Engineering, Institute for Polymer Research (IPR), University of Waterloo, Waterloo, ON N2L 3G1, Canada
* Correspondence: najafi.m@ut.ac.ir (M.N.); greg.russell@canterbury.ac.nz (G.T.R.); mohammadi@npc-rt.ir (Y.M.); penlidis@uwaterloo.ca (A.P.); Tel.: +519-888-4567 (ext. 36634) (A.P.)

Received: 11 September 2020; Accepted: 3 October 2020; Published: 15 October 2020

Abstract: A Monte Carlo simulation algorithm is developed to visualize the impact of various initiator feeding policies on the kinetics of free radical polymerization. Three cases are studied: (1) general free radical polymerization using typical rate constants; (2) diffusion-controlled styrene free radical polymerization in a relatively small amount of solvent; and (3) methyl methacrylate free radical polymerization in solution. The number- and weight-average chain lengths, molecular weight distribution (MWD), and polymerization time were computed for each initiator feeding policy. The results show that a higher number of initiator shots throughout polymerization at a fixed amount of initiator significantly increases average molecular weight and broadens MWD. Similar results are also observed when most of the initiator is added at higher conversions. It is demonstrated that one can double the molecular weight of polystyrene and increase its dispersity by 50% through a four-shot instead of a single shot feeding policy. Similar behavior occurs in the case of methyl methacrylate, while the total time drops by about 5%. In addition, policies injecting initiator at high monomer conversions result in a higher unreacted initiator content in the final product. Lastly, simulation conversion-time profiles are in agreement with benchmark literature information for methyl methacrylate, which essentially validates the highly effective and flexible Monte Carlo algorithm developed in this work.

Keywords: initiator feeding policies; styrene; methyl methacrylate; Monte Carlo simulation; polymer microstructure

1. Introduction

Free radical polymerization is one of the most important polymerization techniques, contributing to the synthesis of about 50% of all the polymers produced worldwide. As a chain polymerization method, it is composed of initiation, propagation, chain transfer and termination steps [1]; this technique enjoys many advantages compared to other polymerization methods, such as ionic polymerization, making it very popular in industry. A number of advantages enhancing its robustness include lower sensitivity to impurities (although sensitive to oxygen, it is tolerant of water), broader range of monomers (almost all vinyl monomers), and milder reaction conditions [2]. Among the drawbacks of this method, one may mention failure to design well-defined macromolecules, which is usually embodied in ill-controlled molecular weight; broad molecular weight distribution; and weakly controlled monomer (sequence) distribution in the case of copolymerization. To overcome these disadvantages, new techniques based on specific feeding policies have been developed [3–7].

For polymerization in a batch reactor, all the monomers and initiators must be added to the reactor in the beginning of the polymerization. Therefore, the desired macromolecular microstructure—i.e., molecular weight average and distribution, branching index and distribution, copolymer composition (distribution), monomer sequence length distribution, etc.—cannot simply be controlled [8]. In particular, polymeric molecules that are formed earlier in a copolymerization may contain a high percentage of the more reactive monomer, while the chains produced later contain a high percentage of the remaining (less reactive) monomer. Since there is no steady state in a batch reactor, and also due to raw material variability, copolymer (and other) properties obtained may differ from batch to batch [9,10]. In a semi-batch reactor, at least one of the reactants is added over time; this mode helps design and produce macromolecules with a well-defined microstructure despite differences in the reactivity of the monomers. Another way of achieving a desired microstructure is to use predefined initiator feed policies [10–12].

In this context, Goto et al. studied the effect of *tert*-butyl hydroperoxide (BHP) initiator on the free radical polymerization of styrene initiated with styroxypiperidine and found that the addition of this initiator tripled the rate of the reaction without a significant influence on the molecular weight distribution and the number of polymer molecules [13]. A monomer feeding policy in a semi-batch atom transfer radical copolymerization (ATRP) showed that the copolymer composition distribution (CCD) could be precisely controlled by slowly feeding the monomers in a planned profile [9,14]. Wang et al. investigated four divinyl monomer feeding policies in the reversible addition-fragmentation (chain) transfer (RAFT) copolymerization of acrylamide and reported the effect of feeding policy on the microstructure and branching of star and hyperbranched chains produced [3]. Furthermore, Diaz-Camacho et al. [15] studied the semi-batch addition of initiator in the nitroxide-mediated polymerization of styrene, and modeled the system using the Predici software, but they slightly over-predicted the effect of semi-continuous addition on polymerization rate [16].

Simulation and modeling techniques have proven to be an extremely useful tool for predicting the microstructural characteristics of polymer chains during polymerization. Modeling of polymerization processes falls into the two categories of deterministic and stochastic modeling. The former involves writing mass balances for all the species present, leading to a set of differential (and algebraic) equations, the numerical solution of which leads to properties such as average molecular weights, average branching indicators, etc. either directly (with the method of moments) or indirectly (when chain lengths are treated individually). In the latter event, there is a very large number of equations that must be solved. This can be a major problem even with increasingly powerful computers used today [17]. A popular variant of the deterministic approach is the use of the method of moments. This method decreases the number of equations by rewriting the kinetically-derived population balances into moment balances [18].

Using mathematical modeling, Arzamendi et al. investigated three different monomer feeding policies in semi-batch emulsion copolymerization in order to optimize control over copolymer composition and minimize polymerization time [19]. In another study, the semi-batch polymerization of styrene, in a monomer and initiator starved-fed reactor, showed that the molecular weight and molecular weight distribution can be effectively controlled in this manner [20]. Pinto et al. optimized a cocktail of initiators in suspension polymerization of vinyl chloride using mathematical modeling and found that by using the optimized cocktail, process time could be significantly decreased compared with a typical single-initiator process [21]. In another work, a kinetic model was developed to consider the effect of a monomer feeding policy on the CCD in the mini-emulsion copolymerization of styrene and butyl acrylate; it was established that the proposed model was in good agreement with polymerization kinetics. Furthermore, several copolymers with a predesigned CCD were successfully synthesized using the presented model [22].

In contrast to deterministic approaches, stochastic approaches do not need to solve a system of differential equations. These approaches are usually based on the master chemical equation and require information about elementary reaction mechanisms and probabilities. This characteristic makes this technique invaluable for modeling complex polymer systems [18,23]. The kinetic Monte Carlo (KMC) algorithm is one of the stochastic approaches that provides an efficient method for investigating the microstructure and spatial topologies of polymer chains. It is also possible to simulate polymerization processes with complex molecular architectures and investigate macromolecular microstructures such as molecular weight distribution, copolymer composition (distribution), monomer sequence length (distribution), long-chain branching, short-chain branching, chain topology, etc. [23–25]

The effect of bi-functional initiators on free radical polymerization of styrene has been studied using Monte Carlo simulation, and it was found that using bi-functional initiators results in increased monomer conversion and molecular weight, and narrower molecular weight distribution compared with mono-functional initiators [26]. Tobita studied long-chain branching (LCB) and the distribution of branching density in poly(vinyl acetate) using this technique [27]. The Monte Carlo method was used in Prescott's work studying the effect of chain-length-dependent termination on reversible deactivation radical polymerization (RDRP) using the reversible addition-fragmentation chain transfer (RAFT) process [28]. Using the Monte Carlo method with the Julia programming language, Pintos et al. simulated the kinetics of RAFT radical polymerization in batch mode, based on the theories of slow fragmentation, intermediate radical termination, and the termination of intermediate radicals with oligomers [29]. The effect of reactivity ratios and initial feed composition on copolymer microstructure was investigated for free radical copolymerization in a comprehensive study using a Monte Carlo simulation method. The model had the ability to illustrate changes in azeotropy and composition drift [30,31]. A high-performance Monte Carlo code was developed for simulating the kinetics of free radical and ATRP of styrene, considering chain-length dependent and diffusion-controlled termination. The obtained results were in close agreement with experimental data [32]. Saeb et al. used Monte-Carlo modeling to study the effect of co-monomer feeding policy on copolymer microstructure in metallocene catalyzed copolymerization of ethylene/1-hexene; this revealed the critical significance of the computerized feeding mode contrasted with the uncontrolled feeding mode [33]. A Monte Carlo method was also used to study the effect of hydrogen and co-catalyst concentration in ethylene polymerization using a Ziegler-Natta catalyst [34]. Finally, Najafi et al. studied four termination modes in styrene ATRP using a Monte Carlo approach [35].

The above is a critical selection of works on the simulation of the kinetics of free radical polymerization, in particular referring to feeding strategies and/or nature of initiation. To the very best of our knowledge, our work that follows is the first ever study using the Monte Carlo simulation method to investigate the effect of initiator feeding policies on a polymerization process and chain properties in semi-batch free radical polymerization. In this regard, our work is also complementary to the studies by Diaz-Camacho et al. [15] and Roa-Luna et al. [16] in that the approach in these two previous references was similar but derived using a commercial package (Predici) for a nitroxide-mediated polymerization system. In this context, we investigate three different cases: a general free radical polymerization with typical kinetic rate coefficient values (Case 1), a diffusion-controlled free radical polymerization of styrene (Case 2, low solvent fraction), and solution free radical polymerization of methyl methacrylate (Case 3, high solvent fraction). Thermal and chemical types of initiation, transfer reactions, and diffusion-control effects were all considered in the different case studies. In order to investigate the effect of initiator feeding policies during the course of the polymerization, three feeding strategies based on the number of shots (NOS), amount of shots (AOS), and time of shots (TOS) were considered in each case, and different indicators of polymerization performance, in terms of polymerization rate, polymer quality, molecular weight averages, and molecular weight distribution, were evaluated. We should state upfront that in these case studies, all being evaluated at the simulation/modeling level, mixing of the initiator shots in the polymerizing mixture is considered perfectly instantaneous and homogeneous.

2. Model Development

2.1. Kinetic Model

Monte Carlo algorithms were developed for three cases, as detailed in Table 1: general free radical polymerization; styrene free radical polymerization in solution (low fraction of solvent); and methyl methacrylate free radical polymerization in solution (high fraction of solvent). Table 1 lists the reactions for each case. In the three cases, the chemical initiation step is related to thermal decomposition of initiator, while thermal monomer self-initiation also contributes in styrene polymerization. Transfer to monomer, transfer to solvent, termination by combination, and termination by disproportionation may occur in each case. The kinetic parameters and constants required for the simulations are listed in Table 2.

Table 1. Elementary reactions in the cases considered in this work: general free radical polymerization (Case 1), styrene free radical polymerization in solution (Case 2) [36], and methyl methacrylate free radical polymerization in solution (Case 3) [37].

Case 1: General Free Radical Polymerization	
Initiator decomposition	$I \xrightarrow{k_d} 2fI^*$
Chain initiation	$I^* + M \xrightarrow{k_i} P_1^*$
Propagation	$P_n^* + M \xrightarrow{k_p} P_{n+1}^*$
Termination by combination	$P_n^* + P_m^* \xrightarrow{k_{tc}} P_{n+m}^*$
Termination by disproportionation	$P_n^* + P_m^* \xrightarrow{k_{td}} P_n^* + P_m^*$
Case 2: Free Radical Polymerization of Styrene	
Chemical Initiation	$I \xrightarrow{k_d} 2fI^*$
	$I^* + M \xrightarrow{4k_p} P_1^*$
Thermal Initiation	$3M \xrightarrow{k_{thermal}} M_1^* + M_2^*$
	$M_1^* + M \xrightarrow{4k_p} P_2^*$
	$M_2^* + M \xrightarrow{4k_p} P_3^*$
Propagation	$P_n^* + M \xrightarrow{k_p} P_{n+1}^*$
Termination by combination	$P_n^* + P_m^* \xrightarrow{k_{tc}} P_{n+m}^*$
Termination by disproportionation	$P_n^* + P_m^* \xrightarrow{k_{td}} P_n^* + P_m^*$
Chain transfer to monomer	$P_n^* + M \xrightarrow{k_{trM}} P_n + P_1^*$
Chain transfer to solvent	$P_n^* + S \xrightarrow{k_{trS}} P_n + P_1^*$
Case 3: Free Radical Polymerization of Methyl Methacrylate	
Chemical Initiation	$I \xrightarrow{k_d} 2fI^*$
	$I^* + M \xrightarrow{k_p} P_1^*$
Propagation	$P_n^* + M \xrightarrow{k_p} P_{n+1}^*$
Termination by disproportionation	$P_n^* + P_m^* \xrightarrow{k_{td}} P_n^* + P_m^*$
Chain transfer to monomer	$P_n^* + M \xrightarrow{k_{trM}} P_n + P_1^*$
Chain transfer to solvent	$P_n^* + S \xrightarrow{k_{trS}} P_n + P_1^*$

Table 2. Rate parameters used in the simulations.

Parameters	Value/Expression	References
Case 1: General Free Radical Polymerization		
f	0.50	This work
k_d (s^{-1})	10^{-3}	This work
k_i (L.mol^{-1}.s^{-1})	10^4	This work
K_p (L.mol^{-1}.s^{-1})	10^4	This work
k_{tc} (L.mol^{-1}.s^{-1})	10^8	This work
k_{td} (L.mol^{-1}.s^{-1})	10^8	This work
Case 2: Free Radical Polymerization of Styrene		
f	0.50	[38,39]
k_d (s^{-1})	$6.78 \times 10^{15} \exp(-17{,}714.0/T)$	[38,39]
$k_{thermal}$ (L^2.mol^{-2}.s^{-1})	$2.19 \times 10^5 \exp(-13{,}800.0/T)$	[40]
K_p (L.mol^{-1}.s^{-1})	$4.266 \times 10^7 \exp(-3910.0/T)$	[41]
$k_{t,0}$ (L.mol^{-1}.s^{-1})	$3.82 \times 10^9 \exp(-958.0/T)$	[42]
K_{trM} (L.mol^{-1}.s^{-1})	$2.31 \times 10^6 \exp(-6377.0/T)$	[40]
K_{trS} (L.mol^{-1}.s^{-1})	1.8	[43]
K_t (L.mol^{-1}.s^{-1})	$k_{t,0} \exp(-0.44 w_p - 6.36 w_p^2 - 0.1704 w_p^3)$	[10]
k_{tc} (L.mol^{-1}.s^{-1})	$0.01 k_t$	[44]
k_{td} (L.mol^{-1}.s^{-1})	$0.99 k_t$	[44]
Note: w_p is the polymer weight fraction.		
Case 3: Free Radical Polymerization of Methyl Methacrylate		
f	0.50	[45]
k_d (s^{-1})	$1.2525 \times 10^{14} \exp(-14{,}770.0/T)$	[43]
K_p (L.mol^{-1}.s^{-1})	$4.92 \times 10^5 \exp(-2191.0/T)$	[45]
k_{trM} (L.mol^{-1}.s^{-1})	$7.177 \times 10^9 \exp(-9036.0/T)$	[43]
k_{trS} (L.mol^{-1}.s^{-1})	$4.673 \times 10^8 \exp(-7902.0/T)$	[43]
k_{td} (L.mol^{-1}.s^{-1})	$9.8 \times 10^7 \exp(-353.0/T)$	[45]

2.2. Monte Carlo Simulation

The kinetic model based on the Monte Carlo algorithm is developed according to the probabilities of reactions. Due to the stochastic nature of Monte Carlo simulation, there may always be differences in the results obtained from each run. Hence, defining a proper simulation volume is vital not only because of its impact on the reliability and accuracy of the simulation results, but also due to its influence on the simulation performance and runtime. A larger simulation volume usually renders more accurate results, but it may lead to longer runtimes. The optimum sample size is specific to each polymerization system since it is greatly influenced by the reaction parameters and initial conditions [23,24]. The Monte Carlo computer code utilized in the current work employs Gillespie's algorithm as its foundation [46]. The concentrations of the species can be related to the number of monomers in the simulation volume according to Equation (1):

$$X_i = C_i \cdot NA \cdot V \tag{1}$$

where X_i is the number of species i and C_i represents the concentration of species i; NA and V stand for Avogadro's number and reaction volume, respectively. In addition, the experimental rate constants (k^{exp}) must be converted to Monte Carlo rate constants (k^{MC}) according to Equations (2)–(5):

Unimolecular reaction:

$$k^{MC} = k^{exp} \tag{2}$$

Bimolecular reaction between different species:

$$k^{MC} = \frac{k^{exp}}{NA \cdot V} \tag{3}$$

Bimolecular reaction between identical species:

$$k^{MC} = \frac{2 \times k^{exp}}{NA \cdot V} \tag{4}$$

Termolecular reaction between identical species:

$$k^{MC} = \frac{6 \times k^{exp}}{NA^2 \cdot V^2} \tag{5}$$

For example, the equation for rate of polymerization, R_p, is:

$$R_p = k_p^{MC} \cdot X_R \cdot X_M \tag{6}$$

where k_p^{MC} stands for the Monte Carlo propagation rate constant and X_R and X_M represent the number of (polymerizing) free radicals and monomer molecules, respectively. The probability of each reaction, P_i, is defined by Equation (7):

$$P_i = \frac{R_i}{R_{total}} = \frac{R_i}{\sum_{j=1}^{n} R_j} \tag{7}$$

where R_i is the rate of reaction i, R_{total} is the total reaction rate of the system, and n denotes the number of reaction channels. Equation 8 is used to determine the reaction channel which should happen at a given time:

$$\sum_{i=1}^{\mu-1} P_i < r_1 \leq \sum_{i=1}^{\mu} P_i \tag{8}$$

where μ is the number of the selected reaction channel and r_1 denotes a random number in the (0,1] interval. An additional random number (r_2) should also be generated to determine the time interval (τ) between two consecutive and sequential reactions (see Equation (9)):

$$\tau = \frac{1}{\sum_{i=1}^{n} R_i} ln\left(\frac{1}{r_2}\right) \tag{9}$$

The improved Mersenne Twister algorithm is embedded in the simulation code to generate random numbers [47]. The simulation code is written in Lazarus 2.0.10 and compiled into 64-bit executable code using FPC 3.2.0. The average runtime of the program is approximately 300 min at a simulation volume of 2.076×10^{-13} lit on a desktop computer equipped with Intel Core i7-3770K (3.50 GHz) and 32 GB of memory (2133 MHz) running Windows 7 Ultimate 64-bit operating system.

3. Results and Discussion

3.1. Initiator Feeding Policies

The simulation is performed in three scenarios according to three different feeding policies. In the first policy, the number of initiator shots injected into the system is varied. Four different feeding policies based on the number of shots (NOS) are simulated. In the second, the amount of initiators per shot is varied; four different feeding policies based on changing the amount of shots (AOS) are chosen. In the third scenario, the time at which the initiator is added to the reactor is varied, thus, four different feeding policies having various times of shots (TOS) are taken into account (Table 3). It is worth mentioning that the total amount of initiator used in each of the three scenarios is the same and equal to $[I]_0$ (of a regular initiator charge). In the NOS policy, the initiator is added to the reactor in one, two, four, and eight steps with equal fractions at distinct reaction conversions. In the AOS policy, the initiator is introduced to the reaction system at four different fractions at distinct conversions. In the feeding policy based on TOS, the initiator is injected at four equal fractions at various conversion levels.

Table 3. Initiator feeding policy configurations with the same total amount of initiator.

Number of Shots	Initiator Fraction
NOS-1	One step (1.000)
NOS-2	Two steps (0.500, 0.500)
NOS-3	Four steps (0.250, 0.250, 0.250, 0.250)
NOS-4	Eight steps (0.125, 0.125, 0.125, 0.125, 0.125, 0.125, 0.125, 0.125)
Amount of Shots	**Initiator Fraction**
AOS-1	(0.10, 0.20, 0.30, 0.40)
AOS-2	(0.40, 0.30, 0.20, 0.10)
AOS-3	(0.10, 0.40, 0.40, 0.10)
AOS-4	(0.40, 0.10, 0.10, 0.40)
Time of Shots	**Conversion of Each Shot (Initiator Fraction is Equal to 0.25 for Each Step)**
TOS-1	(0.00, 0.05, 0.15, 0.25)
TOS-2	(0.00, 0.30, 0.40, 0.50)
TOS-3	(0.00, 0.55, 0.65, 0.75)
TOS-4	(0.00, random, random, random)

3.2. General Free Radical Polymerization

The results of varying the number of initiator shots are shown in Figure 1. In the initiator feeding policy for NOS-1 (A_1), the entire amount of initiator is charged to the reactor at the beginning of the polymerization. Thus, the rate of monomer consumption in NOS-1 (A_2) should be increased in comparison with NOS-2 (B_2), NOS-3 (C_2), and NOS-4 (D_2). Increasing the number of shots, which approaches asymptotically the case of continuous feeding, results in a smaller slope in conversion-monomer curves (NOS-4 (D_2)) in spite of decreasing the amount of initiator in each shot (NOS-1 (A_2)). This indicates slower polymerization, and consequently, larger polymerization times. On the other hand, by distributing the concentration of the initiator at different conversions, because of a drop in the concentration of the initiator at the early stages of the polymerization, the molecular weight increases by changing feeding policy from NOS-1 (A_3) to NOS-4 (D_3). The same change causes a wider molecular weight distribution with a longer tail towards higher molecular weights. Therefore, the molecular weight distribution curves are shifted to the right by increasing the number of shots from NOS-1 (A_4) to NOS-4 (D_4). In fact, increasing the number of shots (NOS) leads to increasing number- and weight-average molecular weights (see third row) and a broader molecular weight distribution (MWD) (fourth row).

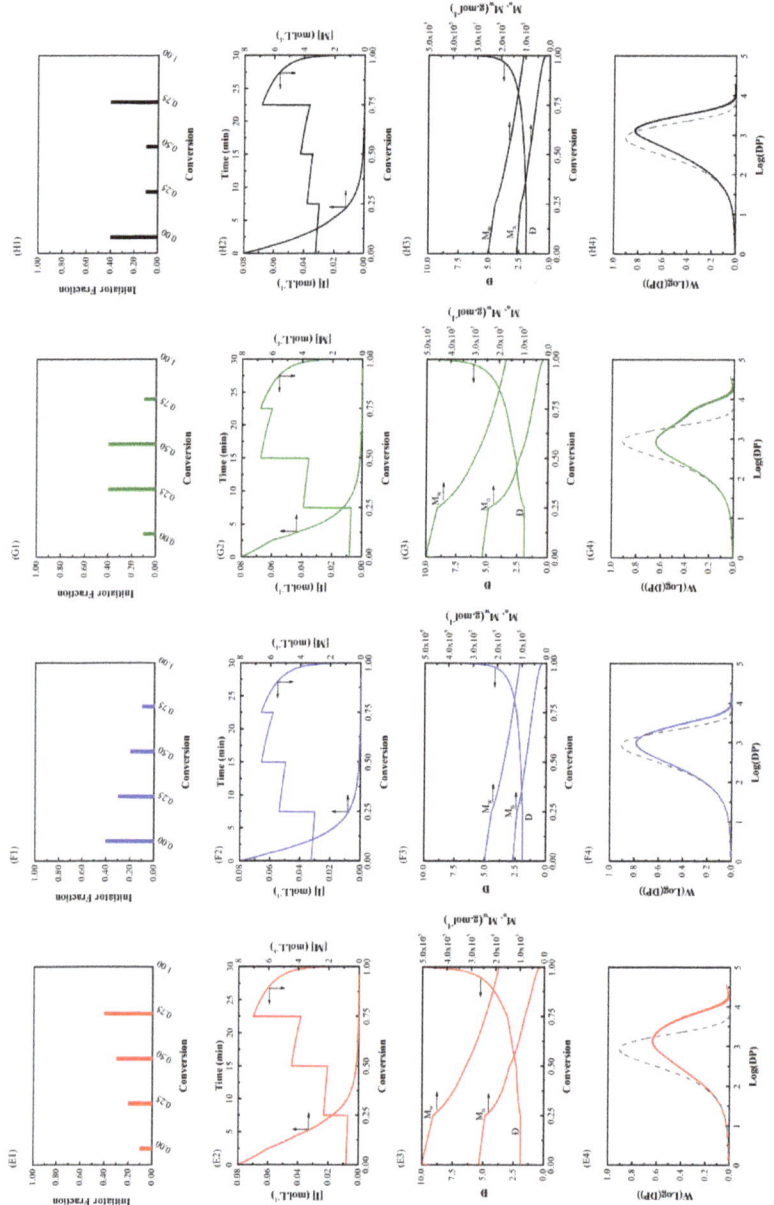

Figure 2. Semi-batch general free radical polymerization (Case 1) using different amount of shots (AOS) feeding policies with initial values $[M]_0 = 8.00$ mol L^{-1} and $[I]_0 = 0.08$ mol L^{-1}. First row: initiator feeding policy for AOS-1 (E$_1$), AOS-2 (F$_1$), AOS-3 (G$_1$), and AOS-4 (H$_1$). Second row: total monomer and initiator consumption versus conversion for AOS-1 (E$_2$), AOS-2 (F$_2$), AOS-3 (G$_2$), and AOS-4 (H$_2$). Third row: number- and weight-average molecular weight and dispersity versus conversion for AOS-1 (E$_3$), AOS-2 (F$_3$), AOS-3 (G$_3$), and AOS-4 (H$_3$). Fourth row: final molecular weight distribution for AOS-1 (E$_4$), AOS-2 (F$_4$), AOS-3 (G$_4$), and AOS-4 (H$_4$) (solid line), where the NOS-1 (A$_4$) result (dashed line from Figure 1) is also presented for comparison.

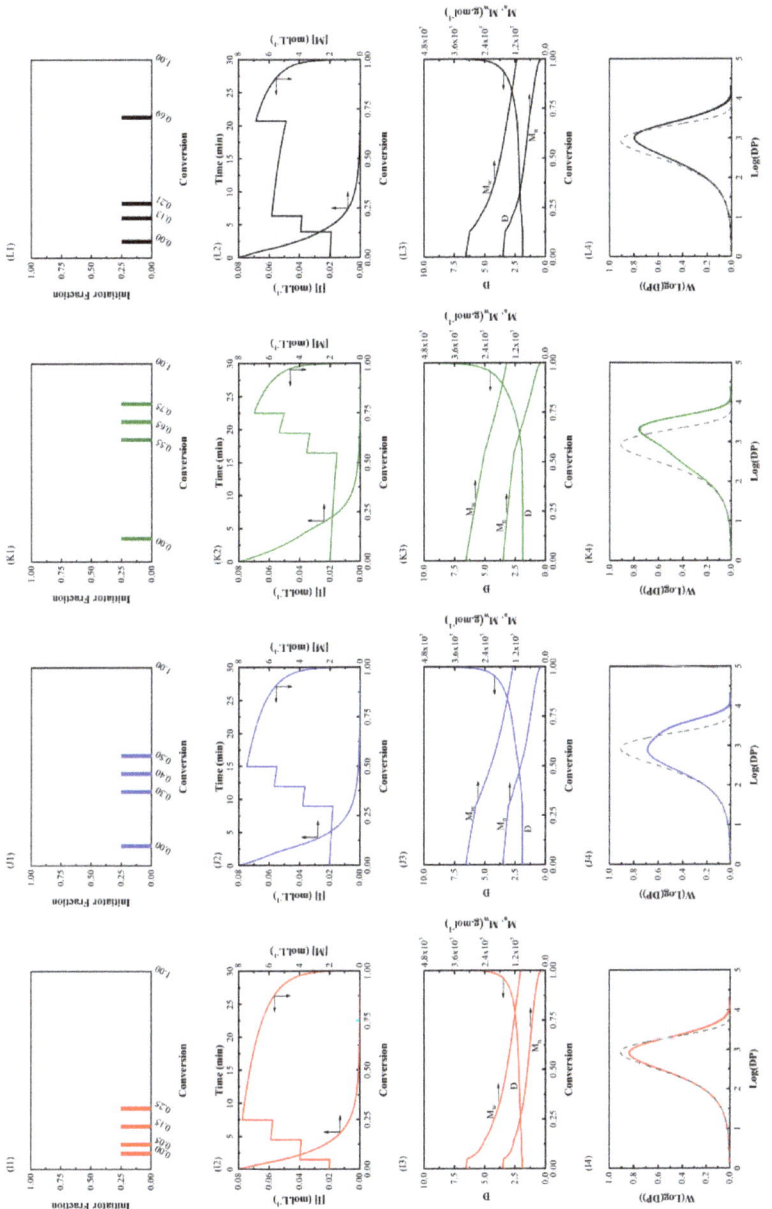

Figure 3. Semi-batch general free radical polymerization (Case 1) using different (TOS) feeding policies with initial values $[M]_0 = 8.00$ mol L^{-1} and $[I]_0 = 0.08$ mol L^{-1}. First row: initiator feeding policy for TOS-1 (I_1), TOS-2 (J_1), TOS-3 (K_1), and TOS-4 (L_1). Second row: total monomer and initiator consumption versus conversion for TOS-1 (I_2), TOS-2 (J_2), TOS-3 (K_2), and TOS-4 (L_2). Third row: number-average and weight-average molecular weight and dispersity versus conversion for TOS-1 (I_3), TOS-2 (J_3), TOS-3 (K_3), and TOS-4 (L_3). Fourth row: molecular weight distribution for TOS-1 (I_4), TOS-2 (J_4), TOS-3 (K_4), and TOS-4 (L_4) (solid line), where the NOS-1 (A_4) result (dashed line from Figure 1) is also presented for comparison.

3.3. Styrene Free Radical Polymerization in Solution

Table 4 presents results obtained from the Monte Carlo simulation of styrene free radical polymerization (Case 2 of Tables 1 and 2; low fraction of solvent). By increasing the number of initiator shots (NOS) from one to eight, the weight-average molecular weight increases by 150%, while the polymerization time drops by 8%. In addition, by meticulously analyzing the results of the simulation, it is found that using the NOS policy not only increases the weight-average molecular weight, but also 'tailors' the molecular weight distribution to some extent, in that one can dial up the dispersity by altering the number of shots. Figure 4 probes this further by presenting the MWD in these scenarios of semi-batch free radical polymerization of styrene for the three different initiator feeding policies. Comparing NOS-1 with TOS-2 (first vs. third column of plots in Figure 4) reveals that changing the policy enables one to increase weight-average molecular weight and broaden the distribution to a moderate extent. Additionally, changing the policy to AOS-1 (middle column of plots in Figure 4) almost doubles the weight-average molecular weight and broadens the chain length distribution (a 150% increase in dispersity, Đ). The AOS-1 and AOS-3 policies provide a wider molecular weight distribution compared to the other policies, which is ascribed to the distribution of the initiator concentration over the course of the reaction. It may be stated that a policy leading to increased weight-average molecular weight usually results in a broadened MWD. In addition, there will be more unreacted initiator in the final polymerization mixture in policies in which an initiator shot is injected at the late stages of the reaction, namely NOS-4, AOS-1, AOS-4, and TOS-3; in all these cases, more than 10% of the initiator remains unreacted. This leads to a decrease in the number of monomer units allotted to each radical in the beginning of the polymerization and, consequently, the emergence of short chains, which is the cause of a widening molecular weight distribution. An increase in the concentration of the residual initiator at the end of the polymerization of styrene means less initiator decomposition over the course of the reaction, which leads to a higher fraction of long chains; as a result, molecular weight distribution widens, and average molecular weight and dispersity deviate from the reference graph (NOS-1 (A_4)). However, unreacted (residual) initiator is undesirable for a number of reasons, both economical and operational (product quality).

Table 4. Simulated data of free radical polymerization of styrene (Case 2) according to various initiator feeding policies (NOS, AOS, and TOS as per Table 3, with an equal total amount of initiator).

Policy	Reaction Time at Conversion 100% (min)	$\bar{M}w$ (g/mol)	Đ	Unreacted Initiator (%)
Number of Shots				
NOS-1	42	22,000	2.1	3.5
NOS-2	39	27,000	2.4	6.7
NOS-3	39	31,000	2.7	9.3
NOS-4	39	34,000	2.9	11
Amount of Shots				
AOS-1	42	41,000	3.4	12.3
AOS-2	39	27,000	2.4	6.6
AOS-3	43	37,000	3.2	8.3
AOS-4	37	30,000	2.6	10.6
Time of Shots				
TOS-1	44	24,000	2.3	3.7
TOS-2	40	30,000	2.7	7.4
TOS-3	41	38,000	3.2	12.8
TOS-4	40	27,000	2.5	6.4

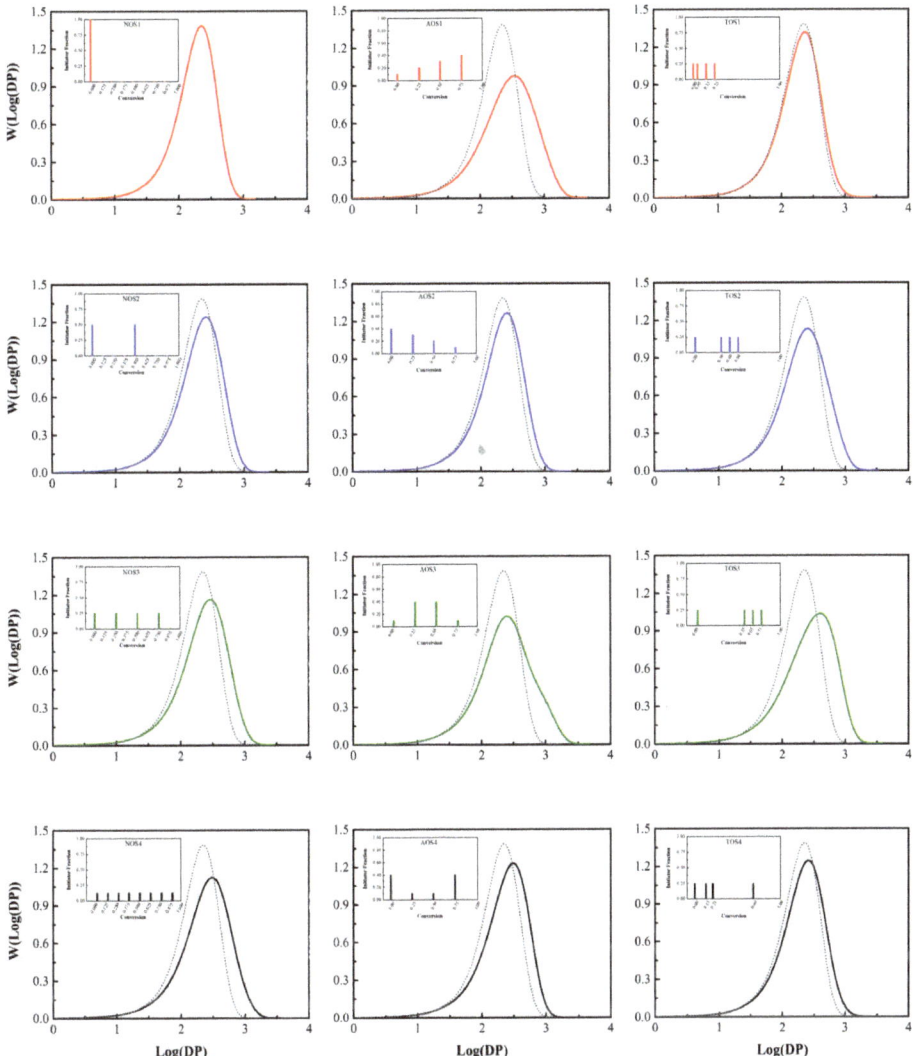

Figure 4. Final molecular weight distributions in semi-batch free radical polymerization of styrene (Case 2) with $[M]_0 = 6.06$ mol L^{-1}, $[S]_0 = 2.54$ mol L^{-1}, $[I]_0 = 0.075$ mol L^{-1}, and a temperature of 138 °C for the three different initiator feeding policies (insets). Solvent mole fraction about 30%; S refers to solvent. Left column: NOS; middle column: AOS; right column: TOS. For comparison, the top left (NOS-1) result is presented in every other plot as a dotted line.

3.4. Free Radical Polymerization of Methyl Methacrylate in Solution

Results of the simulation of methyl methacrylate (MMA) free radical polymerization in solution (Case 3 of Tables 1 and 2) are presented in Table 5 and Figure 5. The simulation captures the fact that the MWDs for MMA polymerization are broader compared with styrene (contrast the values of dispersity in Tables 4 and 5, respectively).

According to the simulation results, increasing the number of shots (NOS) from 1 to 8 (first column of plots in Figure 5, from top to bottom) causes a 75% increase in weight-average molecular weight and a 65% increase in dispersity. The wider distributions (exhibiting a shoulder at high molecular weights) previously observed with the AOS-3 and TOS-3 policies (in Figure 4) are also present in MMA polymerization. Here, by changing the initiator feeding policy from NOS-1 to AOS-1, both weight-average molecular weight and dispersity are increased by a factor of 2. Additionally, the AOS-2 and AOS-4 policies provide a narrower MWD due to the higher fraction of initiator at lower conversions and the lower fraction of initiator at higher conversions. Within these policies, the molecular weight distribution is closer to the reference graph (NOS-1), and the time to achieve complete conversion is reduced. In NOS-4, AOS-1, and TOS-3 feeding policies, due to the higher fraction of initiator concentration at high conversions, a larger amount of initiator remains unreacted in comparison with the other policies. Therefore, the number of decomposed initiator molecules (produced chains) falls, thereby raising molecular weight and widening molecular weight distribution compared with the other policies. Overall, the later the initiator is introduced to the reaction system, the more heterogeneous the microstructure of the produced chains becomes. In fact, a lower fraction of initiator at the beginning of the polymerization generates fewer, but longer chains, thereby broadening chain length distribution, as one can observe from Table 5 and Figure 5.

Finally, in order to 'benchmark' the simulation algorithm and coding developed herein, simulation data for MMA polymerization at four different temperatures were compared with ASPEN modeling data reported elsewhere [48], noting that the ASPEN modeling was already confirmed using experimental data [37]. As can be seen in Figure 6, the conversion-time results for NOS-1 are in complete agreement with the ASPEN data at all temperatures. Furthermore, as expected, raising the temperature increases the rate of monomer consumption.

Table 5. Simulated data of free radical polymerization of methyl methacrylate in solution (Case 3) according to the various initiator feeding policies (NOS, AOS, and TOS of Table 3, with an equal total amount of initiator).

Policy	Reaction Time at Conversion 100% (min)	$\overline{M_w}$ (g/mol)	Đ	Unreacted Initiator (%)
Number of Shots				
NOS-1	1700	19,000	4.2	7.4
NOS-2	1650	24,500	5.3	9.5
NOS-3	1600	30,000	6.2	11.2
NOS-4	1600	33,000	7.0	12.5
Amount of Shots				
AOS-1	1600	40,000	8.4	13.3
AOS-2	1600	25,000	5.4	9.5
AOS-3	1650	37,000	7.9	11.0
AOS-4	1600	27,000	5.7	11.6
Time of Shots				
TOS-1	1700	22,000	4.9	8.1
TOS-2	1650	29,000	6.3	9.9
TOS-3	1600	34,000	7.2	13.1
TOS-4	1630	25,500	5.5	9.7

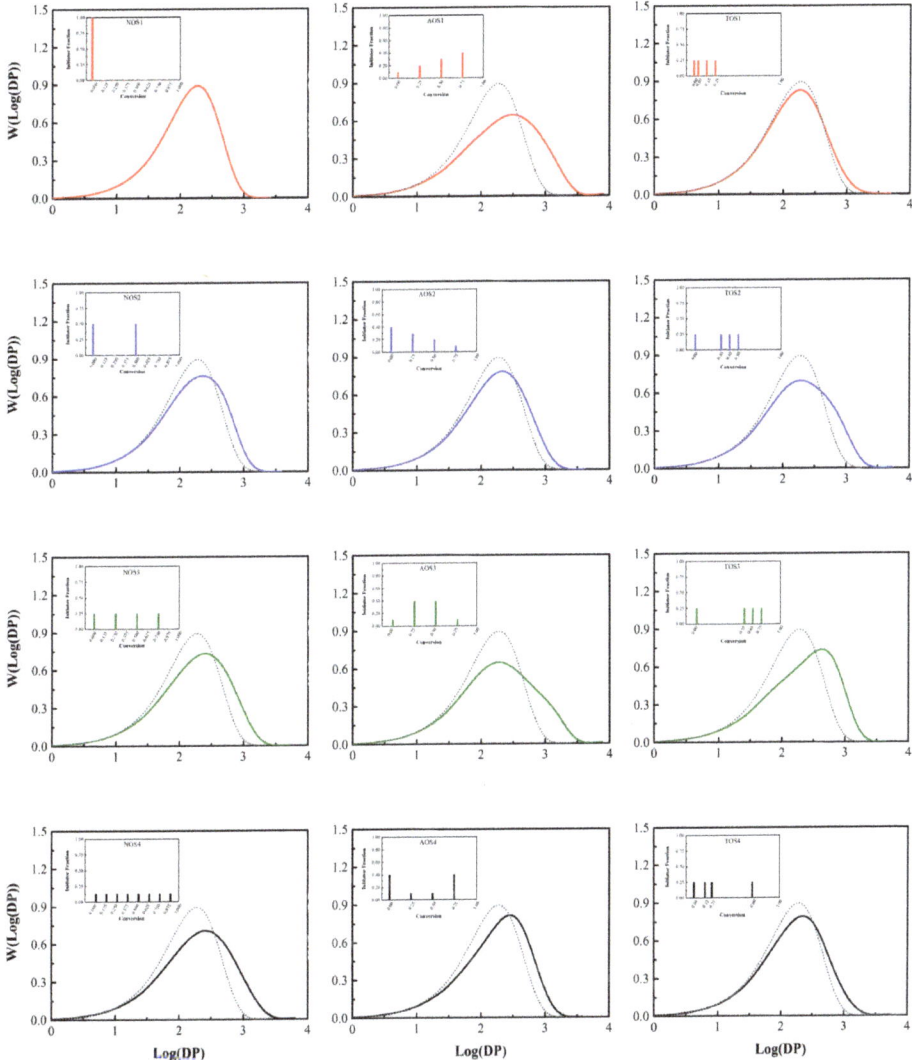

Figure 5. Final molecular weight distributions in semi-batch solution free radical polymerization of methyl methacrylate (Case 3) with $[M]_0$ = 1.0 mol L^{-1}, $[S]_0$ = 3.40 mol L^{-1}, $[I]_0$ = 0.02288 mol L^{-1}, and a temperature of 70 °C for the three different initiator feeding policies (insets). Solvent mole fraction about 75%; S refers to solvent. Left column: NOS; middle column: AOS; right column: TOS. For comparison, the top left (NOS-1) result is presented in every other plot as a dotted line.

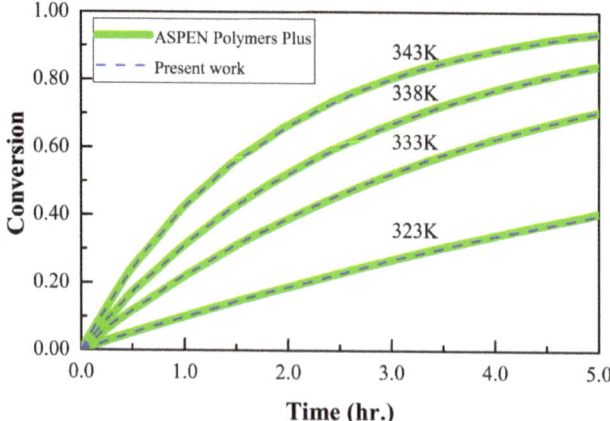

Figure 6. 'Benchmarking' of simulation data by comparison with ASPEN modeling: conversion versus time data for MMA polymerization with NOS-1 initiation policy.

4. Conclusions

This study introduces a new and remarkably simple way of tailoring the rate of polymer synthesis and the obtained MWD. Various initiator feeding policies are defined and evaluated to tune the chain microstructure in a free radical polymerization system. A general free radical polymerization with typical rate constant values (Case 1), free radical polymerization of styrene in solution (Case 2) with a lower solvent fraction, and free radical polymerization of methyl methacrylate in solution (Case 3, at a much higher solvent fraction) are simulated using a Monte Carlo method. By distributing the concentration of initiator at different conversion levels, because of the decrease in the concentration of the initiator at the beginning of the reaction, the molecular weight changes dramatically. Changing initiator feeding policy usually generates a wider molecular weight distribution with a longer tail at higher molecular weights. In addition, the policies in which lower amounts of initiator are injected at the beginning of the reaction produce a large number of long chains and create a tail at the high molecular weight portion of the MWDs. Similar results are achieved by the policies in which most of the initiator is added at high monomer conversions. The policies injecting higher initiator concentrations at lower conversions reach completion much faster. Therefore, there is a compromise between polymerization performance (rate of monomer consumption) and polymer quality, i.e., average molecular weight and molecular weight distribution. As a result, we can exploit an initiator feeding policy tuned for a specific polymerization system to strike a balance between productivity and quality/properties/polymer microstructure.

Author Contributions: A.S., M.N. and Y.M. conceived the original idea (with discussions with G.T.R. and A.P.), designed and executed the computer programming, performed the simulations, analyzed the data, and wrote the original draft of the paper; G.R.T., A.P. and E.V.-L. read and corrected different versions of the manuscript, and provided extra references and discussion points.All authors have read and agreed to the published version of the manuscript.

Funding: This research received no external funding.

Conflicts of Interest: The authors declare no conflict of interest.

References

1. Chiefari, J.; Chong, Y.K.B.K.B.; Ercole, F.; Krstina, J.; Jeffery, J.; Le, T.P.T.P.T.; Mayadunne, R.T.A.T.A.; Meijs, G.F.F.; Moad, C.L.L.; Moad, G.; et al. Living Free-Radical Polymerization by Reversible Addition—Fragmentation Chain Transfer: The RAFT Process. *Macromolecules* **1998**, *31*, 5559–5562. [CrossRef]

2. Noble, B.B.; Coote, M.L. First principles modelling of free-radical polymerisation kinetics. *Int. Rev. Phys. Chem.* **2013**, *32*, 467–513. [CrossRef]
3. Wang, D.; Wang, W.-J.; Li, B.-G.; Zhu, S. Semibatch RAFT polymerization for branched polyacrylamide production: Effect of divinyl monomer feeding policies. *AIChE J.* **2013**, *59*, 1322–1333. [CrossRef]
4. Lowe, A.B.; McCormick, C.L. Homogeneous controlled free radical polymerization in aqueous media. *Aust. J. Chem.* **2002**, *55*, 367–379. [CrossRef]
5. Aydin, M.; Arsu, N.; Yagci, Y.; Jockusch, S.; Turro, N.J. Mechanistic study of photoinitiated free radical polymerization using thioxanthone thioacetic acid as one-component type II photoinitiator. *Macromolecules* **2005**, *38*, 4133–4138. [CrossRef]
6. Crowley, T.J.; Choi, K.Y. Experimental studies on optimal molecular weight distribution control in a batch-free radical polymerization process. *Chem. Eng. Sci.* **1998**, *53*, 2769–2790. [CrossRef]
7. Crowley, T.J.; Choi, K.Y. Discrete optimal control of molecular weight distribution in a batch free radical polymerization process. *Ind. Eng. Chem. Res.* **1997**, *36*, 3676–3684. [CrossRef]
8. Kiparissides, C.; Krallis, A.; Meimaroglou, D.; Pladis, P.; Baltsas, A. From Molecular to Plant-Scale Modeling of Polymerization Processes: A Digital High-Pressure Low-Density Polyethylene Production Paradigm. *Chem. Eng. Technol.* **2010**, *33*, 1754–1766. [CrossRef]
9. Wang, R.; Luo, Y.; Li, B.; Sun, X.; Zhu, S. Design and Control of Copolymer Composition Distribution in Living Radical Polymerization Using Semi-Batch Feeding Policies: A Model Simulation. *Macromol. Theory Simul.* **2006**, *15*, 356–368. [CrossRef]
10. Zhang, M.; Ray, W.H. Modeling of "living" free-radical polymerization processes.I. Batch, semibatch, and continuous tank reactors. *J. Appl. Polym. Sci.* **2002**, *86*, 1630–1662. [CrossRef]
11. Colombani, D. Chain-growth control in free radical polymerization. *Prog. Polym. Sci.* **1997**, *22*, 1649–1720. [CrossRef]
12. Hamielec, A.E.; MacGregor, J.F.; Penlidis, A. Multicomponent free-radical polymerization in batch, semi-batch and continuous reactors. In *Makromolekulare Chemie. Macromolecular Symposia*; Hüthig & Wepf Verlag: Basel, Switzerland, 1987; Volume 10, pp. 521–570.
13. Goto, A.; Fukuda, T. Effects of radical initiator on polymerization rate and polydispersity in nitroxide-controlled free radical polymerization. *Macromolecules* **1997**, *30*, 4272–4277. [CrossRef]
14. Wang, R.; Luo, Y.; Li, B.-G.; Zhu, S. Control of gradient copolymer composition in ATRP using semibatch feeding policy. *AIChE J.* **2007**, *53*, 174–186. [CrossRef]
15. Diaz-Camacho, F.; Lopez-Morales, S.; Vivaldo-Lima, E.; Saldivar-Guerra, E.; Vera-Graziano, R.; Alexandrova, L. Effect of Regime of Addition of Initiator on TEMPO-Mediated Polymerization of Styrene. *Polym. Bull.* **2004**, *52*, 339. [CrossRef]
16. Roa-Luna, M.; Diaz-Barber, M.P.; Vivaldo-Lima, E.; Lona, L.M.F.; McManus, N.T.; Penlidis, A.; Macromol, J. Assessing the Importance of Diffusion-Controlled Effects on Polymerization Rate and Molecular Weight Development in Nitroxide-Mediated Radical Polymerization of Styrene. *Sci. Pure Appl. Chem.* **2007**, *44*, 192. [CrossRef]
17. Mastan, E.; Zhu, S. Method of moments: A versatile tool for deterministic modeling of polymerization kinetics. *Eur. Polym. J.* **2015**, *68*, 139–160. [CrossRef]
18. Mastan, E.; Li, X.; Zhu, S. Modeling and theoretical development in controlled radical polymerization. *Prog. Polym. Sci.* **2015**, *45*, 71–101. [CrossRef]
19. Arzamendi, G.; Asua, J.M. Monomer addition policies for copolymer composition control in semicontinuous emulsion copolymerization. *J. Appl. Polym. Sci.* **1989**, *38*, 2019–2036. [CrossRef]
20. Guiping, C.; Zhongnan, Z.; Huihui, L.; Minghua, Z. Molecular weight distribution of polystyrene produced in a starved feed reactor. *Chinese J. Chem. Eng.* **1999**, *7*, 205–213.
21. Pinto, J.M.; Giudici, R. Optimization of a cocktail of initiators for suspension polymerization of vinyl chloride in batch reactors. *Chem. Eng. Sci.* **2001**, *56*, 1021–1028. [CrossRef]
22. Li, X.; Wang, W.-J.; Weng, F.; Li, B.-G.; Zhu, S. Targeting copolymer composition distribution via model-based monomer feeding policy in semibatch RAFT mini-emulsion copolymerization of styrene and butyl acrylate. *Ind. Eng. Chem. Res.* **2014**, *53*, 7321–7332. [CrossRef]
23. Meimaroglou, D.; Kiparissides, C. Review of Monte Carlo methods for the prediction of distributed molecular and morphological polymer properties. *Ind. Eng. Chem. Res.* **2014**, *53*, 8963–8979. [CrossRef]

24. Brandão, A.L.T.; Soares, J.B.P.; Pinto, J.C.; Alberton, A.L. When polymer reaction engineers play dice: Applications of Monte Carlo models in PRE. *Macromol. React. Eng.* **2015**, *9*, 141–185. [CrossRef]
25. Bruns, W.; Motoc, I.; O'Driscoll, K.F. *Monte Carlo Applications in Polymer Science*; Springer Science & Business Media: Berlin/Heidelberg, Germany, 2012; Volume 27.
26. Khubi-Arani, Z.; Salami-Kalajahi, M.; Najafi, M.; Roghani-Mamaqani, H.; Haddadi-Asl, V.; Ghafelebashi-Zarand, S.M. Simulation of styrene free radical polymerization over bi-functional initiators using Monte Carlo simulation method and comparison with mono-functional initiators. *Polym. Sci. Ser. B* **2010**, *52*, 184–192. [CrossRef]
27. Tobita, H.; Hatanaka, K. Branched structure formation in free radical polymerization of vinyl acetate. *J. Polym. Sci. Part B Polym. Phys.* **1996**, *34*, 671–681. [CrossRef]
28. Prescott, S.W. Chain-length dependence in living/controlled free-radical polymerizations: Physical manifestation and Monte Carlo simulation of reversible transfer agents. *Macromolecules* **2003**, *36*, 9608–9621. [CrossRef]
29. Pintos, E.; Sarmoria, C.; Brandolin, A.; Asteasuain, M. Modeling of RAFT Polymerization Processes Using an Efficient Monte Carlo Algorithm in Julia. *Ind. Eng. Chem. Res.* **2016**, *55*, 8534–8547. [CrossRef]
30. Mohammadi, Y.; Najafi, M.; Haddadi-Asl, V. Comprehensive study of free radical copolymerization using a Monte Carlo simulation method, 1. *Macromol. Theory Simul.* **2005**, *14*, 325–336. [CrossRef]
31. Najafi, M.; Haddadi-Asl, V.; Mohammadi, Y. Application of the Monte Carlo simulation method to the investigation of peculiar free-radical copolymerization reactions: Systems with both reactivity ratios greater than unity (rA > 1 and rB > 1). *J. Appl. Polym. Sci.* **2007**, *106*, 4138. [CrossRef]
32. Najafi, M.; Roghani-Mamaqani, H.; Haddadi-Asl, V.; Salami-Kalajahi, M. A simulation of kinetics and chain length distribution of styrene FRP and ATRP: Chain-length-dependent termination. *Adv. Polym. Technol.* **2011**, *30*, 257–268. [CrossRef]
33. Saeb, M.R.; Mohammadi, Y.; Ahmadi, M.; Khorasani, M.M.; Stadler, F.J. A Monte Carlo-based feeding policy for tailoring microstructure of copolymer chains: Reconsidering the conventional metallocene catalyzed polymerization of α-olefins. *Chem. Eng. J.* **2015**, *274*, 169–180. [CrossRef]
34. Salami-Kalajahi, M.; Najafi, M.; Haddadi-Asl, V. Application of Monte Carlo simulation method to polymerization kinetics over Ziegler–Natta catalysts. *Int. J. Chem. Kinet.* **2009**, *41*, 45–56. [CrossRef]
35. Najafi, M.; Roghani-Mamaqani, H.; Salami-Kalajahi, M.; Haddadi-Asl, V. A comprehensive Monte Carlo simulation of styrene atom transfer radical polymerization. *Chinese J. Polym. Sci.* **2010**, *28*, 483–497. [CrossRef]
36. Fu, Y.; Cunningham, M.F.; Hutchinson, R.A. Modeling of Nitroxide-Mediated Semibatch Radical Polymerization. *Macromol. React. Eng.* **2007**, *1*, 243–252. [CrossRef]
37. Ellis, M.F.; Taylor, T.W.; Gonzalez, V.; Jensen, K.F. Estimation of the molecular weight distribution in batch polymerization. *AIChE J.* **1988**, *34*, 1341–1353. [CrossRef]
38. Buback, M.; Sandmann, J. Pressure and temperature dependence of the decomposition rate of aliphatic tert-butyl peroxyesters. *Zeitschrift für Phys. Chemie* **2000**, *214*, 583. [CrossRef]
39. Buback, M.; Wittkowski, L. Acid-Induced Decomposition of Di-tert-butyl Peroxide in n-Heptane Solution up to High Temperatures and Pressures. *Zeitschrift für Phys. Chemie* **1999**, *210*, 61–81. [CrossRef]
40. Hui, A.W.; Hamielec, A.E. Thermal polymerization of styrene at high conversions and temperatures. An experimental study. *J. Appl. Polym. Sci.* **1972**, *16*, 749–769. [CrossRef]
41. Buback, M.; Gilbert, R.G.; Hutchinson, R.A.; Klumperman, B.; Kuchta, F.-D.; Manders, B.G.; O'Driscoll, K.F.; Russell, G.T.; Schweer, J. Critically evaluated rate coefficients for free-radical polymerization, 1. Propagation rate coefficient for styrene. *Macromol. Chem. Phys.* **1995**, *196*, 3267–3280. [CrossRef]
42. Buback, M.; Kuchta, F.-D. Variation of the propagation rate coefficient with pressure and temperature in the free-radical bulk polymerization of styrene. *Macromol. Chem. Phys.* **1995**, *196*, 1887–1898. [CrossRef]
43. Brandrup, J.; Immergut, E.H.; Grulke, E.A. *Polymer Handbook*, 4th ed.; Wiley: New York, NY, USA, 1999; ISBN 0471166286.
44. Chen, C.-Y.; Wu, Z.-Z.; Kuo, J.-F. Determination of the mode of termination in free-radical copolymerization. *Polym. Eng. Sci.* **1987**, *27*, 553–557. [CrossRef]
45. Schmidt, A.D.; Ray, W.H. The dynamic behavior of continuous polymerization reactors—I: Isothermal solution polymerization in a CSTR. *Chem. Eng. Sci.* **1981**, *36*, 1401–1410. [CrossRef]

46. Gillespie, D.T. Exact stochastic simulation of coupled chemical reactions. *J. Phys. Chem.* **1977**, *81*, 2340–2361. [CrossRef]
47. Matsumoto, M.; Nishimura, T. Mersenne twister: A 623-dimensionally equidistributed uniform pseudo-random number generator. *ACM Trans. Model. Comput. Simul.* **1998**, *8*, 3–30. [CrossRef]
48. AspenTech Methyl Methacrylate Polymerization in Ethyl Acetate. Aspen Polymer Plus User Guide. Examples Application Case B (3rd ed.). 1981. Available online: https://sites.chemengr.ucsb.edu/~ceweb/courses/che184b/aspenplus/PolymersPlusUserGuideVolume2.pdf (accessed on 10 September 2020).

Publisher's Note: MDPI stays neutral with regard to jurisdictional claims in published maps and institutional affiliations.

© 2020 by the authors. Licensee MDPI, Basel, Switzerland. This article is an open access article distributed under the terms and conditions of the Creative Commons Attribution (CC BY) license (http://creativecommons.org/licenses/by/4.0/).

Article

Solution Polymerization of Acrylic Acid Initiated by Redox Couple Na-PS/Na-MBS: Kinetic Model and Transition to Continuous Process

Federico Florit, Paola Rodrigues Bassam, Alberto Cesana and Giuseppe Storti *

Politecnico di Milano, Dipartimento di Chimica, Materiali e Ingegneria Chimica "G. Natta", Piazza Leonardo da Vinci, 32, 20133 Milano, Italy; federico.florit@polimi.it (F.F.); paola.rodrigues@mail.polimi.it (P.R.B.); alberto.cesana@polimi.it (A.C.)
* Correspondence: giuseppe.storti@polimi.it

Received: 28 May 2020; Accepted: 14 July 2020; Published: 16 July 2020

Abstract: This work aims at modeling in detail the polymerization of non-ionized acrylic acid in aqueous solution. The population balances required to evaluate the main average properties of molecular weight were solved by the method of moments. The polymerization process considered is initiated by a persulfate/metabisulfate redox couple and, in particular, the kinetic scheme considers the possible formation of mid-chain radicals and transfer reactions. The proposed model is validated using experimental data collected in a laboratory-scale discontinuous reactor. The developed kinetic model is then used to intensify the discontinuous process by shifting it to a continuous one based on a tubular reactor with intermediate feeds. One of the experimental runs is selected to show how the proposed model can be used to assess the transition from batch to continuous process and allow faster scale-up to industrial scale using a literature approach.

Keywords: Poly(acrylic acid); free-radical polymerization; reaction model; process intensification; semi-batch to continuous

1. Introduction

Poly(acrylic acid) (PAA) is a widely produced polymer with a growing, global annual production of 1.58×10^9 kg as of 2008 [1] and with applications in many industrial sectors. The application is determined according to the molecular weight of the produced polymer: low molecular mass (less than 20 kDa) PAA is used as sequestrant, mid-low mass (20–80 kDa) PAA is adopted in paints, mid-high mass (0.1–1 MDa) PAA is useful in the textile and paper industry, and high weight (larger than 1 MDa) PAA is a flocculating agent and absorbent [1]. The accurate control of the average molecular weight and its distribution are thus critical for efficient production.

The synthesis of PAA is almost exclusively run in aqueous solutions through free-radical polymerization of acrylic-acid (AA) at 40–90 °C in batch (BR) or semi-batch (SBR) chemical reactors. The reaction is highly exothermic (reaction heat of 63 kJ mol^{-1} [2]) and fast, posing a relevant hazard for its production in discontinuous BR systems. The SBR is then preferred as a more efficient thermal control can be achieved with respect to its BR counterpart. A SBR starved process is generally adopted to achieve such control in a monomer-depleted environment. This way AA is consumed at a faster rate than its dosing and no dangerous accumulation of monomer is taking place. On the other hand, this kind of process may require long reaction times and proper disposal of the high amount of water used.

A way to enhance the performance of the PAA production is to shift the discontinuous process to a continuous one by means, for example, of tubular reactors in the general framework of process intensification [3]. Tubular reactors are more efficient in the thermal control with respect to both BRs and SBRs [4] thanks to the higher heat-exchange-area-to-volume ratio. This way, more severe conditions can be used with reduction of both reaction times and downstream utilities. Several methods were developed to shift polymerization processes from batch to continuous, by means of modular reactors (given a particular process) [5,6] or tubular reactors with distributed and localized side injections (for any homogeneous system regardless the chemical kinetics) [7,8].

Typically, the process conditions are designed according to the selected initiating system and the desired product quality (its molecular weight distribution). Commonly used initiators are azo compunds, e.g., azobis-isobutyronitrile (AIBN) and 2,2′-azobis-(2-methyl propionamidine) dihydrochloride (V-50), peroxides, e.g., benzoyl peroxide, and redox initiators, such as persulfate/metabisulfate (PS/MBS) couples [9]. A detailed description of the chemical transformations happening in the reactor is thus paramount to develop processes (both discontinuous and continuous) able to produce the desired product in the most efficient way. A large body of work was carried out (in discontinuous reactors) to accurately determine the reaction rates of AA polymerization. Much effort was spent towards the description of propagation reactions, taking into account many effects such as the initial presence of monomer in the reaction environment [10,11], pH, and the degree of AA dissociation [12,13]. AA radicals also undergo backbiting, leading to the formation of mid-chain radicals (MCR) from which branching occurs [14]. According to the initiating system used, other reactions can occur such as radical transfer to the co-initiator in a redox system [15].

The aim of this work is the development of a comprehensive chemical reaction model for the polymerization of fully-undissociated AA in aqueous solution using the couple PS/MBS as redox initiator. The model is validated with experimental measurements on the discontinuous system. The developed model is then used for process intensification, with emphasis on the transition from discontinuous to continuous production processes.

2. Materials and Methods

2.1. Experimental Tests

2.1.1. Chemicals and Analytical Technique

The experimental tests were run using acrylic acid (99 %), sodium persulfate (Na-PS, $\geq$98 %), sodium metabisulfate (Na-MBS, $\geq$99 %), hydroquinone ($\geq$99 %), sodium nitrate ($\geq$99 %), monobasic sodium phosphate monohydrate ($\geq$98 %), and sodium azide ($\geq$99 %). All chemicals were purchased from Sigma-Aldrich and were used as received without further purification.

For each reaction condition several samples were taken over time. The reaction was inhibited by adding the sample to cold hydroquinone. The samples were then analyzed using gel permeation chromatography (GPC) for the determination of the polymer molecular weight distributions. GPC measurements were carried out at 30 °C using a pH 7 buffer mixture of aqueous solutions of sodium nitrate (0.2 M), monobasic sodium phosphate (0.01 M), and sodium azide (3.1 mM) as eluent with a flowrate of 1 mL min^{-1}. The calibration of the instrument was made using PAA standards with certified peak molecular weights from 1250 to 392,600 g mol^{-1} (all details as Supplementary Materials), thus enabling especially accurate molecular weight evaluations in the range of interest.

Residual monomer was detected by high-performance liquid chromatography (HPLC) equipped with UV detector (wavelength 205 nm) and a column suitable for organic acids. In all cases, the eluent was an aqueous solution of sulfuric acid 0.05 M and the flowrate 0.5 mL min^{-1}.

2.1.2. Discontinuous Reactor

The laboratory SBR is a jacketed glass reactor with a maximum volume of 1 L, equipped with mechanical stirring and reflux condenser. The reagents were fed using three Metrohm 876 Dosimat plus dosing units. PS and MBS are fed into the reactor as aqueous solutions, while the monomer is fed as a pure stream. A schematic representation of the experimental setup is reported in Figure 1.

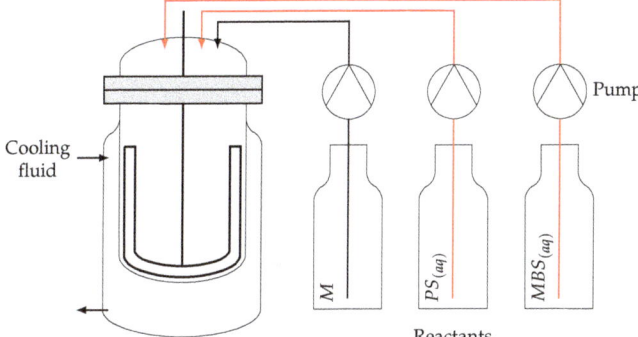

Figure 1. Laboratory SBR scheme. Jacketed reactor equipped with three pumps for the feed of pure monomer and aqueous solutions of PS and MBS.

Species are fed into the reactor with a constant feedrate and different feed durations. In particular, PS and MBS are fed together from the beginning of the process into the reactor, which was initially charged with pure water. After a constant delay, Δt_M^0, the monomer feed is started. Once the entire mass of monomer, m_M^F, is fed over up to the dosing time t_M^F, the PS and MBS aqueous solutions are fed for some more time, Δt_{PS}^F and Δt_{MBS}^F respectively. The total quantities of salts fed into the reactor are m_{PS}^F and m_{MBS}^F for PS and MBS, respectively. Water is also fed into the reactor, as the salts are diluted in aqueous solution; the quantity of water used to dilute the salts is equal to $m_{W,PS}^F$ and $m_{W,MBS}^F$ for PS and MBS, respectively. All species were fed into the reactor at a constant rate. The experimental values of the aforementioned quantities can be found in Table 1 with the exception of the Δt mentioned above. Such values cannot be disclosed because they are proprietary data.

Table 1. Settings of the SBR tests and available number of specimen over time for each set.

Test	m_M^F [g]	m_{PS}^F [g]	m_{MBS}^F [g]	$m_{W,PS}^F$ [g]	$m_{W,MBS}^F$ [g]	m_W^0 [g]	T [°C]	t_M^F [h]	N° Data
SBR1	280	7.8	44.44	37.8	66.66	137.8	90	5	6
SBR2	280	7.8	44.44	37.8	66.66	137.8	90	0.5	4
SBR3	252	7.0	40.00	34.0	60.00	124.0	75	5	6
SBR4	280	7.8	22.24	37.8	33.36	137.8	70	5	6
SBR5	280	7.8	44.44	37.8	66.66	137.8	50	5	6
SBR6	280	7.8	22.24	37.8	33.36	137.8	90	5	6
SBR7	28	0.78	4.444	3.78	6.666	231.8	90	5	6
SBR8	315	8.8	50.00	42.5	75.00	155.0	90	2.5	6
SBR9	252	7.0	40.00	34.0	60.00	124.0	90	1.5	6
SBR10	280	7.8	44.44	37.8	66.66	137.8	90	5	6
SBR11	280	7.8	44.44	37.8	66.66	137.8	90	2.5	6
SBR12	280	7.8	44.44	37.8	66.66	137.8	90	1	6
SBR13	280	7.8	11.12	37.8	16.68	137.8	90	5	6
SBR14	28	0.78	0.00	3.78	0.00	231.8	90	5	6
SBR15	280	7.8	0.00	37.8	0.00	275.6	70	5	4
SBR16	280	7.8	11.12	37.8	16.68	137.8	70	5	6
SBR17	280	7.8	11.12	37.8	16.68	137.8	60	5	6

3. Model Development

3.1. Kinetic Scheme

The considered kinetic scheme extends the basic structure of a polymerization process (initiation, propagation, and termination by both combination and disproportionantion) by adding the possibility of backbiting (formation of MCRs; such radicals are identified by the superscript s while regular terminal radicals by the superscript t) and transfer reactions. The monomer is assumed to be non-ionized, as the pK_a of AA is equal to 4.25 at ambient temperature [16], thus corresponding to a degree of ionization lower than 1%. All reactions are treated as irreversible, elementary reactions, thus their reaction rate is given by a rate constant multiplied by the product of the molar concentrations (indicated in square brackets in the following) of the reactants involved.

The initiation step is due to a redox reaction between PS and MBS [17,18] as well as to the thermal decomposition reaction of PS [19], both reactions producing two radicals:

$$\underbrace{S_2O_8^{2-}}_{PS} + \underbrace{S_2O_3^{2-}}_{MBS} \rightarrow \underbrace{SO_4^{\bullet -} + {}^{\bullet}S_2O_3^{-}}_{2R_0} + SO_4^{2-} \tag{1}$$

$$PS \rightarrow \underbrace{2SO_4^{\bullet -}}_{2R_0} \tag{2}$$

No distinction was done between the radicals formed in the initiation step, by assuming they have the same reactivity towards all other species, and they will be simply indicated as R_0.

Na-MBS is a known chain-transfer agent [15], thus the kinetic scheme also considers the transfer reactions to monomer and MBS. This is a particular reaction arising from the use of MBS and is generally not present in classical kinetic schemes. Transfer to the monomer can also occur [14], while the transfer reaction to dead chains (polymers) was observed to be negligible under the investigated conditions and it is generally not accounted for [14]. Transfer to water was considered negligible, compared to other transfer mechanisms [10].

Termination reactions can occur between any kind of radical, both of the same type (ss, tt) or of different types (st). The distinction between termination by combination or by disproportionation is quantified with a proper constant κ according to the type of radicals involved in the termination reaction.

The complete set of chemical reactions is reported in Table 2 along with each reaction rate computed from the molar concentration of species (reported in square brackets). The monomer is indicated as M, terminal radicals of length n as R_n^s, MCRs of length n as R_n^t, while polymers of length n as P_n (regardless their branching structure). It is assumed terminal radicals and MCR have different reactivity as reported in Table 2.

Table 2. Chemical reaction mechanism and reaction rates for the dissociated AA polymerization in water initiated by PS/MBS redox couple.

Name	Reaction	Rate
Initiation	$PS + MBS \rightarrow 2R_0 + SO_4^{2-}$	$k_1[PS][MBS]$
	$PS \rightarrow 2R_0$	$k_d[PS]$, efficiency f
Propagation	$R_n^s + M \rightarrow R_{n+1}^s$	$k_p^s[M][R_n^s]$
	$R_n^t + M \rightarrow R_{n+1}^s$	$k_p^t[M][R_n^t]$
Backbiting	$R_n^s \rightarrow R_n^t$	$k_{bb}[R_n^s]$
Transfer to monomer	$R_n^s + M \rightarrow P_n + R_1^s$	$k_{trM}^s[M][R_n^s]$
	$R_n^t + M \rightarrow P_n + R_1^s$	$k_{trM}^t[M][R_n^t]$
Transfer to MBS	$R_n^s + MBS \rightarrow P_n + R_0$	$k_{trMBS}^s[MBS][R_n^s]$
	$R_n^t + MBS \rightarrow P_n + R_0$	$k_{trMBS}^t[MBS][R_n^t]$
Termination by combination	$R_n^s + R_m^s \rightarrow P_{n+m}$	$k_t^{ss}(1-\kappa^{ss})[R_n^s][R_m^s]$
	$R_n^s + R_m^t \rightarrow P_{n+m}$	$k_t^{st}(1-\kappa^{st})[R_n^s][R_m^t]$
	$R_n^t + R_m^t \rightarrow P_{n+m}$	$k_t^{tt}(1-\kappa^{tt})[R_n^t][R_m^t]$
Termination by disproportionation	$R_n^s + R_m^s \rightarrow P_n + P_m$	$k_t^{ss}\kappa^{ss}[R_n^s][R_m^s]$
	$R_n^s + R_m^t \rightarrow P_n + P_m$	$k_t^{st}\kappa^{st}[R_n^s][R_m^t]$
	$R_n^t + R_m^t \rightarrow P_n + P_m$	$k_t^{tt}\kappa^{tt}[R_n^t][R_m^t]$

3.2. Governing Equations

It is assumed that all reactors are isothermal, then characterized by constant and uniform temperature value. Mixing is assumed to be perfect in the SBR while plug-flow behavior is assumed in the tubular reactor. AA monomer will be indicated as M, water as W, Na-PS as PS, Na-MBS as MBS, and polymer (PAA) as P.

The model for the SBR is initially developed and applied to validate the proposed kinetic scheme using experimental data. Then, the model for the continuous system is presented, with reference to tubular reactors. Finally, a kinetics-free procedure, which details can be found elsewhere [7,8], is applied to adapt the recipe already effective in SBR to a continuous production process based on a series of tubular reactors with intermediate feed streams.

3.2.1. SBR Model

For every species i it is possible to write a molar conservation equation:

$$\frac{dn_i}{dt} = F_i + R_i V \tag{3}$$

for $i = 1 \ldots NC$, where NC is the number of components (species), n_i the number of moles of species i in the SBR at a given time, t, F_i the (possibly-non-constant) molar feed-rate of species i at a given time, R_i the production rate of species i as a function of temperature, T, and composition, and V the reaction volume at a given time. This equation is equipped with the initial condition:

$$n_i(0) = \frac{m_i^0}{MW_i} \tag{4}$$

where m_i^0 is the mass of species i which is initially charged into the reactor and MW_i is species i molecular weight. For the test run in this work, only water is initially charged into the reactor, thus $m_i^0 = 0$ for all species except water.

The molar flowrate changes in time according to:

$$F_M = \frac{m_M^F}{MW_M(t_M^F - \Delta t_M^0)} H(t - \Delta t_M^0) H(t_M^F - t) \tag{5}$$

$$F_{PS} = \frac{m_{PS}^F}{MW_{PS}(t_M^F + \Delta t_{PS}^F)} H(t_M^F + \Delta t_{PS}^F - t) \tag{6}$$

$$F_{MBS} = \frac{m_{MBS}^F}{MW_{MBS}(t_M^F + \Delta t_{MBS}^F)} H(t_M^F + \Delta t_{MBS}^F - t) \tag{7}$$

$$F_W = \frac{m_{W,PS}^F}{MW_W(t_M^F + \Delta t_{PS}^F)} H(t_M^F + \Delta t_{PS}^F - t) + \frac{m_{W,MBS}^F}{MW_W(t_M^F + \Delta t_{MBS}^F)} H(t_M^F + \Delta t_{MBS}^F - t) \tag{8}$$

where $H(t)$ the Heaviside function. These functions take into account all the delays happening in the experimental runs. All other species (namely, radicals and polymers) have a null feedrate. The species production rate is given by:

$$R_i = \sum_n^{NR} \nu_{in} r_n \tag{9}$$

where NR is the number of reactions happening in the system, ν_{in} the stoichiometric coefficient of species i in reaction n, and r_n the n-th reaction rate, as from Table 2.

Under the assumption of ideal mixtures, species volumes can be considered to be additive, meaning:

$$V = \sum_{i}^{NC} \frac{m_i}{\rho_i} \approx \frac{m_W + m_{PS} + m_{MBS}}{\rho_W} + \frac{m_M}{\rho_M} + \frac{m_P}{\rho_P} \qquad (10)$$

where $m_i = n_i MW_i$ is the mass of species i at a given time, and ρ_i the density of pure species i at a given temperature. The pure species density is provided in Table 3 as a polynomial function of temperature.

Table 3. Pure species density as a function of temperature [14,20]. Temperature in [°C].

Species	Density [g cm^{-3}]
W	$0.9999 + 2.3109 \times 10^{-5} T - 5.448{,}07 \times 10^{-6} T^2$
M	$1.0731 - 1.0826 \times 10^{-3} T - 7.2379 \times 10^{-7} T^2$
P	$1.7 - 6.0 \times 10^{-4} T$

To solve the polymer population balances in an effective way, the method of moments (MoM) is used. By definition, the moments of a generic order k are:

$$\lambda_k^s = \sum_{j=1}^{\infty} j^k n_{R_j^s} \qquad (11)$$

$$\lambda_k^t = \sum_{j=1}^{\infty} j^k n_{R_j^t} \qquad (12)$$

$$\mu_k = \sum_{j=1}^{\infty} j^k n_{P_j} \qquad (13)$$

respectively, the k-th order moment of the secondary radicals, tertiary radicals, and dead polymers. $n_{R_j^s}$ is the number of moles of the secondary radicals of length j, $n_{R_j^t}$ for tertiary radicals, and n_{P_j} for dead chains. The moments behave as pseudo-species, representing the ensemble of radicals and polymers, thus they change in time as reaction proceeds.

This way the main average properties of the molecular weight distribution can be easily computed. In particular, the number-average chain length (number-average degree of polymerization) is:

$$DP_n = \frac{\sum_{j=1}^{\infty} j n_{P_j}}{\sum_{j=1}^{\infty} n_{P_j}} = \frac{\mu_1}{\mu_0} \qquad (14)$$

the weight-average chain length (weight-average degree of polymerization) is:

$$DP_w = \frac{\sum_{j=1}^{\infty} j^2 n_{P_j}}{\sum_{j=1}^{\infty} j n_{P_j}} = \frac{\mu_2}{\mu_1} \qquad (15)$$

and the polydispersity is:

$$Đ = \frac{DP_w}{DP_n} = \frac{\mu_2 \mu_0}{\mu_1^2} \qquad (16)$$

Furthermore, the average radical chain length can be computed as:

$$DP_n^R = \frac{\lambda_1^s + \lambda_1^t}{\lambda_0^s + \lambda_0^t} \qquad (17)$$

The minimum number of moments to be computed is therefore three for the polymer (μ_0, μ_1, μ_2) and two for each type of radicals ($\lambda_0^s, \lambda_0^t, \lambda_1^s, \lambda_1^t$). If of interest, additional moments could be used to get more detailed information on the molecular weight distribution.

Using the definitions of moments, it is possible to write equations describing their change in time, exploiting the fact that for radicals and polymers no convective term is present (the feedrate is null). From Equation (3):

$$\frac{d\lambda_k^s}{dt} = \sum_{j=1}^{\infty} j^k \frac{dn_{R_j^s}}{dt} = \sum_{j=1}^{\infty} j^k R_{R_j^s} V = R_{\lambda_k^s} V \qquad (18)$$

$$\frac{d\lambda_k^t}{dt} = \sum_{j=1}^{\infty} j^k \frac{dn_{R_j^t}}{dt} = \sum_{j=1}^{\infty} j^k R_{R_j^t} V = R_{\lambda_k^t} V \qquad (19)$$

$$\frac{d\mu_k}{dt} = \sum_{j=1}^{\infty} j^k \frac{dn_{P_j}}{dt} = \sum_{j=1}^{\infty} j^k R_{P_j} V = R_{\mu_k} V \qquad (20)$$

Furthermore, as no radical nor polymer is present at the beginning of the reaction, the initial quantity of all moments is null, meaning $\lambda_k^s(0) = \lambda_k^t(0) = \mu_k(0) = 0$.

Using the reaction rates reported in Table 2, one can write the production rates of the molecular species, as well as those of the moments:

$$R_M = -k_p^s[M][R_0] - (k_p^s + k_{trM}^s)[M][\lambda_0^s] - (k_p^t + k_{trM}^t)[M][\lambda_0^t] \qquad (21)$$

$$R_W = 0 \qquad (22)$$

$$R_{PS} = -k_1[PS][MBS] - k_d[PS] \qquad (23)$$

$$R_{MBS} = -k_1[PS][MBS] - (k_{trMBS}^s[\lambda_0^s] + k_{trMBS}^t[\lambda_0^t])[MBS] \qquad (24)$$

$$R_{\lambda_k^s} = k_p^s[M][R_0] + k_p^s[M]\sum_{l=0}^{k}\binom{k}{l}[\lambda_l^s] - k_p^s[M][\lambda_k^s] - k_{bb}[\lambda_k^s] + k_p^t[M]\sum_{l=0}^{k}\binom{k}{l}[\lambda_l^t] + \\
+ k_{trM}^s[M]([\lambda_0^s] - [\lambda_k^s]) + k_{trM}^t[M][\lambda_0^t] - k_{trMBS}^s[\lambda_k^s][MBS] - k_t^{ss}[\lambda_0^s][\lambda_k^s] - k_t^{st}[\lambda_k^s][\lambda_0^t] \qquad (25)$$

$$R_{\lambda_k^t} = k_{bb}[\lambda_k^s] - k_p^t[M][\lambda_k^t] - k_{trMBS}^t[\lambda_k^t][MBS] - k_t^{st}[\lambda_0^s][\lambda_k^t] - k_t^{tt}[\lambda_0^t][\lambda_k^t] \qquad (26)$$

$$R_{\mu_k} = k_{trM}^s[M][\lambda_k^s] + k_{trM}^t[M][\lambda_k^t] + \left(k_{trMBS}^s[\lambda_k^s] + k_{trMBS}^t[\lambda_k^t]\right)[MBS] + \\
+ \frac{1}{2}\left(k_t^{ss}(1-\kappa^{ss})\sum_{l=0}^{k}\binom{k}{l}[\lambda_l^s][\lambda_{k-l}^s] + k_t^{st}(1-\kappa^{st})\sum_{l=0}^{k}\binom{k}{l}[\lambda_l^s][\lambda_{k-l}^t] + \\
+ k_t^{tt}(1-\kappa^{tt})\sum_{l=0}^{k}\binom{k}{l}[\lambda_l^t][\lambda_{k-l}^t]\right) + k_t^{ss}\kappa^{ss}[\lambda_0^s][\lambda_k^s] + k_t^{st}\kappa^{st}\left([\lambda_k^s][\lambda_0^t] + [\lambda_0^s][\lambda_k^t]\right) + \\
+ k_t^{tt}\kappa^{tt}[\lambda_0^t][\lambda_k^t] \qquad (27)$$

The molar concentration of species i is given by the ratio n_i/V, while the concentration of moments is given by the ratio between moment and volume.

Assuming quasi-steady-state conditions for the radicals produced during initiation, the consumption rate is approximated by the propagation rate only (being this term much higher than any other consumption term), meaning:

$$R_{R_0} \approx 2fk_d[PS] + 2k_1[PS][MBS] + (k_{trMBS}^s[\lambda_0^s] + k_{trMBS}^t[\lambda_0^t])[MBS] - k_p^S[M][R_0] = 0 \qquad (28)$$

from which it is possible to compute analytically the concentration of radicals (of length zero):

$$[R_0] = \frac{2fk_d[PS] + 2k_1[PS][MBS] + (k_{trMBS}^s[\lambda_0^s] + k_{trMBS}^t[\lambda_0^t])[MBS]}{k_p^S[M]} \qquad (29)$$

The mass fraction of species i can be expressed as:

$$\omega_i = \frac{m_i}{\sum_i^{NC}\left(m_i^0 + MW_i \int_0^t F_i(\tau)\,d\tau\right)} \quad (30)$$

where τ is an auxiliary variable of integration. The initial mass fraction of monomer is $\omega_M^0 = m_M^0/\sum_i^{NC} m_i^0$.

The monomer overall conversion at a given time is defined, for the SBR, as

$$\chi = \frac{m_M^0 + MW_M \int_0^t F_M(\tau)\,d\tau - MW_M n_M}{m_M^0 + m_M^F} \quad (31)$$

while the instantaneous conversion is

$$\mathscr{I} = 1 - \frac{MW_M n_M}{m_M^0 + MW_M \int_0^t F_M(\tau)\,d\tau} \quad (32)$$

3.2.2. Tubular Reactor Model

In a similar manner, a tubular reactor model can be written under the assumption of uniform distribution of species over the cross section of the reactor. This condition can be fulfilled in different ways, e.g., by using static mixers. Accordingly the tubular reactor can be modeled as a plug-flow reactor (PFR) [21]. A series of PFRs with intermediate feedings (depicted in Figure 2) can be modeled by noting that the inlet stream to one reactor is equal to the outlet stream from the previous one with the addition of the intermediate stream. The model is therefore able to describe the single PFR and proper boundary conditions are used to link the PFRs in the series.

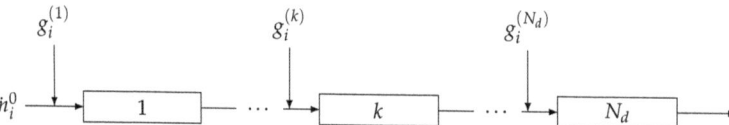

Figure 2. Series of N_d PFRs with intermediate injections.

For each species i the molar conservation equation along the axial coordinate of the reactor, x, in terms of molar flowrates reads:

$$\frac{d\dot{n}_i}{dx} = R_i \mathscr{A} \quad (33)$$

where $\dot{n}_i$ is species i molar flowrate at a given axial position, and $\mathscr{A}$ is the cross-sectional area of the reactor (assumed uniform along the axial coordinate). R_i is computed as from Equation (9). The boundary condition at the entrance of the reactor is:

$$\dot{n}_i(0) = \frac{\dot{m}_i^{prev} + g_i}{MW_i} \quad (34)$$

where $\dot{m}_i^{prev}$ is the outlet mass flowrate of species i from the previous reactor and g_i the intermediate feedrate of species i pertaining to the considered PFR of the series (note that g_i can be non-uniform along the series of PFRs). For the first reactor in the series $\dot{m}_i^{prev} = \dot{m}_i^0$ is the mass flowrate of species i fed at the entrance of the series of PFRs. The values of the inlet and intermediate feedrates, along with their composition, are computed through a kinetics free procedure according to the number of reactors in the series, N_d, starting from a given SBR recipe [7,8]. All reactors in the series are assumed to have the same geometrical parameters (length and cross-sectional area).

Under the assumption of ideal mixtures, the volumetric flowrate can be computed in a given position as:

$$\dot{V} = \sum_i^{NC} \frac{\dot{m}_i}{\rho_i} \approx \frac{\dot{m}_W + \dot{m}_{PS} + \dot{m}_{MBS}}{\rho_W} + \frac{\dot{m}_M}{\rho_M} + \frac{\dot{m}_P}{\rho_P} \qquad (35)$$

where $\dot{m}_i = \dot{n}_i MW_i$ is the mass flowrate of species i in the same position.

The MoM is again applied with the following definitions:

$$\lambda_k^s = \sum_{j=1}^{\infty} j^k \dot{n}_{R_j^s} \qquad (36)$$

$$\lambda_k^t = \sum_{j=1}^{\infty} j^k \dot{n}_{R_j^t} \qquad (37)$$

$$\mu_k = \sum_{j=1}^{\infty} j^k \dot{n}_{P_j} \qquad (38)$$

and Equations (14)–(17) still apply. The change of the moments along the axial coordinate is described by:

$$\frac{d\lambda_k^s}{dx} = \sum_{j=1}^{\infty} j^k \frac{d\dot{n}_{R_j^s}}{dx} = \sum_{j=1}^{\infty} j^k R_{R_j^s} \mathscr{A} = R_{\lambda_k^s} \mathscr{A} \qquad (39)$$

$$\frac{d\lambda_k^t}{dx} = \sum_{j=1}^{\infty} j^k \frac{d\dot{n}_{R_j^t}}{dx} = \sum_{j=1}^{\infty} j^k R_{R_j^t} \mathscr{A} = R_{\lambda_k^t} \mathscr{A} \qquad (40)$$

$$\frac{d\mu_k}{dx} = \sum_{j=1}^{\infty} j^k \frac{d\dot{n}_{P_j}}{dx} = \sum_{j=1}^{\infty} j^k R_{P_j} \mathscr{A} = R_{\mu_k} \mathscr{A} \qquad (41)$$

with the boundary conditions $\lambda_k^s(0) = \lambda_k^{s,prev}$, $\lambda_k^t(0) = \lambda_k^{t,prev}$, and $\mu_k(0) = \mu_k^{prev}$, as the outlet of a reactor is entirely fed to the following in the series and no radical nor polymer is assumed to be fed through the intermediate stream. The values used for the boundary conditions at the first reactor of the series are $\lambda_k^{s,prev} = \lambda_k^{t,prev} = \mu_k^{prev} = 0$ being that neither radicals nor polymer are assumed to be present in the inlet stream.

The production rates of all species and moments are computed in the same manner of the SBR using Equations (21)–(29), with the definition of molar concentration given by $\dot{n}_i/\dot{V}$ for the different species, and the ratio between the moment and the volumetric flowrate for the moment concentration.

The i-th species mass fraction is given by:

$$\omega_i = \frac{\dot{m}_i}{\sum_i^{NC}(\dot{m}_i^{prev} + g_i)} \qquad (42)$$

and the inlet monomer mass fraction to the series of PFRs with intermediate feed streams is defined as $\omega_M^0 = \dot{m}_M^0 / \sum_i^{NC} \dot{m}_i^0$.

The monomer overall conversion for the series of PFRs at a given axial position is equal to:

$$\chi = \frac{\dot{m}_M^0 + G_M - MW_M \dot{n}_M}{\dot{m}_M^0 + G_M^{tot}} \qquad (43)$$

where G_M is the sum of the monomer intermediate feedrates, g_M, from the first up to the considered reactor of the series and G_M^{tot} is the sum of all the monomer intermediate feedrates, corresponding to the value of G_M for the last reactor of the series. The instantaneous conversion is instead equal to:

$$\mathscr{I} = 1 - \frac{MW_M \dot{n}_M}{\dot{m}_M^0 + G_M} \qquad (44)$$

3.2.3. Parameter Evaluation

Table 4 reports the expressions required to calculate every reaction rate constant, as an Arrhenius-like function of temperature, T, and of other process parameters, namely species i mass fraction, ω_i, initial/inlet monomer mass fraction, ω_M^0, and the average radical chain length, DP_n^R. Table 5 lists auxiliary variables necessary to compute the reaction rate constants available in literature. It can be noted that only one parameter is unknown from literature, namely E_1. Generally, this parameter ranges between 4811 and 7217 K for redox initiators [18]. The actual value of this parameter will be found fitting the model predictions to the experimental data.

Namely, parameter E_1 was evaluated considering a subset of SBR runs (SBR1 to SBR5 in Table 1). The remaining experiments will be instead used for model validation. The optimal value of the parameter E_1 was determined by minimizing the sum of the square errors of the model predictions with respect to the experimental data of weight-average chain length of the polymer at a given monomer conversion. The problem to be solved is therefore the minimization of an objective function, $\mathscr{O}$:

$$\min_{E_1} \mathscr{O} = \min_{E_1} \sum_{l=1}^{5} \sum_{m}^{\exp} \left(DP_{w,m}^{EXPl} - DP_{w,m}^{MODl} \right)^2 \qquad (45)$$

where $DP_{w,m}^{MODl}$ is the calculated weight-average chain length at a given time for the m-th sample of the l-th SBR test, when $DP_{w,m}^{EXPl}$ is measured experimentally. The total number of experimental values used for the fitting procedure is 28, as reported in Table 1.

Table 4. Chemical reaction rate constants and sources. Temperature in [K], first-order constants in $[s^{-1}]$, second-order constants in $[L\,mol^{-1}\,s^{-1}]$.

Rate Constant	Expression	Source
k_1	$k_1^{45} \exp\left[-E_1\left(\frac{1}{T} - \frac{1}{318.15}\right)\right]$	This study
k_d	$1.11 \times 10^{12} \exp\left(-\frac{13110}{T}\right)$	[19]
k_p^s	$3.2 \times 10^7 \eta_p \exp\left(-\frac{1564}{T}\right)$	[14]
k_p^t	$0.755 k_p^s \exp\left(-\frac{2464}{T}\right)$	[14]
k_{bb}	$9.94 \times 10^8 \exp\left(-\frac{4576}{T}\right)$	[14]
k_{trM}^s	$7.5 \times 10^{-5} k_p^s$	[14]
k_{trM}^t	$7.5 \times 10^{-5} k_p^t$	[14]
k_{trMBS}^s	$0.039 k_p^s$	[22]
k_{trMBS}^t	$0.039 k_p^t$	[22]
k_t^{ss}	$k_t^{1,1,ss} N^*$	[11,14]
k_t^{st}	$0.3 k_t^{ss}$	[11,14]
k_t^{tt}	$0.01 k_t^{ss}$	[11,14]

Table 5. Auxiliary variables for the computation of the rate constants.

Variable	Expression	Source
f	0.5	[22,23]
k_1^{45}	2.033×10^{-2}	[22]
η_p	$0.11 + (1 - 0.11)\exp(-3\omega'_M)$	[14]
ω'_M	$\omega_M/(\omega_M + \omega_W + \omega_{PS} + \omega_{MBS})$	[14]
$k_t^{1,1,ss}$	$9.78 \times 10^{11} \exp\left(-\dfrac{1860}{T}\right)\zeta$	[14]
ζ	$1.56 - 1.77\omega_M^0 - 1.2(\omega_M^0)^2 + (\omega_M^0)^3$	[11,14]
N^*	$\begin{cases}(DP_n^R)^{-0.66} & DP_n^R \leq 30 \\ 30^{-0.44}(DP_n^R)^{-0.16} & DP_n^R > 30\end{cases}$	[11,14]
κ^{ss}	0.05	[14,23]
κ^{st}	0.4	[14,23]
κ^{tt}	0.8	[14,23]

4. Results and Discussion

The fitting procedure was carried out by evaluating the objective function at different values according to the range proposed in [18]. The objective function, $\mathcal{O}$, behavior at changing values of the fitting parameter E_1 is reported in Figure 3. The optimum value of the parameter is found to be 5000 K. This value will be used in the following for the validation procedure.

An example among those used for the fitting (test SBR4) is shown in Figure 4, where the weight-average chain length and polydispersity are plotted as a function of the overall monomer conversion for both the model predictions and the experimental results. A very good agreement is achieved for the polymer properties (in terms of degree of polymerization and polydispersity) during the reaction process. This means the model is consistent at any degree of monomer conversion. All other cases used in the fitting procedure can be found in the Supplementary Materials.

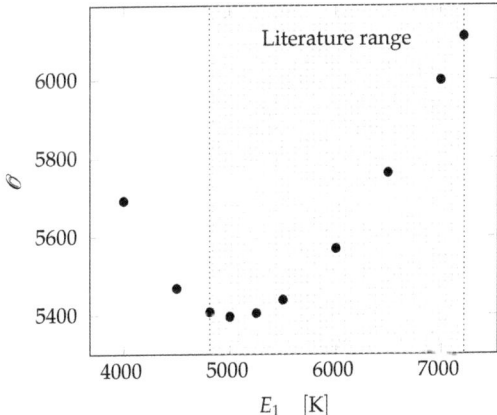

Figure 3. Minimization of the objective function $\mathcal{O}$ to find fitting parameter E_1. The range indicated by [18] is highlighted.

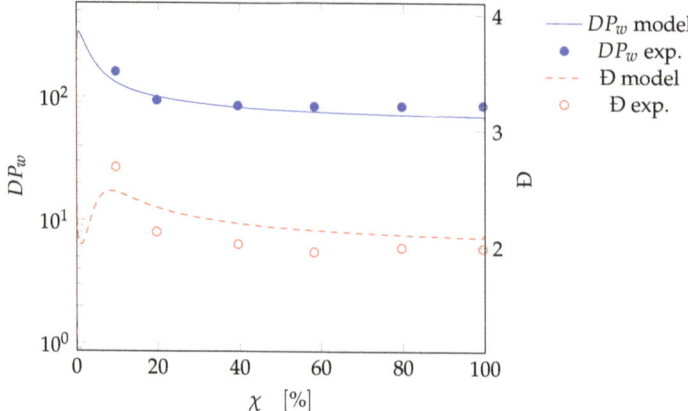

Figure 4. Fitting test SBR4: weight-average degree of polymerization (DP_w) and polydispersity (Đ) as a function of the monomer overall conversion. Curves: model results; symbols: experimental data.

The kinetic model can now be used to simulate all other tests for the SBR (namely tests SBR6-SBR17) reported in Table 1. An example of the results is depicted in Figure 5 (test SBR6). A good agreement between the experiments and the model predictions is verified, both on qualitative and quantitative standpoint. Both the weight-average chain length and the polydispersity are accurately predicted as a function of conversion. The instantaneous conversion profile, for this case, is typical of starved processes. The monomer is immediately consumed as soon as it is fed into the reactor, leading to almost complete conversion at every time. Another case is reported in Figure 6, where again the prediction of DP_w and Đ is reasonably accurate. In this case the instantaneous conversion profile is not typical of a starved process, being not complete at each instant. The model is therefore capable of predicting both starved and non-starved systems for the PAA synthesis in SBR. The results for all other cases can be found in the Supplementary Materials. It was observed that the contribution of transfer reactions and backbiting was crucial for the model predictions. In fact, the contributions of these reaction rates are shown in Figure 7 as ratio with respect to the propagation rate along the reaction coordinate for the case SBR7. The relevance of the backbiting reaction is confirmed, while chain transfer to MBS is indeed more effective to determine the molecular weight than chain transfer to monomer, as expected from [22].

All SBR tests are summarized in parity plots, reporting the predicted weight-average chain length versus the corresponding experimental values (Figure 8a), and similar plots for the number-average chain length (Figure 8b), the polydispersity (Figure 8c), and the overall conversion (Figure 8d). The cases used in the fitting procedure are denoted as crosses, while the others are indicated as circles. For each test all data in time are reported. Notably, the values of chain length are predicted accurately up to a factor of 2 (reported in Figure 8a,b as dotted lines). The polydispersity is predicted fairly well, with most of the experimental data within an interval of ±0.5 (dotted lines in Figure 8c). The conversion is always slightly underestimated with differences of up to 4% (dotted lines in Figure 8d). These deviations are deemed acceptable and the model is considered to be validated in both starved and non-starved conditions.

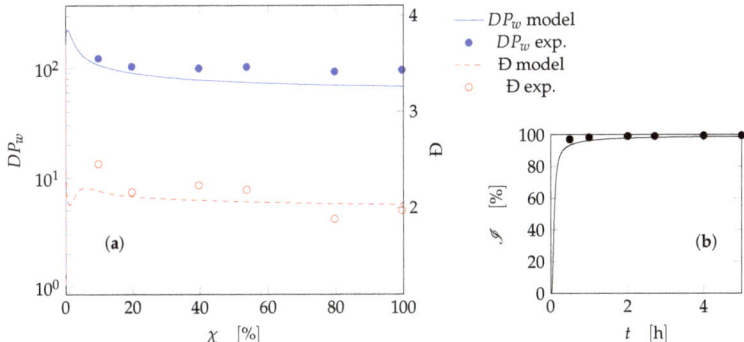

Figure 5. Test SBR6: (**a**) weight-average degree of polymerization (DP_w) and polydispersity (Đ) as a function of the monomer overall conversion; (**b**) instantaneous conversion as a function of time. Curves: model results; symbols: experimental data.

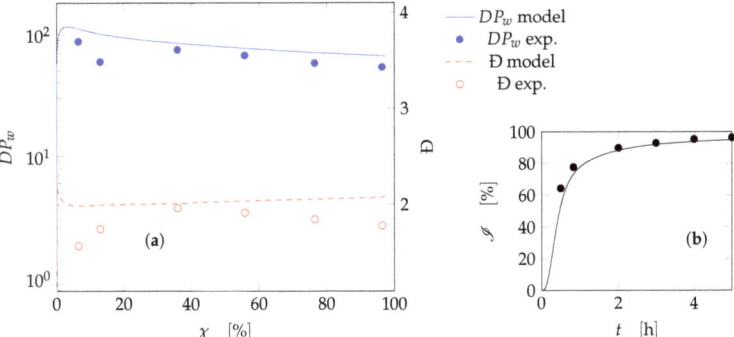

Figure 6. Test SBR7: (**a**) weight-average degree of polymerization (DP_w) and polydispersity (Đ) as a function of the monomer overall conversion; (**b**) instantaneous conversion as a function of time. Curves: model results; symbols: experimental data.

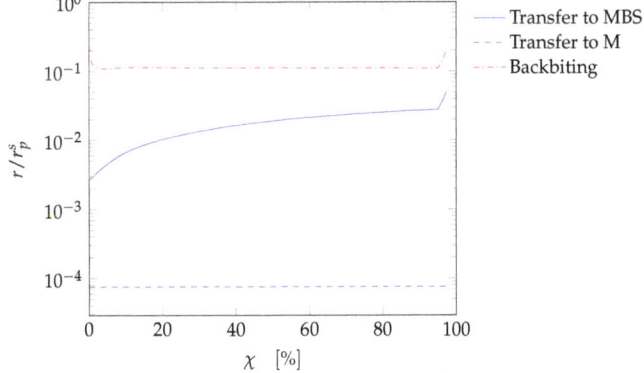

Figure 7. Ratio of selected reaction rates to the propagation rate of terminal radicals as a function of conversion for case SBR7.

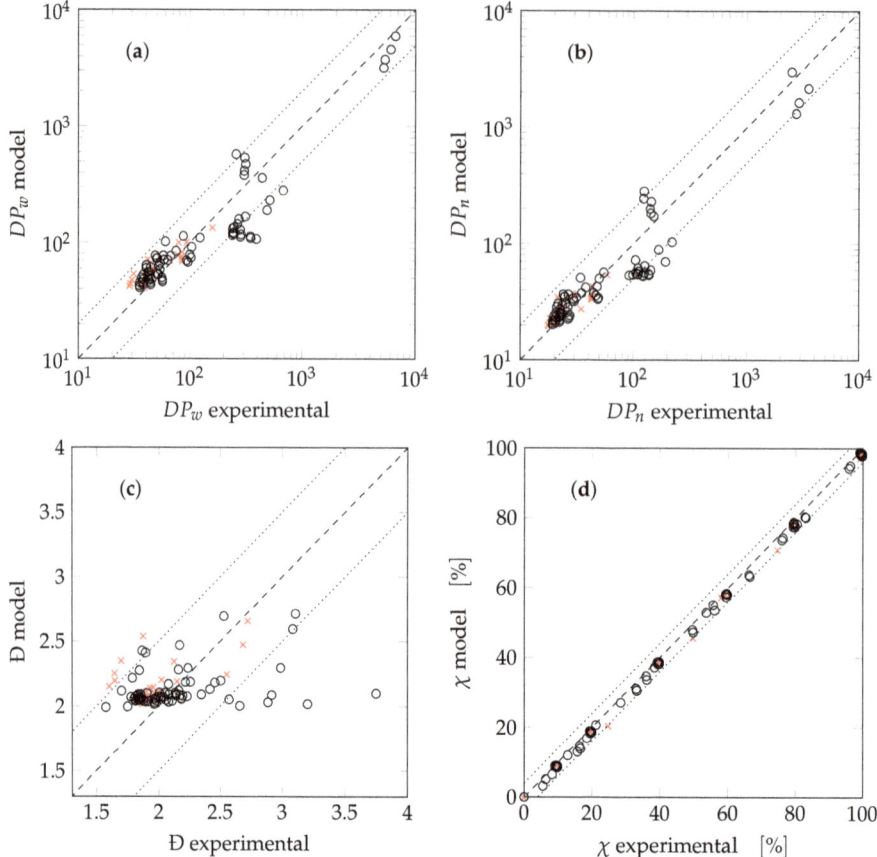

Figure 8. Parity plots for the SBR tests: (**a**) weight-average chain length; (**b**) number-average chain length; (**c**) polydispersity; (**d**) overall conversion. Crosses indicate data used for the fitting procedure.

The validated kinetic model can now be applied to the tubular reactor model. The recipe of tests SBR7 was chosen for the transition to a continuous process. The kinetics-free procedure for isothermal systems detailed in [7] was used to determine the process and geometrical parameters of the continuous series of PFRs at different number of reactors in the series, N_d. The procedure allows to choose arbitrarily the total inlet mass flowrate and the cross-sectional area, which were set to $1.00\,\mathrm{g\,s^{-1}}$ and $20.3\,\mathrm{cm^2}$, respectively. The total length of the series of PFRs, L, using these conditions is equal to $10.3\,\mathrm{m}$, thus every single reactor has a length equal to L/N_d. The inlet feed stream is, in this case, pure water. The intermediate feed policy for each fed species can be computed through this procedure and an example result for $N_d = 10$ is shown in Figure 9 as a function of the location along the series of reactors.

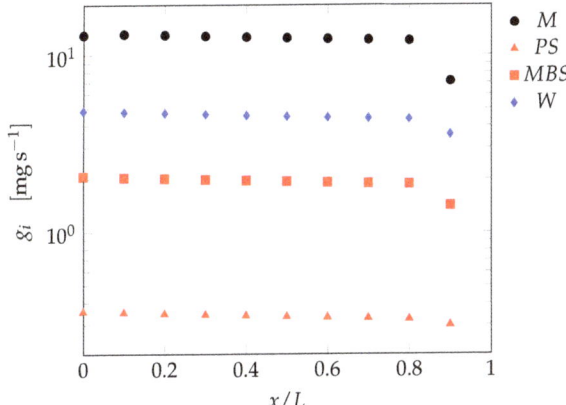

Figure 9. Intermediate species feedrates positioned along the dimensionless axial coordinate of a series of $N_d = 10$ PFRs as computed from the recipe of test SBR7.

The kinetic model can be used to optimize the number of reactors in the series such that the polymer produced in the continuous reactor has similar properties to the one produced in the discontinuous system. The aforementioned procedure guarantees perfect reproduction of the discontinuous reactor product when an infinite number of PFRs in the series is applied. This is of course only an idealized situation and a finite (possibly low) number of reactors is desirable. Such finite value has been identified by parametrically changing N_d and evaluating the difference in performance between the SBR final product and the continuous reactor product in terms of molecular weight and polydispersity. The performance of the continuous system is evaluated through the following quantities:

$$DP_w^R = \frac{DP_w^{cont.}}{DP_w^{SBR}} \qquad (46)$$

$$\Delta\mathit{Đ} = \mathit{Đ}^{cont.} - \mathit{Đ}^{SBR} \qquad (47)$$

where *cont.* refers to value of the quantity at the outlet of the last PFR in the series, while *SBR* refers to the value at the end of the process in the SBR. It is desired to have a DP_w^R as close to 1 as possible and a $\Delta\mathit{Đ}$ as close to 0 as possible. These quantities are reported in Figure 10 as a function of N_d. Low values of N_d will have higher deviations from the desired properties as these configurations are more similar to a batch reactor rather than a SBR [7]. It can be seen that for a high number of reactors in series (above 20 elements), the continuous system perfectly reproduces the SBR product both in terms of average molecular weight and polydispersity. During the validation procedure, a difference in dispersity equal to 0.5 was deemed acceptable, as well as variations of up to two times in terms of molecular weight. To fulfill the same criteria at least 7 reactors in series are needed. For the sake of example, the trends of DP_w and $\mathit{Đ}$ as a function of χ, and $\mathscr{I}$ along the axial coordinate are reported in Figure 11 for the case of $N_d = 10$. While the discontinuities in the plots are related to the addition of species between the reactors, the reproduction of polymer quality is indeed satisfactory.

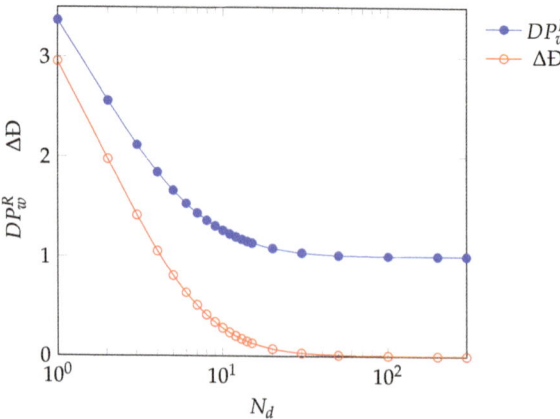

Figure 10. Quality measures for the reproducibility of the product of tests SBR7 in a continuous series of PFRs according to the number of reactors in the series.

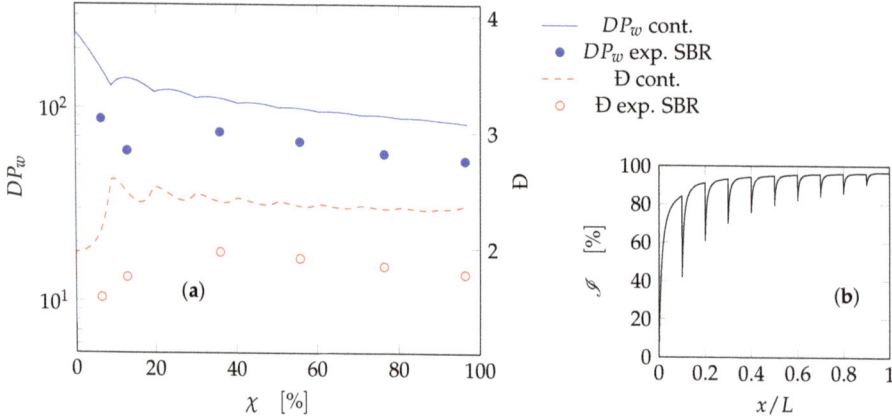

Figure 11. Series of $N_d = 10$ PFRs with intermediate feed streams: (**a**) weight-average degree of polymerization (DP_w) and polydispersity (Đ) as a function of the monomer overall conversion; (**b**) instantaneous conversion along the dimensionless axial coordinate. Curves: model results for the continuous reactor; symbols: experimental data for the SBR.

On the other hand, the major advantage of this continuous system is that the productivity of the system can be scaled arbitrarily by adding more reactors in parallel. This can be readily implemented when a compact configuration of heat-exchanger reactors is considered, that is multiple parallel tubes with especially large surface-to-volume ratio. Furthermore, to achieve the same productivity (also considering dead times), an industrial SBR would require a reaction volume that is 30 % larger than the continuous reactor. Therefore, the continuous process is intrinsically safer, since smaller holdups of the different species are involved at constant productivity. The transition from the SBR to the continuous system achieves higher degree of process intensification and enables less demanding scale-up procedure from laboratory to industrial scale. This is due to the arbitrary value of the inlet flowrate to the continuous reactor, which allows an arbitrary change in the productivity together with the possibility to have a simple configuration with reactors in parallel, exploiting a scale-out procedure. If the parallel configuration is adopted (e.g., heat-exchanger reactor), then a single tube is representative of the whole ensemble and can be used as a pilot for testing.

It is industrial practice to adopt a series of mixed reactors to run continuous polymerizations. The proposed method could be in principle applied also to a series of continuous stirred-tank reactors (CSTRs) with intermediate injections, replacing each single PFR with a CSTR [24]. Nevertheless from a practical standpoint, the tubular reactor retains the aforementioned advantages with respect to a mixed vessel, especially in terms of compactness, safety, and ease of scaling with a parallel configuration. Rescaling a series of CSTRs could require extensive plant changes or pose safety issues given the fact that industrial CSTRs have a lower thermal efficiency than tubular reactors.

5. Conclusions

A detailed kinetic model for the polymerization of non-ionized acrylic acid initiated with a redox couple (Na-PS/Na-MBS) in aqueous solution was presented and validated using experimental data obtained from a laboratory scale SBR under isothermal conditions. The model takes into account the formation of MCRs and the transfer reactions to one of the species in the redox couple and to the monomer. The population balances of the resulting model were solved through the MoM and the governing equations are generalized to consider any number of moments. The kinetic model was validated over temperatures ranging from 50 to 90 °C.

Afterwards, the kinetic model was used to intensify one of the test cases by shifting it to the continuous mode using a literature approach. The continuous reactor, a series of tubular reactors with intermediate feed streams, enabled a scale-up of the recipe which would considerably reduce the volume of an equivalent (in terms of productivity) industrial SBR. In particular, a volume reduction up to 30% was found for the examined test case. Therefore, this model is a valuable tool to drive the transition of discontinuous polymerization processes to continuous ones in the framework of process intensification.

Supplementary Materials: The following are available online at http://www.mdpi.com/2227-9717/8/7/850/s1: Table S1: PAA standards used for the calibration of GPC. Peak molecular weight and polydispersity; Figures S1–S17: Tests SBR1-SBR17: weight-average degree of polymerization (DP_w) and polydispersity (Đ) as a function of the monomer overall conversion; instantaneous conversion along the dimensionless axial coordinate; model results and experimental data.

Author Contributions: Conceptualization, F.F. and G.S.; methodology, F.F. and G.S.; software, F.F.; validation, P.R.B. and A.C.; formal analysis, F.F.; investigation, P.R.B. and A.C.; resources, G.S.; data curation, F.F. and P.R.B.; writing—original draft preparation, F.F.; writing—review and editing, G.S.; visualization, F.F.; supervision, G.S.; project administration, G.S.; funding acquisition, G.S. All authors have read and agreed to the published version of the manuscript.

Funding: Financial support of Innovhub for one of us (F.F.) is gratefully acknowledged.

Acknowledgments: The authors gratefully thank an industrial partner, who wishes to remain anonymous, for the support of the experimental work.

Conflicts of Interest: The authors declare no conflict of interest.

Abbreviations

The following abbreviations are used in this manuscript:

AA	Acrylic acid
BR	Batch reactor
GPC	Gel permeation chromatography
HPLC	High-performance liquid chromatography
MBS	Metabisulfate
MCR	Mid-chain radicals
MoM	Method of moments
PAA	Poly(acrylic acid)
PFR	Plug-flow reactor
PS	Persulfate
SBR	Semi-batch reactor

References

1. Herth, G.; Schornick, G.; Buchholz, F.L. Polyacrylamides and Poly(Acrylic Acids). In *Ullmann's Encyclopedia of Industrial Chemistry*; American Cancer Society: Atlanta, GA, USA, 2015; pp. 1–16.
2. Roberts, D. Heats of Polymerization. A Summary of Published Values and Their Relation to Structure. *J. Res. Natl. Bur. Stand.* **1950**, *44*, 221–232. [CrossRef]
3. Stankiewicz, A.I.; Yan, P. 110th Anniversary: The Missing Link Unearthed: Materials and Process Intensification. *Ind. Eng. Chem. Res.* **2019**, *58*, 9212–9222. [CrossRef]
4. Anxionnaz, Z.; Cabassud, M.; Gourdon, C.; Tochon, P. Transposition of an Exothermic Reaction From a Batch Reactor to an Intensified Continuous One. *Heat Transf. Eng.* **2010**, *31*, 788–797. [CrossRef]
5. Goerke, T.; Kohlmann, D.; Engell, S. Transfer of Semibatch Processes to Continuous Processes with Side Injections—Opportunities and Limitations. *Macromol. React. Eng.* **2016**, *10*, 364–388. [CrossRef]
6. Kohlmann, D.; Chevrel, M.C.; Hoppe, S.; Meimaroglou, D.; Chapron, D.; Bourson, P.; Schwede, C.; Loth, W.; Stammer, A.; Wilson, J.; et al. Modular, Flexible, and Continuous Plant for Radical Polymerization in Aqueous Solution. *Macromol. React. Eng.* **2016**, *10*, 339–353. [CrossRef]
7. Florit, F.; Busini, V.; Storti, G.; Rota, R. From semi-batch to continuous tubular reactors: A kinetics-free approach. *Chem. Eng. J.* **2018**, *354*, 1007–1017. [CrossRef]
8. Florit, F.; Busini, V.; Storti, G.; Rota, R. Kinetics-free transformation from non-isothermal discontinuous to continuous tubular reactors. *Chem. Eng. J.* **2019**, *373*, 792–802. [CrossRef]
9. Omidian, H.; Zohuriaan-Mehr, M.; Bouhendi, H. Aqueous solution polymerization of neutralized acrylic acid using $Na_2S_2O_5/(NH_4)_2S_2O_8$ redox pair system under atmospheric conditions. *Int. J. Polym. Mater. Polym. Biomater.* **2003**, *52*, 307–321. [CrossRef]
10. Ebdon, J.; Huckerby, T.; Hunter, T. Free-radical aqueous slurry polymerizations of acrylonitrile: 1. End-groups and other minor structures in polyacrylonitriles initiated by ammonium persulfate/sodium metabisulfite. *Polymer* **1994**, *35*, 250–256. [CrossRef]
11. Barth, J.; Meiser, W.; Buback, M. SP-PLP-EPR Study into Termination and Transfer Kinetics of Non-Ionized Acrylic Acid Polymerized in Aqueous Solution. *Macromolecules* **2012**, *45*, 1339–1345. [CrossRef]
12. Anseth, K.S.; Scott, R.A.; Peppas, N.A. Effects of Ionization on the Reaction Behavior and Kinetics of Acrylic Acid Polymerizations. *Macromolecules* **1996**, *29*, 8308–8312. [CrossRef]
13. Çatalgil Giz, H.; Giz, A.; Alb, A.M.; Reed, W.F. Absolute online monitoring of acrylic acid polymerization and the effect of salt and pH on reaction kinetics. *J. Appl. Polym. Sci.* **2004**, *91*, 1352–1359. [CrossRef]
14. Wittenberg, N.F.G.; Preusser, C.; Kattner, H.; Stach, M.; Lacík, I.; Hutchinson, R.A.; Buback, M. Modeling Acrylic Acid Radical Polymerization in Aqueous Solution. *Macromol. React. Eng.* **2016**, *10*, 95–107. [CrossRef]
15. Ebdon, J.; Huckerby, T.; Hunter, T. Free-radical aqueous slurry polymerizations of acrylonitrile: 2. End-groups and other minor structures in polyacrylonitriles initiated by potassium persulfate/sodium bisulfite. *Polymer* **1994**, *35*, 4659–4664. [CrossRef]
16. Riddick, J.A.; Bunger, W.B.; Sakano, T.K. *Organic Solvents: Physical Properties and Methods of Purification*, 4th ed.; John Wiley & Sons: Hoboken, NJ, USA, 1986; p. 376.
17. Misra, G.; Bajpai, U. Redox polymerization. *Prog. Polym. Sci.* **1982**, *8*, 61–131. [CrossRef]
18. Odian, G. Radical Chain Polymerization. In *Principles of Polymerization*; John Wiley & Sons, Ltd.: Hoboken, NJ, USA, 2004; Chapter 3, pp. 198–349.
19. Costa, C.; Santos, V.; Araujo, P.; Sayer, C.; Santos, A.; Fortuny, M. Microwave-assisted rapid decomposition of persulfate. *Eur. Polym. J.* **2009**, *45*, 2011–2016. [CrossRef]
20. Kuchta, F.D.; van Herk, A.M.; German, A.L. Propagation Kinetics of Acrylic and Methacrylic Acid in Water and Organic Solvents Studied by Pulsed-Laser Polymerization. *Macromolecules* **2000**, *33*, 3641–3649. [CrossRef]
21. Thakur, R.; Vial, C.; Nigam, K.; Nauman, E.; Djelveh, G. Static mixers in the process industries—A review. *Chem. Eng. Res. Des.* **2003**, *81*, 787–826. [CrossRef]
22. Gutierrez, C.G.; Cáceres Montenegro, G.; Minari, R.J.; Vega, J.R.; Gugliotta, L.M. Scale Inhibitor and Dispersant Based on Poly(Acrylic Acid) Obtained by Redox-Initiated Polymerization. *Macromol. React. Eng.* **2019**, *13*, 1900007. [CrossRef]

23. Minari, R.J.; Caceres, G.; Mandelli, P.; Yossen, M.M.; Gonzalez-Sierra, M.; Vega, J.R.; Gugliotta, L.M. Semibatch Aqueous-Solution Polymerization of Acrylic Acid: Simultaneous Control of Molar Masses and Reaction Temperature. *Macromol. React. Eng.* **2011**, *5*, 223–231. [CrossRef]
24. Florit, F.; Busini, V.; Rota, R. Kinetics-free process intensification: From semi-batch to series of continuous chemical reactors. *Chem. Eng. Process. Process. Intensif.* **2020**, *154*, 108014. [CrossRef]

© 2020 by the authors. Licensee MDPI, Basel, Switzerland. This article is an open access article distributed under the terms and conditions of the Creative Commons Attribution (CC BY) license (http://creativecommons.org/licenses/by/4.0/).

Article

Thermal Pyrolysis of Polystyrene Aided by a Nitroxide End-Functionality. Experiments and Modeling

Almendra Ordaz-Quintero, Antonio Monroy-Alonso and Enrique Saldívar-Guerra *

Centro de Investigación en Química Aplicada, Blvd. Enrique Reyna 140, Saltillo Coahuila 25294, Mexico; almendra.ordaz@ciqa.edu.mx (A.O.-Q.); antonio.monroy.alonso@hotmail.com (A.M.-A.)
* Correspondence: enrique.saldivar@ciqa.edu.mx; Tel.: +52-844-439-4408

Received: 27 February 2020; Accepted: 31 March 2020; Published: 5 April 2020

Abstract: The thermal pyrolysis of polystyrene (PS) is gaining importance as the social pressure for achieving a circular economy is growing; moreover, the recovery of styrene monomer in such a process is especially relevant. In this study, a simple thermal pyrolysis process in the temperature range of 390–450 °C is developed. A working hypothesis is that by using a nitroxide-end functionalized PS (PS-T or dormant polymer), the initiation process for the production of monomer (unzipping) during the PS pyrolysis could be enhanced due to the tendency of the PS-T to activate at the nitroxide end. Two types of PS were used in this work, the first one was synthesized by free-radical polymerization (FRP-dead polymer) and the second by nitroxide-mediated polymerization (NMP) using three levels of nitroxide to initiator ratio: 1.3, 1.1, and 0.9. Analysis of the recovered products of the pyrolysis by gas-mass spectroscopy shows that the yield of styrene increases from ~33% in the case of dead polymer to ~38.5% for PS-T. A kinetic and mathematical model for the pyrolysis of dead and dormant polymer is proposed and solved by the method of moments. After a parameter sensitivity study and data fitting, the model is capable of explaining the main experimental trends observed.

Keywords: polystyrene; thermal pyrolysis; nitroxide mediated polymerization; mathematical modeling

1. Introduction

Plastics are widely used materials due to their physical and chemical properties; [1] additionally, they are cheap, light, long-lasting [2], and relatively easy to produce [3]. These outstanding properties are the ones that have led to their excessive production and irresponsible consumption, originating severe ecological problems that could cause irreversible damage to the environment [4–6].

Essentially there are three traditional solutions for plastic waste management, each one presenting significant drawbacks: Landfill disposal, recycling, and combustion [6]. In landfill disposal, none of the material resources used to produce plastics are recovered. With respect to the second alternative, the recycling capacity of polymers is inherently limited since after a determined number of recycling processes the materials exhibit reduced mechanical properties and the plastic ends up being thrown away. Regarding the third option, recovering energy from combustion is feasible, but it causes negative effects on the environment and human health.

Given the increasing social pressure to take care of the environment and the present trends towards circular economy, a better option to manage plastic waste would be to transform it into defined chemical compounds, especially the source monomer, with the required purity and quality to be used again as raw material in the same polymerization or in other processes. One such technique that permits the recovery of high-value compounds is pyrolysis, which can be finely tuned to favor some product distribution over others. The practical implementation of a chemical recycling strategy requires a concerted effort among different sectors of the society (the general population, municipal governments, polymer producers, and recycling companies), but there are already some successful examples of this model.

Pyrolysis of plastics can be carried out via thermal or catalytic routes, and the distribution of obtained products is influenced by a wide range of parameters, including the type of waste used, the type of reaction system, the residence time, the pressure range, the presence or absence of catalysts, the presence of hydrogen donor compounds, and several others that affect the composition and yield of the obtained products, which consist of complex mixtures that require further separation if individual products are desired.

Purely thermal pyrolysis processes temperatures can reach up to 900 °C which enhance the rate of secondary decomposition reactions and reduce the liquid yield, producing low-quality products with no predictable composition of a broad range of compounds [7]. Catalytic pyrolysis, on the other hand, promotes the polymer decomposition at lower temperatures and shorter times, reducing the energy consumption and enhancing the selectivity of the obtained products, but implying extra costs of the catalyst and its regeneration. In the case of the present proposed process, despite the reactions taking place purely thermally (with temperatures above, but close to the ceiling point of polystyrene), the obtained distribution of products shows a selectivity similar to that of a catalytic process.

Submitting polymers to heat deeply alters their main chain structure, their substituent atoms, and their side chains, provoking random scission at weak links in the main chain, at a chain-end, or in labile structures and, depending on the site of reaction in the polymer structure and the conditions chosen for the particular system, the resulting radicals follow different paths of reaction, from depolymerization of the main chain to inter- or intramolecular transfer reactions [8].

The pyrolysis of pyrolysis of polystyrene (PS) has been a topic of interest for many years, and it is not surprising that renewed interest is rising considering all of the above trends and given that PS is a thermoplastic used throughout the world in a variety of applications.

Thermal degradation of PS is considered to be a chain radical process [9–11], but despite numerous studies and experiments performed to understand its mechanism, many aspects of the problem remain unresolved, making it still difficult to accurately predict the product distribution at a given set of conditions. Additionally, as the process system temperature increases, so does its complexity [8].

The recovery of styrene monomer in the highest possible yield is a desired output of the pyrolysis process of PS, since this would contribute to the consolidation of a circular economy. Therefore, any effort directed towards this goal should be welcome. It is our hypothesis that modifying PS with a nitroxide moiety at a chain-end could enhance the generation of monomer when this material is subjected to thermal pyrolysis at a temperature higher than the ceiling temperature of PS. In this paper, empirical evidence that supports this hypothesis is presented and a mechanistic and mathematical model that is consistent with the experimental observations is also presented and discussed. Before that, background information that supports and motivates this hypothesis is introduced. In the next paragraphs, a brief literature review is made, including a description of the reaction mechanism and previous experimental work. Some of the studies are of academic nature, while many others are associated with patents.

Reaction Mechanism. It has been reported that the molecular weight of polystyrene decreases on heating in vacuum at 250 °C, and when heated above 300 °C volatile products are formed [8,10]. With regard to the initiation process, it is accepted that the radicals are formed by different mechanisms such as chain-end initiation, random mid-chain scission, and scission at irregular structures or "weak links" distributed along the polymer chain such as head-to-head linkages, chain branches, and unsaturated bonds [9,10,12,13]. Lehrle et al. [13] concluded that between 450 to 480 °C initiation occurs both by random mid-chain scission and at chain-ends, and Madorsky [14] pointed out that between 300 to 400 °C it degrades preferably at the ends and that simultaneously some random degradation takes place.

Propagation reactions include intramolecular transfer, which involves the transfer of a hydrogen atom within a single polymer chain and intermolecular hydrogen transfer, comprising the transfer of a hydrogen atom between polymer chains. Also included in this category is the depropagation reaction, where no hydrogen transfer occurs; this reaction is essentially the reverse of the polymerization propagation, also called unzipping or depolymerization, and is mainly responsible for the production of styrene monomer [10,15].

It has been claimed that the most frequent propagation reaction taking place in the 280–350 °C temperature range is β-scission [9], either at a reaction end or at a mid-chain position, a reversible reaction that becomes faster as the ceiling temperature is approached (310–390 °C depending on the source [16,17]); below 300 °C the production of styrene is hardly observed. The second most important reaction is the hydrogen abstraction from the main chain by any radical, producing a saturated chain-end from the attacking radical, and a new radical that undergoes β-scission [9]. At higher temperatures (600 and 700 °C), increased production of benzene, toluene, ethylbenzene, and α-methyl styrene and decreased production of styrene have been observed, probably due to the predominance of intramolecular hydrogen transfer reactions at this range of temperatures [18]. Concerning termination, some authors propose a mechanism involving the recombination [9,11], or the disproportionation reaction [19] between two radicals; while others suggest the occurrence of depropagation until the end of the polymer molecule [13].

Published studies related to the thermal degradation and pyrolysis of polystyrene [13,20–25] have been performed in a variety of reaction systems and sets of conditions, including reaction temperatures, reaction times, and molecular weights of the polystyrene sample, resulting in a broad collection of reaction yields and product distribution. The available reported experimental data fall into three types: chemical nature of the products (that help to elucidate the degradation mechanism), rate of evolution of products, and the change of molecular weight in the residue.

Jellinek [22] reported experiments carried out in long open-to-air tubes containing 0.1 to 0.5 g of material, to show the influence of oxygen on the thermal degradation of polystyrene. The reactions were described as too violent above 220 °C, many side reactions set in and the material quickly became yellow and brown. They assumed that the first stage of the oxidation consisted of the formation of hydro peroxide groups, which led to chain scission, and that at later stages, the reaction slowed down due to the formation of antioxidants such as benzaldehyde. For practical application, to avoid the presence of oxygen in the resulting products, pyrolysis processes are designed to occur in an inert atmosphere. Ebert et al. [23] pyrolyzed samples of polydisperse polystyrene with molecular weights between 60,000 and 220,000 Da in ampoules free of oxygen at temperatures between 270 and 336 °C, assessing the sample weight loss as a function of degradation time. They also performed simulations of the thermal degradation assuming that scission and unzipping occur independently, and attributed scission as the main cause for the decrease in molecular weight, while unzipping as the sole cause of volatiles production. Additionally, in oxygen-free glass ampoules, Wegner and Patat [26] analyzed n-butyl lithium initiated polystyrene samples with molecular weights between 110,000 to 900,000 Da, submitting them to temperatures between 200 to 325 °C, concluding that the polystyrene molecule does not contain thermolabile bonds corresponding to the weak-links theory.

Reactions conducted in a fluidized-bed reactor with operation temperatures in the range 450–700 °C, nitrogen atmosphere, heated by an external electric furnace, permitted liquid recovery above 90%, with the main components being styrene monomer, dimer and trimer, benzene, toluene, ethylbenzene, and α-methyl styrene [20].

Simard et al. [21] reported PS pyrolysis experiments using flat bottom vessels and temperatures between 370–420 °C, with a maximum reaction time of 45 min, observing that 85 to 95% of the obtained liquid fraction are styrene, dimer, and trimer. Anderson and Freeman [25] assessed the weight loss of samples heated from 246 to 430 °C, observing decomposition at 320 °C and 99% of weight loss at 430 °C.

Small scale PS pyrolysis experiments have also been performed. Richards and Salter [24] intentionally synthesized polystyrene containing weak links in its structure that differ from the usual head–tail links of the polymer backbone and should, therefore, have different thermal properties. The pyrolysis experiments were performed at 276–329 °C, and the main product from both polymers was styrene, although significant quantities of dimer and trimer were also formed, the main difference being that the molecular weight of the polymer with weak links decreased more rapidly.

Several patents regarding polystyrene pyrolysis have also been filed [27–32] reporting the use of temperatures varying in a wide range, 330–870 °C, styrene monomer recovery from 33 to 80%, and diverse processes, including the use of a fluidized bed reactor [31] and the use of solvents like toluene [32].

In general, pyrolysis temperature is a very important factor affecting the product distribution [20]; in all the studies both higher temperatures and longer reaction times resulted in an increase in the amount of the liquid yield [21] and a decrease in the molecular weight of the residue. Many of these experiments were done with very small samples (micrograms and milligrams) and none of them had an initial charge greater than 25 g. At increasing temperatures more monomer is formed reaching a maximum at 600 °C where secondary reactions take place.

As can be seen, the use of solvents and catalytic systems allows high yield of styrene recovery, although some drawbacks associated with these processes are evident, such as high operating temperatures and a wide range of obtained products by purely thermal pyrolysis.

On the other hand, nitroxide-mediated polymerization (NMP) is among the most popular techniques in the category of RDRP (reversible deactivation radical polymerization) that provide living character to free radical polymerizations and control the structure of the resulting polymers, enabling the production of polymers with narrow molecular weight distributions and the formation of well-defined polymer architectures such as block copolymers [33]. Roughly speaking, these techniques simultaneously provide some of the advantages of living polymerizations (structure control) and conventional radical polymerization (robustness to impurities and to protic media). NMP, which is especially suited for styrene polymerization, is based on a reversible termination mechanism between the propagating growing species and a stable free radical (nitroxide radical), that acts as a control agent to generate a macro alkoxyamine (dormant polymer) as predominant species. The dormant polymer eventually regenerates a propagating radical and a nitroxide radical through a homolytic breakage as the dormant and the radical species are in dynamic equilibrium resulting from the reversible deactivation-activation reactions just described (see Figure 1). As a result of the NMP process, the final product is mainly constituted by dormant polymer end-functionalized with a nitroxide moiety. For most practical purposes this polymer can be prepared to be very similar to the product produced via conventional free-radical polymerization (except for its dispersity). However, its nitroxide-end functionality should make it more favorable to depolymerization in a certain range of temperatures (via unzipping) than the conventional PS, due to the increased possibility of chain-end initiation of the depolymerization reaction via rupture of the oxygen–carbon bond between the nitroxide moiety and the rest of the polymeric chain.

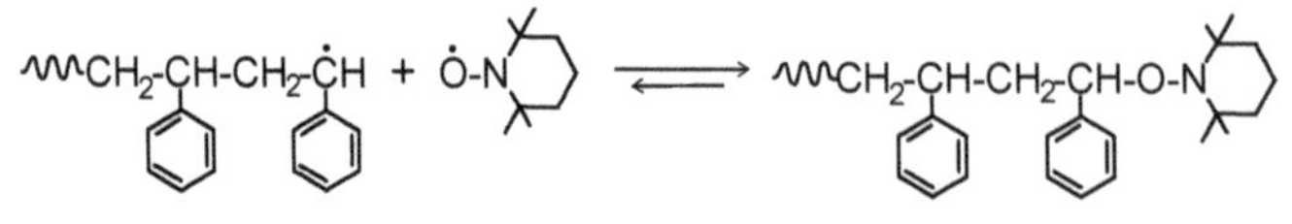

Figure 1. Depolymerization initiation reaction via oxygen–carbon bond rupture (nitroxide radical activation/deactivation).

With respect to the thermal pyrolysis process, it is known that chain-end initiation in mild temperature conditions or moderately above the polystyrene ceiling temperature (310 °C) leads to unzipping and increased production of styrene monomer over other products of this process.

Roland and Schmidt-Naake [34] studied the polymerization of styrene with TEMPO and benzoyl peroxide, claiming that the reversible capping with TEMPO can introduce a weak link at the end of the polymer, such as the bond between polymer and nitroxide. They conclude that polymer degradation for this material occurs in the same temperature range as for non-nitroxide polystyrene (400 °C). However the PS-T thermal degradation curve shows an additional step of mass loss at temperatures below 300 °C, suggesting two reactions, one apparently being the cleavage of the polystyrene–nitroxide bond followed by depolymerization, and the other being the breakage of the N-O bond in the TEMPO moiety, the importance of both reactions changing with increasing temperature.

The lack of knowledge of the position and nature of the initial scission of the polystyrene thermal degradation has restricted its quantitative analysis.

Taking into account this background, the goals of this work are twofold: (i) First, to define a process to achieve the thermal decomposition of conventional (free-radical) polystyrene in one step, and (ii) to compare the developed process for the thermal pyrolysis of conventional PS with another in which PS possessing a nitroxide end-functionality is used. This functionality is introduced in the polymer previously polymerizing styrene in the presence of the TEMPO nitroxide (PS-T) (TEMPO is (2,2,6,6-tetramethylpiperidin-1-yl) oxyl). For the comparison, precise pyrolysis conditions to induce depolymerization reactions that generate styrene monomer as main product and chemicals of high energetic value in the absence of solvents and catalysts at relatively mild conditions of temperature and pressure are first determined. TEMPO is used to provide living character to the free radical polymerization producing polystyrene with an extreme ended in a nitroxide moiety (PS-T) that, from the origin, has characteristics that favor depolymerization. As mentioned above, the hypothesis behind this work is that PS-T will depolymerize in a certain range of temperatures (via unzipping) by chain-end initiation of the reaction at the nitroxide functionality, promoting depropagation reactions in relatively mild temperature conditions or moderately above the polystyrene ceiling temperature. It is believed that above the T_c of polystyrene the uncapping (activation) reaction of the nitroxide moiety at the end of the polystyrene chain will leave a polystyryl radical that will undergo unzipping. The experiments seem to validate this hypothesis to some extent. On the other hand, a mathematical model of the pyrolysis reaction is developed to explain major effects observed and especially the effect of the end-nitroxide group in the PS-T chain.

2. Materials and Methods

2.1. Materials

Styrene (99%) was purified using an inhibitor removal column (4-tert-butylcatechol). Benzoyl peroxide (98%), (2, 2, 6, 6-tetramethyl-1-piperidinyl-1-yl) oxyl (TEMPO), all reactants from Aldrich (St. Louis Missouri, USA) (Manufacturer name, city, state if US or Canada, country) and industrial grade methanol (90%) (Proquisa, Saltillo, Mexico) were used as received.

2.2. Reaction System

The depolymerization system used in this work is shown in Figure 2 and consists of a batch pressurized reactor and a condenser.

The 50 mL reactor vase is made of stainless steel, and it comprises a heating source in the form of an electric mantle which provides homogeneous temperature increase inside the reactor and keeps the reaction temperature between 300 and 500 °C with the help of a thermocouple situated inside the reaction mixture and a digital controller that manipulates the current sent to the heating mantle. The reactor is also fitted with a manometer, inlet and outlet valves (to feed inert gas to the system and generate a free oxygen atmosphere), and a third outlet valve that connects the reactor to a double tube condenser which uses a low temperature fluid (−5 °C) that allows the rapid condensation of the generated vapors.

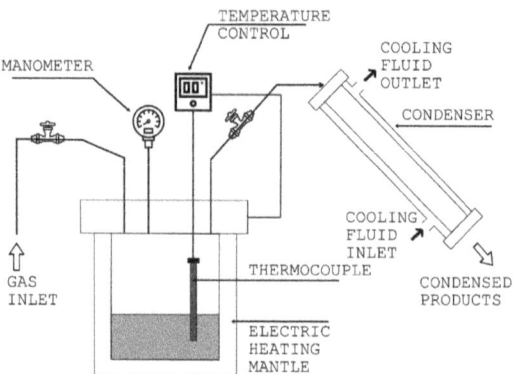

Figure 2. Depolymerization reaction system.

2.3. Polystyrene Synthesis Methods

Polystyrene that would be subjected to the pyrolysis experiments was synthesized by two different methods: by conventional free radical polymerization, and by nitroxide mediated radical polymerization (NMP).

In the conventional free radical polymerization (FRP), the reaction was performed in a batch bulk process at constant temperature (90 °C) in a 150 mL glass reactor, inert atmosphere, and continuous agitation, using benzoyl peroxide as initiator. The agitation was implemented via a mechanical agitator that was used until the viscosity of the reaction media made it not possible to continue in this way. After this point, reached at conversions above 80%, the reaction was left without agitation until conversions of 99% were reached and then stopped by lowering the reaction temperature to ambient temperature. The polymer was subsequently recovered by adding acetone to it (around 10–15% wt. with respect to the total mass of polymer) to allow the material to flow, followed by precipitation of the reaction mixture in methanol. The precipitate was later filtered and dried, obtaining white polymer dust as product.

Styrene polymerizations by the NMP method were performed using BPO as initiator and TEMPO as the stable free radical, varying the nitroxide (N) to initiator (I) ratio; three levels were established (N/I: 1.3, 1.1, and 0.9), they were all performed in a batch reactor at constant temperature (130 °C) with continuous agitation and under inert atmosphere. When conversions of 99% were reached, the reactions were stopped by submerging the reactor in a cooling ice bath, lowering the system temperature. The polymer from the reaction mixture was recovered following the same procedure described before for the PS produced by conventional FRP. The product was obtained as a white dust.

The polymers obtained by FRP and NMP were analyzed by size exclusion chromatography (SEC) using an Agilent Technologies Mod G7810A chromatograph (Agilent Technologies, Santa Clara, CA, USA) at 40 °C, with a sample concentration of 1 mg mL^{-1}, solvent flow of 1 mL/min and a polystyrene standard (Agilent Technologies, Santa Clara, CA, USA); the results are summarized in Table 1. The experiments were designed to obtain approximately the same number average molecular weight in all the cases (40–50,000 Da).

Table 1. Properties of synthesized polystyrene.

Sample	M_n	M_w	M_w/M_n
NMP, N/I = 1.3	46,100	56,800	1.23
NMP, N/I = 1.1	44,800	56,900	1.27
NMP, N/I = 0.9	42,500	55,500	1.3
FRP	50,040	96,700	1.9

2.4. Polystyrene Thermal Pyrolysis Experiments

The process to depolymerize all the polystyrene samples was the same and started by introducing 10 g of the synthesized plastic material inside the reactor and displacing the oxygen atmosphere with nitrogen to guarantee that the experiment was developed under inert conditions. Once this was achieved, all the valves of the reactor were closed and the temperature was rapidly increased, provoking a pressure rise; the time needed to reach the defined temperature set point was between 5 to 7 min. When the pressure reached from 7 to 12 psig, the valve connecting the reactor to the condenser was turned open, allowing the vapor products to migrate from the reactor to the condenser, after which the products were recovered as a liquid mixture. After reaching the temperature set point, the reaction proceeded until the liquid product was no longer obtained (about 10 min). The total reaction time was considered from the point at which the heating started to the moment at which the liquid outflow stopped. Experiments were carried out varying the pyrolysis temperature: 390, 420, and 450 °C and the source polystyrene (see Table 2), to investigate the effect of this variable on the prepared samples. After the reaction was finished and the reactor was cooled, two types of samples were obtained from each experiment: a liquid mixture from the condensed vapors and the residue left in the reactor in the form of a dark-colored oil with some solid degradation fragments (in the following discussion this will be referred as the solid fraction). Only the condensed liquid mixture was analyzed since it contains the most valuable products (including the monomer).

Table 2. Pyrolysis experiments performed.

Sample	Pyrolysis Temperature		
NMP, N/I = 1.3	390 °C * (Tri)	420 °C (Tri)	450 °C (Tri)
NMP, N/I = 1.1	390 °C	420 °C	450 °C
NMP, N/I = 0.9	390 °C	420 °C	450 °C
FRP	390 °C (Tri)	420 °C (Tri)	450 °C (Tri)

* (Tri) = Run and analyzed by triplicate.

3. Results

3.1. Product Analysis

The obtained products from the polystyrene pyrolysis experiments were analyzed to determine the most abundant components present in the liquid mixture. A gas chromatograph (Thermo Finnigan by Thermo Fisher Scientific Waltham, MA, USA) equipped with a selective mass detector (DSQ Trace 2000) (Thermo Electron Corporation, Austin, TX, USA) and a Thermo TG-5ms column were used. All the samples were analyzed under the same conditions with helium as the carrier gas. The temperature program was as follows: the temperature was held at 60 °C for 3 min, then programmed to reach 300 °C at a heating rate of 10 °C min^{-1} and held at that temperature for 10 min. The injector temperature was set at 250 °C. The transfer temperature line of the mass detector was set at 280 °C and the mass range was from 32 to 650 amu. The products were identified according to their fragmentation patterns using a library included in the software of the chromatograph.

3.2. Kinetic Model

As described before, the radical process of conventional polystyrene pyrolysis comprises the typical steps of initiation, propagation, transfer to polymer, and termination: Although initiation can occur due to several causes, there seems to be general agreement that consists of the formation of free radicals after bond breakage by the action of heat mainly either at a chain-end (which results in the production of monomer) or at a mid-chain position along the polymer backbone. Propagation covers the competition of three different reaction mechanisms: unzipping, intramolecular, and intermolecular hydrogen transfer, the first one is sometimes also called depolymerization and is taken to be the reverse of chain growth, while transfer to polymer consists in the abstraction of a hydrogen from the same or another molecule. The termination step can occur via recombination and disproportionation reactions.

Our proposed model, which aims to describe both conventional and nitroxide-modified polystyrene pyrolysis behavior, contains the following reactions: mid-chain random scission, end chain scission or activation, transfer to polymer, β-scission, depropagation, and termination by combination and by disproportionation, as can be seen in Table 3. P_n denotes living polymer with length n, while R_n, S_n, D_n, M, and $M\cdot$ denote live polymer of length n with nitroxide in one end, dormant polymer of length n, dead polymer of length n, monomer and monomer radical, respectively. The kinetic constants are denoted as: k_b (mid-chain random scission), k_{be} (end-chain scission), $k_{tr\beta}$ (transfer to polymer + β-scission, simplified mechanism), k_{rev} (depropagation), k_{tc} (termination by combination), and k_{td} (termination by disproportionation). To simplify the description, some assumptions and approximations are made:

Table 3. Polystyrene depolymerization kinetic mechanism including the presence of nitroxide-end polystyrene.

Mechanism	Reaction
Mid chain random scission, dormant polymer	$S_n \xrightarrow{k_b} P_n\cdot + \cdot R_{n-r}$
Mid chain random scission, dead polymer	$D_n \xrightarrow{k_b} P_r\cdot + \cdot P_{n-r}$
End chain scission, dead polymer	$D_n \xrightarrow{k_{be}} M\cdot + P_{n-1}$
* End chain scission or activation, dormant polymer	$S_n \underset{k_d}{\overset{k_a}{\rightleftarrows}} N + P_n$
Transfer to polymer + β-scission (1)	$P_n\cdot + S_m \xrightarrow{k_{tr\beta}} D_n \; {}^{1/2}\!\!\nearrow P_{m-r} + \cdot \; {}^{1/2}\!\!\searrow R_{m-r}\cdot + S_r \; + D_r$
Transfer to polymer + β-scission (2)	$R_n\cdot + S_m \xrightarrow{k_{tr\beta}} S_n \; {}^{1/2}\!\!\nearrow P_{m-r} + \cdot \; {}^{1/2}\!\!\searrow R_{m-r}\cdot + S_r \; + D_r$
Transfer to polymer + β-scission (3)	$P_n\cdot + D_m \xrightarrow{k_{tr\beta}} D_n + P_{m-r} + \cdot D_r$

Table 3. Cont.

Mechanism	Reaction
Transfer to polymer + β-scission (4)	$\sim\!\!\sim^\bullet R_n + \sim\!\!\sim D_m \xrightarrow{k_{trp}} \sim\!\!\sim S_n + \sim\!\!\sim^\bullet P_{m-r} + \sim\!\!\sim D_r$
De-propagation	$\sim\!\!\sim^\bullet P_m \xrightarrow{k_{rev}} \sim\!\!\sim^\bullet P_{m-1} + M$ $\sim\!\!\sim^\bullet R_m \xrightarrow{k_{rev}} \sim\!\!\sim^\bullet R_{m-1} + M$
Termination by combination	$\sim\!\!\sim^\bullet P_n + {}^\bullet\!\!\sim\!\!\sim P_m \xrightarrow{k_{tc}} \sim\!\!\sim D_{n+m}$ $\sim\!\!\sim^\bullet P_n + {}^\bullet\!\!\sim\!\!\sim R_m \xrightarrow{k_{tc}} \sim\!\!\sim S_{n+m}$ $\sim\!\!\sim^\bullet R_n + {}^\bullet\!\!\sim\!\!\sim R_m \xrightarrow{k_{tc}} \sim\!\!\sim S_{n+m}$
Termination by disproportionation	$\sim\!\!\sim^\bullet P_n + {}^\bullet\!\!\sim\!\!\sim P_m \xrightarrow{k_{td}} \sim\!\!\sim D_n + \sim\!\!\sim D_m$ $\sim\!\!\sim^\bullet P_n + {}^\bullet\!\!\sim\!\!\sim R_m \xrightarrow{k_{td}} \sim\!\!\sim D_n + \sim\!\!\sim S_m$ $\sim\!\!\sim^\bullet R_n + {}^\bullet\!\!\sim\!\!\sim R_m \xrightarrow{k_{td}} \sim\!\!\sim S_n + \sim\!\!\sim S_m$
Termination with monomeric radicals	$\sim\!\!\sim^\bullet P_n + M^\bullet \begin{array}{l} \xrightarrow{k_{tc}} \sim\!\!\sim D_{m+1} \\ \xrightarrow{k_{td}} M + \sim\!\!\sim D_m \\ \xrightarrow{1/2} Et\text{-}B + \sim\!\!\sim D_m \end{array}$ $\sim\!\!\sim^\bullet R_n + M^\bullet \begin{array}{l} \xrightarrow{k_{tc}} \sim\!\!\sim S_{m+1} \\ \xrightarrow{k_{td}} M + \sim\!\!\sim S_m \\ \xrightarrow{1/2} Et\text{-}B + \sim\!\!\sim S_m \end{array}$

* Notice that this is the same reaction illustrated in Figure 2.

Dormant polymer with one or two nitroxide-functionalized ends will be lumped together into a single quantity. Upon mid-chain random scission of these species to form two radicals, it will be assumed that only one radical contains a nitroxide-functionalized end. Notice that as the number of chain ends increases in the system due to the pyrolysis process, the probability of a chain ending in a nitroxide moiety decreases. This introduces some error, but it will be neglected for simplicity since it is assumed that on the average the error introduced is small.

Initiation includes the possible formation of free radicals either in the terminal or in a mid-chain random position. The corresponding kinetic coefficients could be different depending on the position (end-chain or mid-chain) since the radicals produced in each case are of different nature, mid-chain scission forms one primary radical and one secondary benzyl radical, while end-chain scission forms a secondary benzyl radical; both reactions can occur on dormant or dead polymer (S_n, D_n) as described in Table 3. Transfer to polymer followed by β-scission can occur between any of two types of living chains (normal, P_n, or with a nitroxide-end functionality, R_n) and one of two types of inactive chains (dormant, S_n, or dead, D_n). In the case of dormant polymer, the initiation can also occur at the nitroxide end via the activation reaction. The rest of the reactions, depropagation and termination, are the same as those appearing in the standard mechanism of free-radical polymerization, with subtle variations depending on the possible presence of a nitroxide-end functionality in one or both reacting chains.

3.3. Mathematical Model

Based on the described reaction scheme, the corresponding mathematical model was formulated writing a mass balance equation for each species. Then, the method of moments was used to keep track of changes of the average molecular weights as a function of time.

Dead polymer, $n = 1, \ldots, \infty$

$$\begin{aligned}
\frac{dD_n}{dt} = &-k_b n D_n \; -k_{be} D_n + k_{tr\beta} P_n \sum_{m=1}^{\infty} m\, D_m - k_{tr\beta}\, n\, D_n \sum_{m=1}^{\infty} P_m \\
&+ k_{tr\beta} \left(\sum_{s=1}^{\infty} P_s \right) \sum_{m=n+1}^{\infty} m\, D_m\left(\tfrac{1}{m}\right) + k_{tr\beta} P_n \sum_{m=1}^{\infty} m\, S_m \\
&+ \frac{k_{tr\beta}}{2} \left(\sum_{r=1}^{\infty} P_r \right) \sum_{m=n+1}^{\infty} m\, S_m\left(\tfrac{1}{m}\right) + \frac{k_{tr\beta}}{2} \left(\sum_{r=1}^{\infty} R_r \right) \sum_{m=n+1}^{\infty} m\, S_m\left(\tfrac{1}{m}\right) \\
&+ k_{tr\beta} \left(\sum_{r=1}^{\infty} R_r \right) \sum_{m=n+1}^{\infty} m\, D_m\left(\tfrac{1}{m}\right) - k_{tr\beta}\, n D_n \sum_{m=1}^{\infty} R_m \\
&+ \frac{k_{tc}}{2} \sum_{m=1}^{n-1} P_m P_{n-m} + k_{td} P_n \sum_{m=1}^{\infty} P_m + k_{td} P_n \sum_{m=1}^{\infty} R_m + k_{tc} P_{n-1} M \\
&+ k_{td} P_n M
\end{aligned} \qquad (1)$$

Dormant polymer, $n = 1, \ldots, \infty$

$$\begin{aligned}
\frac{dS_n}{dt} = &-k_b n S_n \; -k_a S_n + k_d M\, P_n - k_{tr\beta}\, n\, S_n \sum_{m=1}^{\infty} P_m + \frac{k_{tr\beta}}{2} \left(\sum_{r=1}^{\infty} P_r \right) \sum_{m=n+1}^{\infty} m\, S_m\left(\tfrac{1}{m}\right) \\
&- k_{tr\beta}\, n\, S_n \sum_{m=1}^{\infty} R_m + k_{tr\beta}\, R_n \sum_{m=1}^{\infty} m\, S_m \\
&+ \frac{k_{tr\beta}}{2} \left(\sum_{r=1}^{\infty} R_r \right) \sum_{m=n+1}^{\infty} m\, S_m\left(\tfrac{1}{m}\right) + k_{tr\beta} R_n \sum_{m=1}^{\infty} m\, D_m \\
&+ \tfrac{1}{2} k_{tc} \sum_{m=1}^{n-1} P_m R_{n-m} + \tfrac{1}{2} k_{tc} \sum_{m=1}^{n-1} R_m R_{n-m} + k_{td} R_n \sum_{m=1}^{\infty} P_m \\
&+ k_{td} R_n \sum_{m=1}^{\infty} R_m + k_{tc} R_{n-1} M \cdot + k_{td} R_n M
\end{aligned} \qquad (2)$$

Live polymer, $n = 1, \ldots, \infty$

$$\begin{aligned}
\frac{dP_n}{dt} = &k_a S_n - k_d N P_n \; + 2k_b \sum_{m=n+1}^{\infty} m\, D_m\left(\tfrac{1}{m}\right) + k_{be} D_{n+1} + k_b \sum_{m=n+1}^{\infty} m\, S_m\left(\tfrac{1}{m}\right) \\
&- k_{rev} P_n + k_{rev} P_{n+1} - k_{tr\beta} P_n \sum_{m=1}^{\infty} m\, S_m \\
&+ \frac{k_{tr\beta}}{2} \left(\sum_{r=1}^{\infty} P_r \right) \sum_{m=n+1}^{\infty} m\, S_m\left(\tfrac{1}{m}\right) + \frac{k_{tr\beta}}{2} \left(\sum_{r=1}^{\infty} R_r \right) \sum_{m=n+1}^{\infty} m\, S_m\left(\tfrac{1}{m}\right) \\
&- k_{tr\beta} P_n \sum_{m=n+1}^{\infty} m\, D_m + k_{tr\beta} \sum_{r=1}^{\infty} P_r \sum_{m=1}^{\infty} m\, D_m\left(\tfrac{1}{m}\right) \\
&+ k_{tr\beta} \sum_{r=1}^{\infty} R_r \sum_{m=n+1}^{\infty} m\, D_m\left(\tfrac{1}{m}\right) - k_t P_n \sum_{m=1}^{\infty} P_m - k_t P_n \sum_{m=1}^{\infty} R_m - k_t P_n M
\end{aligned} \qquad (3)$$

Live polymer with a nitroxide at the chain-end, $n = 1, \ldots, \infty$

$$\frac{dR_n}{dt} = k_b \sum_{m=n+1}^{\infty} m\, S_m\left(\frac{1}{m}\right) + \frac{k_{tr\beta}}{2}\left(\sum_{m=1}^{\infty} P_m\right) \sum_{r=n+1}^{\infty} r\, S_r\left(\frac{1}{r}\right) - k_{tr\beta} R_n \sum_{m=1}^{\infty} m\, S_m$$
$$+ \frac{k_{tr\beta}}{2}\left(\sum_{r=1}^{\infty} R_r\right) \sum_{m=n+1}^{\infty} m\, S_m\left(\frac{1}{m}\right) - k_{tr\beta} R_n \sum_{m=1}^{\infty} m\, D_m + k_{rev} R_{n+1} \quad (4)$$
$$- k_{rev} R_n - k_t R_n \sum_{m=1}^{\infty} P_m - k_t R_n \sum_{m=1}^{\infty} R_m - k_t R_n M$$

Monomer

$$\frac{dM}{dt} = k_{rev} \sum_{m=1}^{\infty} P_m + \frac{1}{2} k_{td} M \cdot \sum_{m=1}^{\infty} P_m + k_{rev} \sum_{m=1}^{\infty} R_m + \frac{1}{2} k_{td} M \cdot \sum_{m=1}^{\infty} R_m \quad (5)$$

Monomeric radicals

$$\frac{dM}{dt} = k_{rev} \sum_{m=1}^{\infty} P_m + \frac{1}{2} k_{td} M \cdot \sum_{m=1}^{\infty} P_m + k_{rev} \sum_{m=1}^{\infty} R_m + \frac{1}{2} k_{td} M \cdot \sum_{m=1}^{\infty} R_m \quad (6)$$

Nitroxide radicals

$$\frac{dN}{dt} = k_a S_n - k_d N \cdot P_n \quad (7)$$

Ethyl benzene

$$\frac{dEt-B}{dt} = \frac{1}{2} k_{td} M \cdot \sum_{m=1}^{\infty} P_m + \frac{1}{2} k_{td} M \cdot \sum_{m=1}^{\infty} R_m \quad (8)$$

Four polymer populations exist in this system: Dead polymer, dormant polymer, growing polymeric radical and polymer radical with a nitroxide end functionality; moments for these species are defined in Equations (9)–(12) respectively:

$$\lambda_i = \sum_r r^i D_r \quad (9)$$

$$\xi_i = \sum_r r^i S_r \quad (10)$$

$$\mu_i = \sum_r r^i P_r \quad (11)$$

$$\psi_i = \sum_r r^i R_r \quad (12)$$

The method of moments applied to the previously described balance equations generates Equations (13)–(15) for the zero-th, first, and second moments of dead polymer, respectively, Equations (16)–(18) for the moments of dormant polymer, Equations (19)–(21) for polymeric radical moments, and Equations (22)–(24) for the moments of polymeric radicals with a nitroxide end functionality.

$$\begin{aligned}\frac{d\lambda_0}{dt} = & -k_b \lambda_1 - k_{be}\lambda_0 + k_{td}\mu_0^2 + \frac{k_{tc}}{2}\mu_0^2 + k_{td}\mu_0\psi_0 + k_{tc}M\cdot\mu_0 + k_{td}M\cdot\mu_0 \\ & + k_{tr\beta}\mu_0\lambda_1 - k_{tr\beta}\mu_0\lambda_1 + k_{tr\beta}(\lambda_1 - \lambda_0)(\mu_0 + \psi_0) + k_{tr\beta}\mu_0\xi_1 \\ & + \frac{k_{tr\beta}}{2}(\mu_0 + \psi_0)B_0(\xi_1 - \xi_0) - k_{tr\beta}\lambda_1\psi_0 \\ & B_0 = 1 \end{aligned} \quad (13)$$

$$\begin{aligned}\frac{d\lambda_1}{dt} = & -k_b\lambda_2 - k_{be}\lambda_1 + k_{td}\mu_1\mu_0 + k_{tc}\mu_1\mu_0 + k_{td}\mu_1\psi_0 + k_{tc}M(\mu_1 + \mu_0) + k_{td}M\cdot\mu_1 \\ & + k_{tr\beta}[\mu_1\lambda_1 - \lambda_2\mu_0 + \frac{(\mu_0+\psi_0)}{2}(\lambda_2 - \lambda_1) + \mu_1\xi_1 \\ & + \frac{1}{4}(\mu_0 + \psi_0)(\xi_2 - \xi_1) - \lambda_2\psi_0] \end{aligned} \quad (14)$$

$$\frac{d\lambda_2}{dt} = -k_b\lambda_3 - k_{be}\lambda_2 + k_{td}\mu_2\mu_0 + k_{tc}\left(\mu_0\mu_2 + \mu_1^2\right) + k_{td}\mu_2\psi_0 + k_{tc}M\cdot(\mu_2 + 2\mu_1 \\
+ \mu_0) + k_{td}M\cdot\mu_2 + k_{tr\beta}[\mu_2\lambda_1 - \lambda_3\mu_0] \\
+ (\mu_0 + \psi_0)\left(\tfrac{1}{3}\lambda_3 - \tfrac{1}{2}\lambda_2 + \tfrac{1}{6}\lambda_1\right) + \mu_2\xi_1 \\
+ \tfrac{1}{2}(\mu_0 + \psi_0)\left(\tfrac{1}{3}\xi_3 - \tfrac{1}{2}\xi_2 + \tfrac{1}{6}\xi_1\right) - \lambda_3\psi_0]$$
(15)

$$\frac{d\xi_0}{dt} = -k_b\xi_1 - k_a\xi_0 + k_dN\mu_0 + \tfrac{k_{tc}}{2}(\mu_0\psi_0 + \psi_0\psi_0) + k_{td}(\psi_0\mu_0) + k_{td}\psi_0^2 + k_{tc}M \\
\cdot\psi_0 + k_{td}\psi_0M - k_{tr\beta}\xi_1\mu_0 - k_{tr\beta}\xi_1\psi_0 + \tfrac{k_{tr\beta}}{2}(\mu_0 + \psi_0)(\xi_1 - \xi_0) \\
+ k_{tr\beta}\psi_0(\xi_1 + \lambda_1)$$
(16)

$$\frac{d\xi_1}{dt} = -k_b\xi_2 - k_a\xi_1 + k_dN\mu_1 + \tfrac{k_{tc}}{2}(\mu_1\psi_0 + \mu_0\psi_1) + k_{tc}\psi_1\psi_0 + k_{td}(\psi_1\mu_0) \\
+ k_{td}\psi_1\psi_0 + k_{tc}M\cdot(\psi_0 + \psi_1) + k_{td}\psi_1M - k_{tr\beta}\xi_2\mu_0 - k_{tr\beta}\xi_2\psi_0 \\
+ \tfrac{k_{tr\beta}}{4}(\mu_0 + \psi_0)(\xi_2 - \xi_1) + k_{tr\beta}\psi_1(\xi_1 + \lambda_1)$$
(17)

$$\frac{d\xi_2}{dt} = -k_b\xi_3 - k_a\xi_2 + k_dN\mu_2 + \tfrac{k_{tc}}{2}(\mu_2\psi_0 + 2\mu_1\psi_1 + \mu_0\psi_2) + k_{tc}\left(\psi_2\psi_0 + \psi_1^2\right) \\
+ k_{td}(\psi_2\mu_0) + k_{td}\psi_2\psi_0 + k_{tc}M(\psi_2 + 2\psi_1 + \psi_0) + k_{td}\psi_2M \\
- k_{tr\beta}\xi_3\mu_0 - k_{tr\beta}\xi_3\psi_0 + \tfrac{k_{tr\beta}}{2}(\mu_0 + \psi_0)\left(\tfrac{1}{3}\xi_3 - \tfrac{1}{2}\xi_2 + \tfrac{1}{6}\xi_1\right) \\
+ k_{tr\beta}\psi_2(\xi_1 + \lambda_1)$$
(18)

$$\frac{d\mu_0}{dt} = k_a\xi_0 - k_dN\mu_0 + k_{be}\lambda_0 - k_t\mu_0^2 - k_t\mu_0\psi_0 - k_t\mu_0M - k_{tr\beta}\mu_0\xi_1 \\
+ k_{tr\beta}(\mu_0 + \psi_0)(\lambda_1 - \lambda_0) - k_{tr\beta}\mu_0\lambda_1 + \tfrac{1}{2}k_{tr\beta}(\mu_0 + \psi_0)(\xi_1 - \xi_0) \\
+ 2k_b[(\lambda_1 - \lambda_0) + (\xi_1 - \xi_0)]$$
(19)

$$\frac{d\mu_1}{dt} = k_a\xi_1 - k_dN\,\mu_1 + k_{be}(\lambda_1 - \lambda_0) - k_{rev}\mu_1 + k_{rev}(\mu_1 - \mu_0) - k_t\mu_1\mu_0 - k_t\mu_1\psi_0 - k_t\mu_1M \\
- k_{tr\beta}\mu_1\lambda_1 + k_{tr\beta}(\mu_0 + \psi_0)\left[\tfrac{1}{2}(\lambda_2 - \lambda_1)\right] - k_{tr\beta}\mu_1\xi_1 + \tfrac{1}{2}k_{tr\beta}(\mu_0 + \psi_0)\left[\tfrac{1}{2}(\xi_2 - \xi_1)\right] + k_b\left[(\lambda_2 - \lambda_1) + \tfrac{1}{2}(\xi_2 - \xi_1)\right]$$
(20)

$$\frac{d\mu_2}{dt} = k_a\xi_2 - k_dN\mu_2 + k_{be}(\lambda_2 - 2\lambda_1 + \lambda_0) - k_{rev}\mu_2 + k_{rev}(\mu_2 - 2\mu_1 + \mu_0) - k_t\mu_2\mu_0 \\
- k_t\mu_2\psi_0 - k_t\mu_2M - k_{tr\beta}\mu_2\lambda_1 + k_{tr\beta}(\mu_0 + \psi_0)\left[\tfrac{1}{3}\lambda_3 - \tfrac{1}{2}\lambda_2 + \tfrac{1}{6}\lambda_1\right] \\
- k_{tr\beta}\mu_2\xi_1 + \tfrac{1}{2}k_{tr\beta}(\mu_0 + \psi_0)\left[\tfrac{1}{3}\xi_3 - \tfrac{1}{2}\xi_2 + \tfrac{1}{6}\xi_1\right] \\
+ 2k_b\left[\tfrac{1}{3}\lambda_3 - \tfrac{1}{2}\lambda_2 + \tfrac{1}{6}\lambda_1 + \tfrac{1}{3}\xi_3 - \tfrac{1}{2}\xi_2 + \tfrac{1}{6}\xi_1\right]$$
(21)

$$\frac{d\psi_0}{dt} = -k_t\psi_0\mu_0 - k_t\psi_0^2 - k_t\psi_0M + k_b(\xi_1 - \xi_0) + \tfrac{1}{2}k_{tr\beta}(\mu_0 + \psi_0)(\xi_1 - \xi_0) \\
- k_{tr\beta}\psi_0(\xi_1 + \lambda_1)$$
(22)

$$\frac{d\psi_1}{dt} = -k_{rev}\psi_0 - k_t\psi_1\mu_0 - k_t\psi_1\psi_0 - k_t\psi_1M + \tfrac{k_b}{2}(\xi_2 - \xi_1) \\
+ \tfrac{1}{2}k_{tr\beta}(\mu_0 + \psi_0)\left[\tfrac{1}{2}(\xi_2 - \xi_1)\right] - k_{tr\beta}\psi_1(\xi_1 + \lambda_1)$$
(23)

$$\frac{d\psi_2}{dt} = k_{rev}(\psi_2 - 2\psi_1 + \psi_0) - k_{rev}\psi_2 - k_t\psi_2\mu_0 - k_t\psi_2\psi_0 - k_t\psi_2M \\
+ k_b\left[\tfrac{1}{3}\xi_3 - \tfrac{1}{2}\xi_2 + \tfrac{1}{6}\xi_1\right] + \tfrac{1}{2}k_{tr\beta}(\mu_0 + \psi_0)\left[\tfrac{1}{3}\xi_3 - \tfrac{1}{2}\xi_2 + \tfrac{1}{6}\xi_1\right] \\
- k_{tr\beta}\psi_2(\xi_1 + \lambda_1)$$
(24)

From the previous equations, the number average molecular weight can be calculated using Equation (25)

$$M_n = \frac{\lambda_1 + \xi_1 + \mu_1 + \psi_1}{\lambda_0 + \xi_0 + \mu_0 + \psi_0}$$
(25)

Notice that the moment equations are not closed since the second moment depends on the third moment. To solve this problem, the well-known expression of Saidel and Katz [35] is used to estimate the third moment in terms of the lower moments.

$$\lambda_3 = \frac{\lambda_2}{\lambda_0\lambda_1}\left(2\lambda_2\lambda_0 - \lambda_1^2\right)$$
(26)

The mathematical model formulation yielded a set of ordinary differential equations (ODE's) that were numerically solved using the FORTRAN programming language with the DDASSL routine for the integration of the equations. Kinetic rate constant values were estimated to match the recorded experimental data as explained below.

3.4. Experimental Results

Since the experiments at the extreme levels of N/I (0 and 1.3) were run by triplicate, it was possible to estimate a pooled variance (σ^2) for the relevant responses (one for each temperature), and therefore error bars of 1σ are included in most of the plots.

Figure 3 shows the relative yields of the solid, liquid, and gaseous fractions obtained in the pyrolysis experiments using a given N/I ratio or by FRP carried at different temperatures. It can be noted that at higher pyrolysis temperatures the liquid fraction has a tendency to increase while the gas and solid fractions diminish. Comparing all the samples, the ones corresponding to the N/I ratio of 1.3 exhibit the higher values of recovered liquid fraction, the maximum value reached being 88% at 420 °C, while the lowest values correspond to the FRP samples. These results point out to a favorable effect of the presence of the nitroxide moiety at the end of the polymer chains on the pyrolysis process; more will be discussed below. All the percentages are referred to the total amount of polymer charged at the beginning of the experiments.

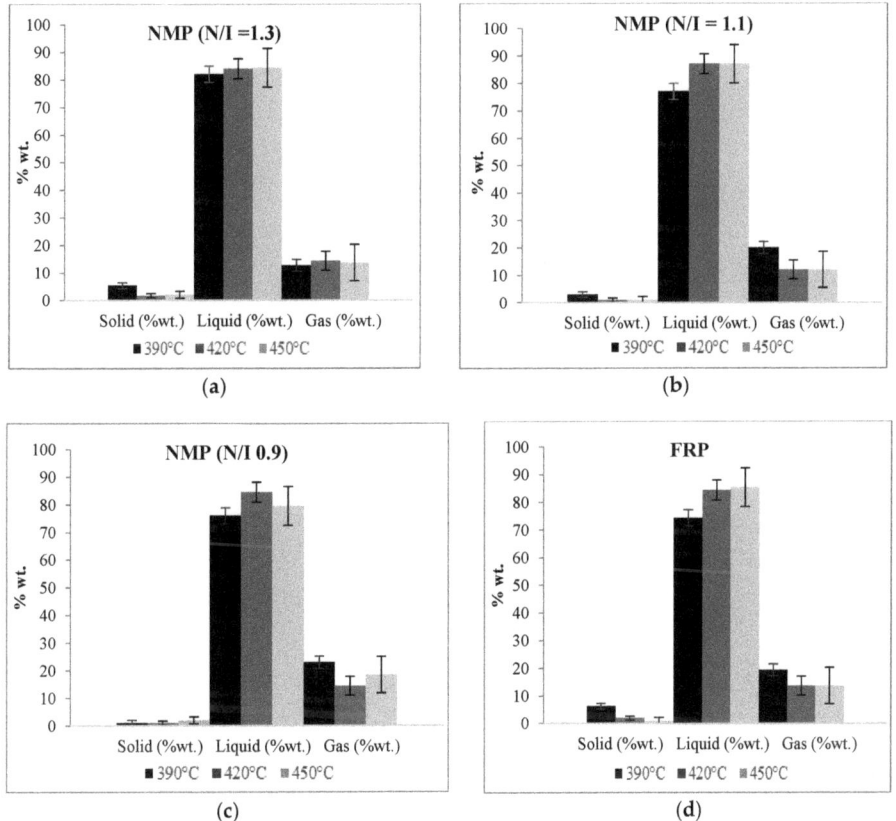

Figure 3. Yields of recovered fractions of the pyrolyzed samples: (**a**) N/I = 1.3, (**b**) N/I = 1.1, (**c**) N/I = 0.9, and (**d**) FRP.

Figure 4 shows the identified products of the PS pyrolysis in the liquid fraction; their relative amounts are plotted for each PS pyrolysis experiment at three reaction temperatures (390, 420, and 450 °C). The main product in all cases is styrene followed by styrene dimer in second place; an extra category is included denoted as "mixture", which corresponds to the sum of all the components that could not be identified, but individually amount to less than 1 wt.%. In the case of the FRP samples, stilbene and benzoic acid are also generated and the compounds in the non-identified mixture reach the highest values of all the samples.

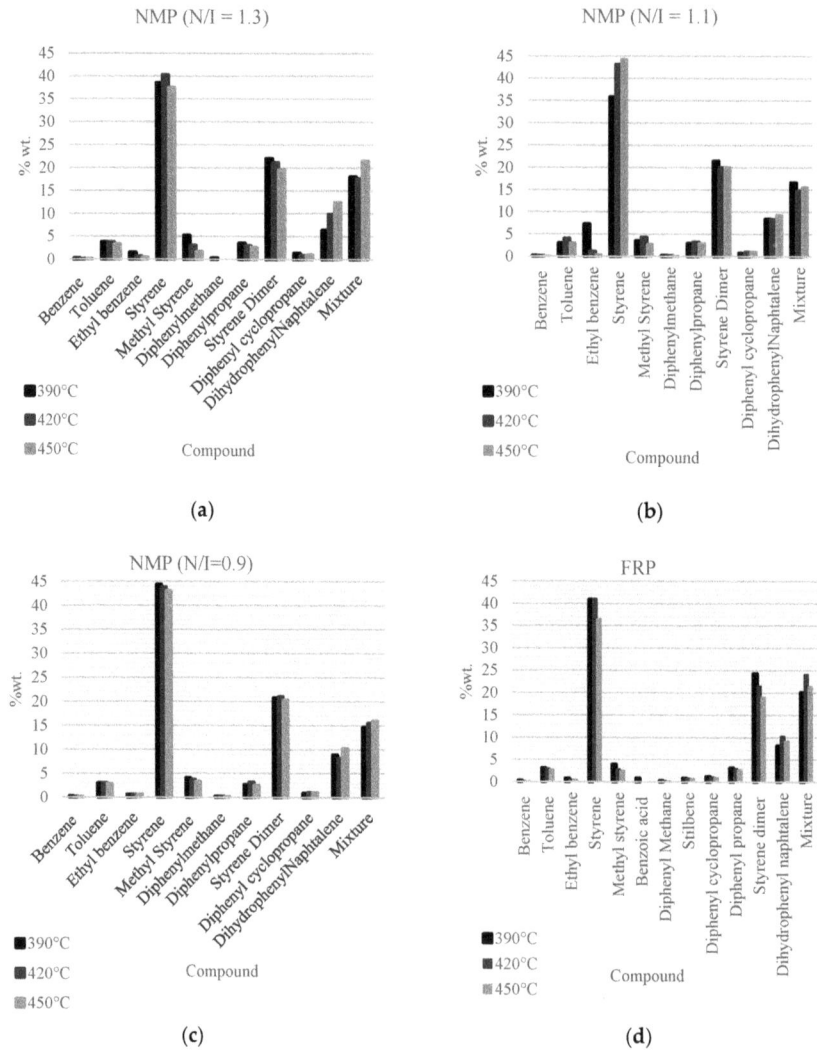

Figure 4. % wt. yield of the polystyrene depolymerization products in the liquid fraction: (a) N/I = 1.3, (b) N/I = 1.1, (c) N/I = 0.9, and (d) FRP.

Figure 5 shows the absolute yields of styrene monomer and styrene dimer of the different samples (with respect to the total load of PS, not only with respect to the liquid fraction recovered), indicating that at higher reaction temperatures the styrene dimer concentration tends to lower for the reaction with N/I = 1.3, but it is the opposite for the reaction with N/I = 0.9. The monomer recovery yield exhibits a more complex behavior with temperature. For samples N/I = 1.1 and N/I = 0.9, the monomer recovery yield increases with higher temperatures, while for the N/I = 1.3 and the FRP samples maxima for monomer yield are exhibited. Clearly, not all the effects observed are significant, but in some cases the differences are evident, although generally moderate.

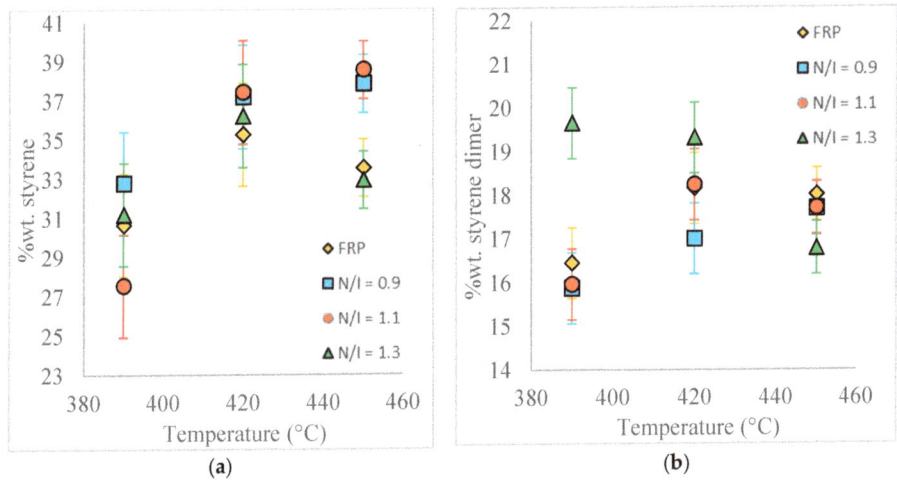

Figure 5. Absolute yields of (**a**) styrene and (**b**) styrene dimer of the analyzed samples.

Comparing all the samples at 450 °C, the highest styrene yield (38.5%) is obtained using N/I = 1.1, and the lowest using N/I = 1.3 (around 33%) or, almost identical, FRP. Interestingly, at 420 °C all the samples with nitroxide show higher styrene yield than the FRP sample.

Pyrolysis reaction time is also affected by the use of PS samples containing nitroxide. Compared with FRP PS, the pyrolysis of these samples results in shorter reaction times at comparable temperatures, except in the case of FRP PS at 450 °C (Figure 6).

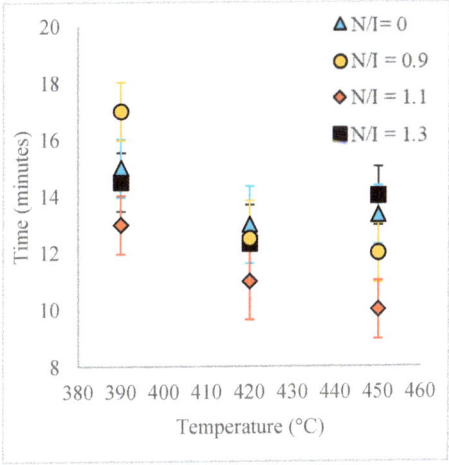

Figure 6. Reaction time (min) vs. temperature (C).

During the experimentation, special emphasis was put in obtaining reproducible results, and therefore the experimental design included replicate runs. Additionally, by performing replicate experiments, we were able to observe clear qualitative differences in the presence and absence of nitroxide in the samples. For example, for the blank reactions, the pressure increase was more abrupt, and the liquid recovery began in all cases at around 380 °C (during the temperature increase ramp). On the other hand, for the dormant polymer samples, especially those of N/I = 1.3, the pressure increase was smoother and more gradual, and the liquid recovery began before, at temperatures between 370–380 °C. These and other observations, as well as some of the quantitative results, reinforce the conclusion that the pyrolysis of PS samples with and without nitroxide ends behave differently.

3.5. Mathematical Modeling Results

The mathematical model developed here was intended as a first approach to qualitatively and quantitatively understand the experimental results observed. At this stage, only main experimental trends, particularly reaction time and the effect of nitroxide chain-ends on reaction time, were taken into account to tune the model parameters, since not much detailed information, such as time evolution of the molecular weight distribution or of the product distribution (in terms of individual chain-lengths), was available. Ongoing work in our group is directed towards getting more detailed experimental information as well as more detailed mathematical models.

With respect to the kinetic parameters, values are reported in the literature for some of them. Table 4 summarizes some rate constants for PS depolymerization available from the literature that were estimated in different reaction systems.

Table 4. Reaction kinetic constants of polystyrene depolymerization from literature sources.

Reference	[36]		[37]		[38]	
Processes	A [a,b]	E [b] (kcal/mol)	A [a,b]	E [b] (kcal/mol)	A [a,b]	E [b] (kcal/mol)
Chain fission (s^{-1})	5.0×10^{13}	63.7	7.98×10^{15}	2.3	1.0×10^{16}	2.3
Chain fission allyl (s^{-1})	5.0×10^{12}	58.6	-	-	5.5×10^{13}	2.3
H- abstraction of end chain $\left(\frac{L}{mol\,seg}\right)$	5.0×10^{7}	13.5	-	-	-	-
H-abstraction of mid chain $\left(\frac{L}{mol\,seg}\right)$	5×10^{7}	16.5	-	-	-	-
H-abstraction	-	-	1.5×10^{8}	13.3	2.1×10^{6}	12
Mid chain β-scission (s^{-1})	1.2×10^{13}	27.0	2.62×10^{12}	12.9	4.1×10^{12}	11.4
End chain β-scission (s^{-1})	1.6×10^{13}	25.8	2.62×10^{12}	12.9	4.1×10^{12}	11.4
Radical recombination $\left(\frac{L}{mol\,seg}\right)$	5.0×10^{6}	14.0	2.2×10^{9}	2.3	1.1×10^{11}	2.3
Disproportionation $\left(\frac{L}{mol\,seg}\right)$	-	-	1.14×10^{8}	2.3	5.5×10^{9}	2.3

[a] Units compatible with those shown in first column. [b] A and E are pre-exponential and activation energy in the corresponding Arrhenius expression for the kinetic constant.

Literature values for the kinetic constants were tested using the mathematical model in terms of moments described by the Equations (5)–(8) and (13)–(24); however, the sets of parameter values available in the literature were either incomplete or corresponded to kinetic models with differences with respect to the one proposed here. When they were tried in the present kinetic model, they yielded inconsistent results in the sense of high stiffness and/or non-convergence of the numerical integration of the differential equations, as well as some results without proper physical meaning.

4. Discussion

4.1. Sensitivity Analysis

Given the lack of useful values from the literature, it was decided to perform a sensitivity analysis to fit a consistent set of kinetic parameter values of the model. The results of the sensitivity analysis are summarized in Figures 7–11, which show essentially the sensitivity of the pyrolysis model in terms of the evolution of the polymer M_n and the monomer generation with the time of reaction upon varying different reaction coefficients. The base set of used kinetic constants at 350 °C in all cases is listed in Table 5. In the value of M_n plotted the definition given by Equation (25) was used, which represents a lumped value of average molecular weight that includes all the present polymer populations (naturally, the terms corresponding to dormant polymer and live polymer with a nitroxide end are zero for the dead polymer case). The criterion used to choose the set of parameters of Table 5, among other viable sets, was that they produced qualitative results that were consistent with the experimental observations and also that their values fell in the vicinity of similar parameters reported in the literature when available (same order of magnitude). Clearly, the proper selection of kinetic parameters for this system is a difficult and still unsolved problem which requires further investigation. Ideally, the kinetic parameters should be evaluated by independent experiments as model-free as possible; however, this goal is quite challenging for systems exhibiting the complexity of the present one. The proposed kinetic parameters are assumed as a good starting point for further research, and are valuable to show the feasibility of the mathematical model presented and for parameter sensitivity studies, but it is important to mention that this parameter set is not unique and there may be other sets of values similarly viable and consistent with the experimental observations.

The discussion of Figures 6–11 is done by groups of plots. In each case analyzed, four plots are presented in each figure: two of them correspond to pyrolysis experiments of dead polymer having no functionality; the other two correspond to the pyrolysis of dormant polymer with a nitroxide functionality at its end. In the first two plots, the evolution (degradation) of M_n of the dead polymer and of the dormant polymer respectively is shown; in the other two plots, the evolution of monomer for the dead and dormant polymer is respectively exhibited.

Table 5. Depolymerization kinetic constants at 350 °C.

Mechanism (L mol^{-1}s^{-1} Unless Indicated)	Variable	Value
End Chain scission (s^{-1})	k_{be}	8.0×10^2
Transfer + β-scission	k_{trb}	1.1×10^2
Depropagation (s^{-1})	k_{rev}	1.1×10^{-1}
Termination by combination	k_{tc}	5.8×10^1
Termination by disproportionation	k_{td}	2×10^1
Mid chain scission (s^{-1})	k_b	1.5×10^{-1}
Deactivation of dormant species (at 130 °C)	k_d	4.78×10^7 *
Activation of dormant species (s^{-1}) (at 130 °C)	k_a	3.16×10^{-3} *

* Arrhenius expressions: $k_d = 5 \times 10^9 \exp\left(-\frac{16,000}{RT}\right)$; $k_a = 4 \times 10^{13} \exp\left(-\frac{124,000}{RT}\right)$; R in J mol^{-1}K^{-1} [39].

Figure 7 shows the effect of the variation of the value of k_{be} (end-chain scission) on the pyrolysis of dead and dormant polymer. Unexpectedly, in both cases, a smaller value of this kinetic coefficient results in a faster decomposition rate and in lower M_n at the end of the reaction time. In the following discussion the case of dead polymer will be taken as an illustration for the explanation, but the case of dormant polymer follows a similar pattern. To understand why this behavior is observed, the evolution of moments 0 and 1 of live and dead polymer were investigated and it was found that almost instantaneously most of the dead chains are activated; that is, they go from a dead state to live polymer (produced by the different available initiation mechanisms). There are however differences in the dynamics of this activation depending on the values of k_{be}. When lower values of this constant are used, the live polymer concentration rises more slowly (and the dead polymer concentration also decreases more slowly), but it reaches a higher maximum and then a higher stationary value, at the same time that the dead polymer concentration is less depleted in this case (higher stationary concentration). In fact, higher remaining concentration of dead polymer in these conditions favors a relatively faster steady formation of live polymer (via chain-end activation) and leads to a higher quasi-stationary concentration of live polymer (roughly proportional to $\mu_0 \approx \left[\frac{k_{be}\lambda_0}{k_t}\right]^{1/2}$, the lower value of k_{be} is more than compensated by a higher λ_0). Finally, an ultimate higher concentration of live polymer leads to faster depropagation and higher monomer production. The effect of k_{be} is more pronounced for the pyrolysis of dead polymer than for dormant polymer. This is expected since a large portion of the initial decomposition step of the dormant polymer should occur at the nitroxide end, making the other contribution less important.

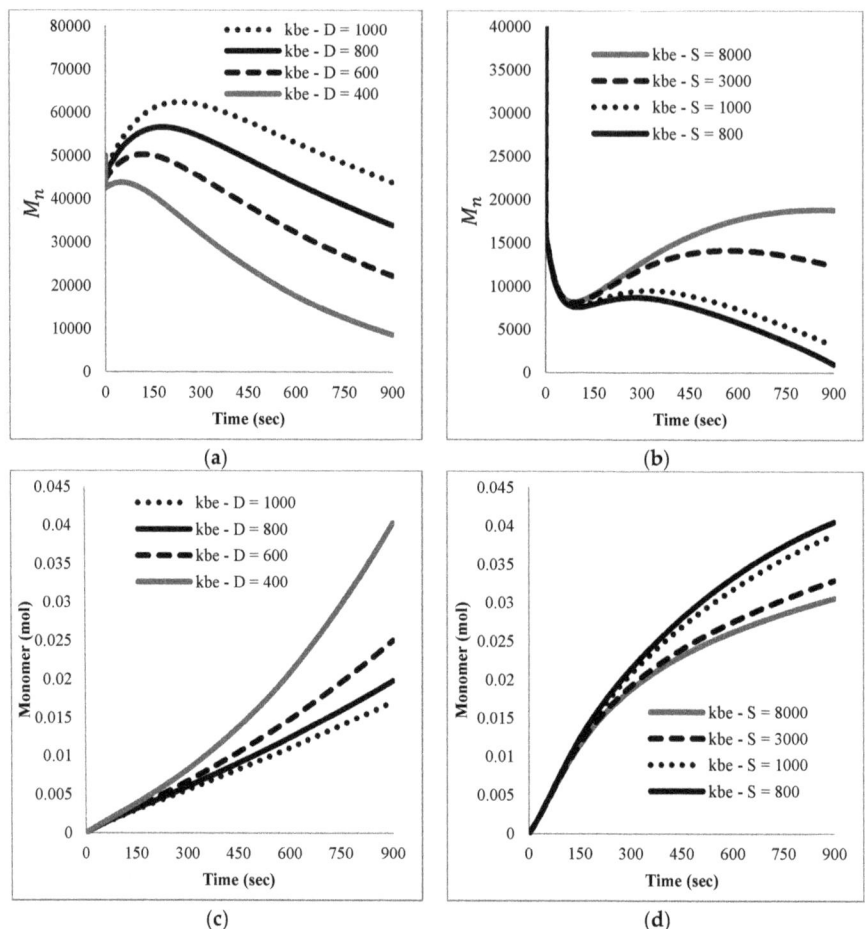

Figure 7. Effect of the variation of the chain-end scission coefficient, k_{be}, value on the pyrolysis M_n evolution of (**a**) dead polymer and (**b**) dormant polymer, and on the monomer generation of (**c**) dead polymer and (**d**) dormant polymer.

Figure 8 shows the sensitivity of the M_n evolution when the value of the transfer to polymer + β-scission coefficient, k_{trb}, is changed. As this value increases, the polymer decomposes faster, but the behavior shows significant quantitative differences for the dead and the dormant polymer. In both cases the value of M_n goes through a minimum, then increases to reach a maximum and finally steadily decreases; however, for the dormant polymer cases the initial minima are much more pronounced and the subsequent maxima are less marked, indicating a more effective decomposition in the case of dormant polymer. This is confirmed by the plots of monomer evolution that grows faster and reaches higher values (roughly double) for the case of dormant polymer pyrolysis compared to that of dead polymer.

Figure 9 shows the effect of the termination by combination value (k_{tc}) over the M_n evolution. For dead polymer the effect is quite complex. At early stages of the reaction (less than 100 s) M_n values reach a maximum, which is higher and is reached faster for higher values of k_{tc} (an expected output); however, as the activation reactions proceed, the faster dynamics apparently promoted by higher values of k_{tc} accelerate the polymer degradation leading faster to lower values of M_n. These shorter chains imply larger number of dead chain-ends that can undergo end-scission, accelerating the degradation of the polymer and the monomer production. Similar effects are present in the case of dormant polymer, but they are significantly attenuated by the competing mechanism of deactivation/degradation at the nitroxide end, which tends to dominate, especially at lower values of k_{tc}.

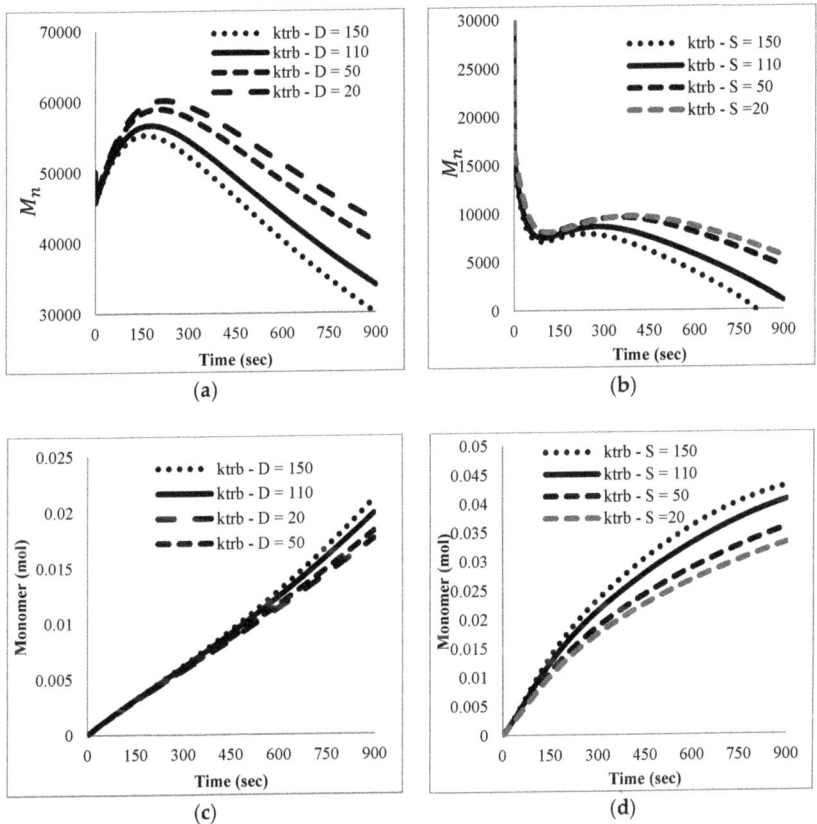

Figure 8. Effect of the variation of transfer to polymer + β-scission (k_{trb}) values on the pyrolysis M_n evolution of (**a**) dead polymer and (**b**) dormant polymer, and on the monomer generation of (**c**) dead polymer and (**d**) dormant polymer.

Remarkably, no reasonable results can be obtained by using typical values of this coefficient at temperatures in the range of 100–200 °C, which are in the order of 10^8 Lmol^{-1}s^{-1} [40] and much lower values had to be used in the simulations; otherwise numerical convergence for the solution of the equations was not possible. Some researchers that have modeled the thermal pyrolysis of PS omit this reaction [41] at all; others, for example Kruse et al. [37,38], use a standard Arrhenius expression for this constant extrapolating its value at the high pyrolysis temperatures; however, they report to have used a gel effect expression modifying the Arrhenius expression which would lower the value of k_{tc} by several orders of magnitude. They do not report the resulting range of values for this constant, but they may lie close to the values used in this work. Since it is not clear that diffusion limitations are present at these conditions, it was preferred here not to speculate and use instead a constant value that was adequate for the numerical solution of the model and resulted in reasonable outputs. In our next study this issue will be explored further.

In Figure 10 the effect of the reverse propagation value (k_{rev}) over the M_n evolution is shown. As for other parameters, for example those shown in Figures 7 and 8, the pyrolysis process for the dormant polymer is faster than for the dead polymer. The dormant polymer swiftly reaches a very low value of M_n, which is lower the higher the value of k_{rev}, while for dead polymer the value of M_n exhibits a gradually decreasing value even at the final time of the process. The generation of monomer in both cases is consistent with the decrease of the molecular weight.

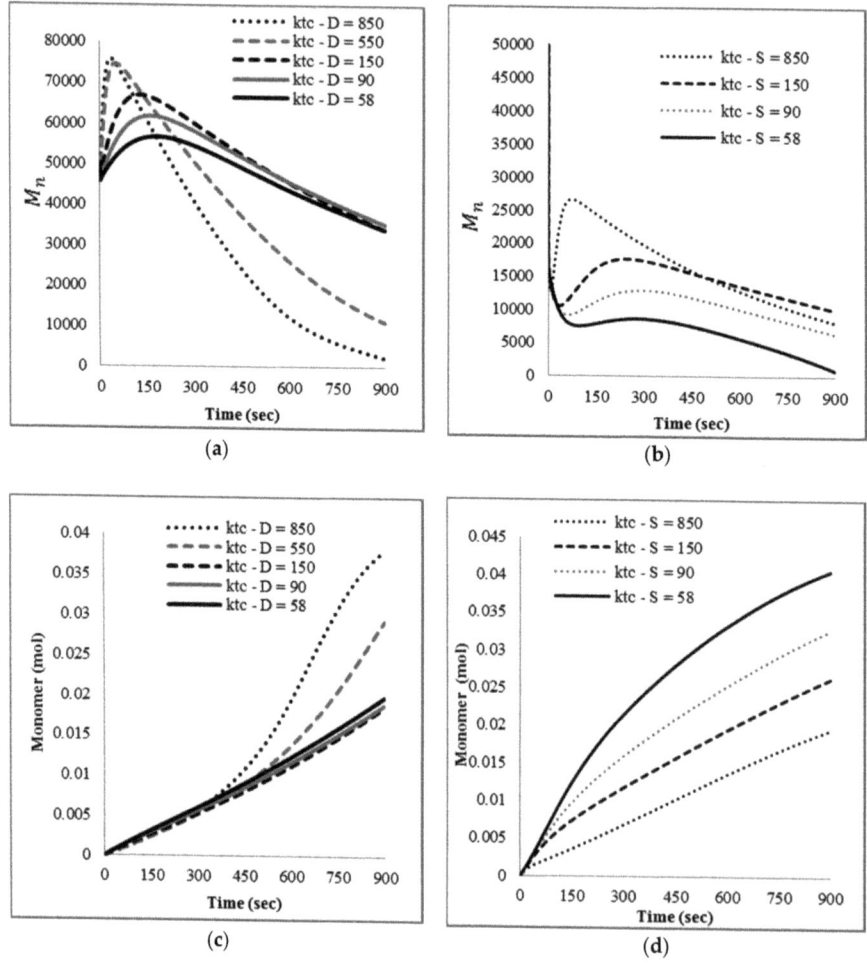

Figure 9. Effect of the variation of the termination by combination coefficient, k_{tc}, value on the pyrolysis M_n evolution of (**a**) dead polymer and (**b**) dormant polymer, and on the monomer generation of (**c**) dead polymer and (**d**) dormant polymer.

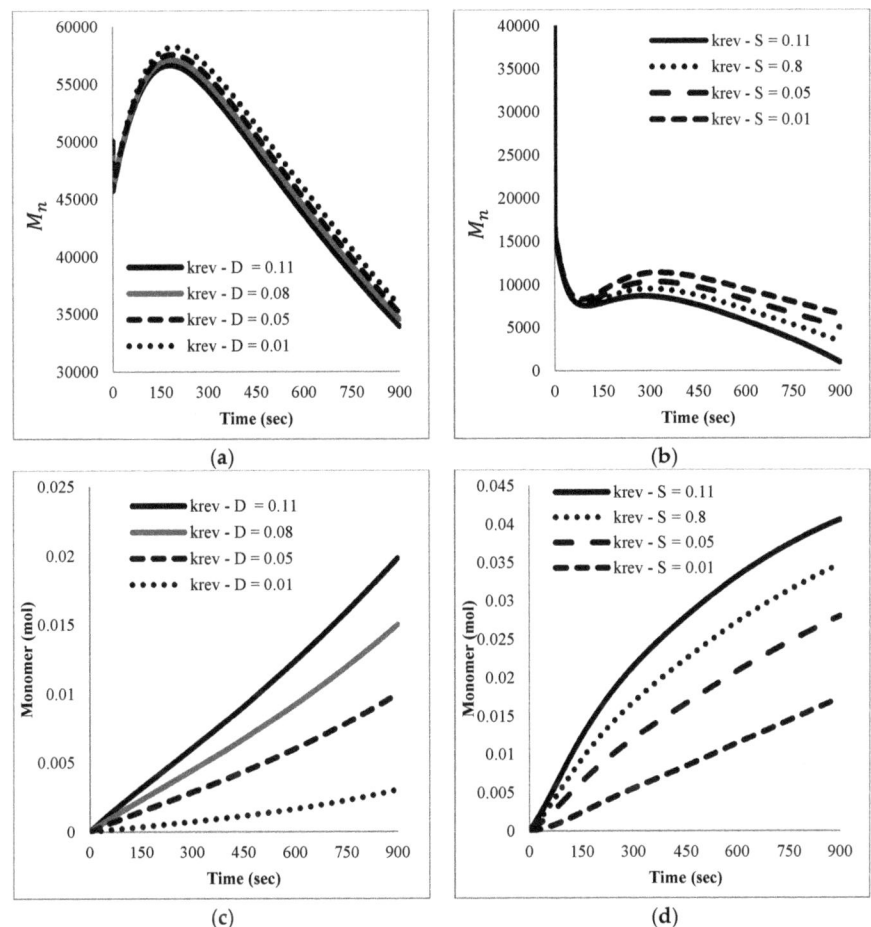

Figure 10. Effect of the variation of the reverse propagation coefficient value, k_{rev}, on the pyrolysis M_n evolution of (**a**) dead polymer and (**b**) dormant polymer, and on the monomer generation of (**c**) dead polymer and (**d**) dormant polymer.

The sensitivity of the outputs to k_{td} is not included here since it results in effects closely related to those observed with k_{tc}.

Figure 11 shows the effect of the mid-chain random scission value (k_b) on the number average molecular weight and monomer generation evolution. Mid-chain β-scission has been identified as a key step in the reduction of molecular weight of polymers, resulting in the formation of an end-chain polymer radical and a stable polymer with a double bond at the chain end and it is considered a major contribution to the formation of styrene monomer from PS. [42] In the sensitivity calculations for the dead polymer case the behavior is somewhat complex and non-linear, but the magnitude of the effects are relatively small. These small effects, combined with the fact that the M_n value combines the populations of live and dead polymer, partly explains the complexity observed. Notice that the use of k_b values, far from those shown in Figure 11, resulted in no convergence of the integration algorithm; therefore, the values studied were restricted to a small range. On the other hand, for the dead polymer case, the monomer generation grows with an increase in the values of k_b, as expected. With respect to the dormant polymer case, the behavior of M_n is simpler, as this output decreases faster the larger the value of k_b, as it could be expected; it seems that the presence of the activation reaction at the nitroxide end helps in removing small non-linear effects. Consistent with this behavior, the generation of monomer increases with larger values of k_b.

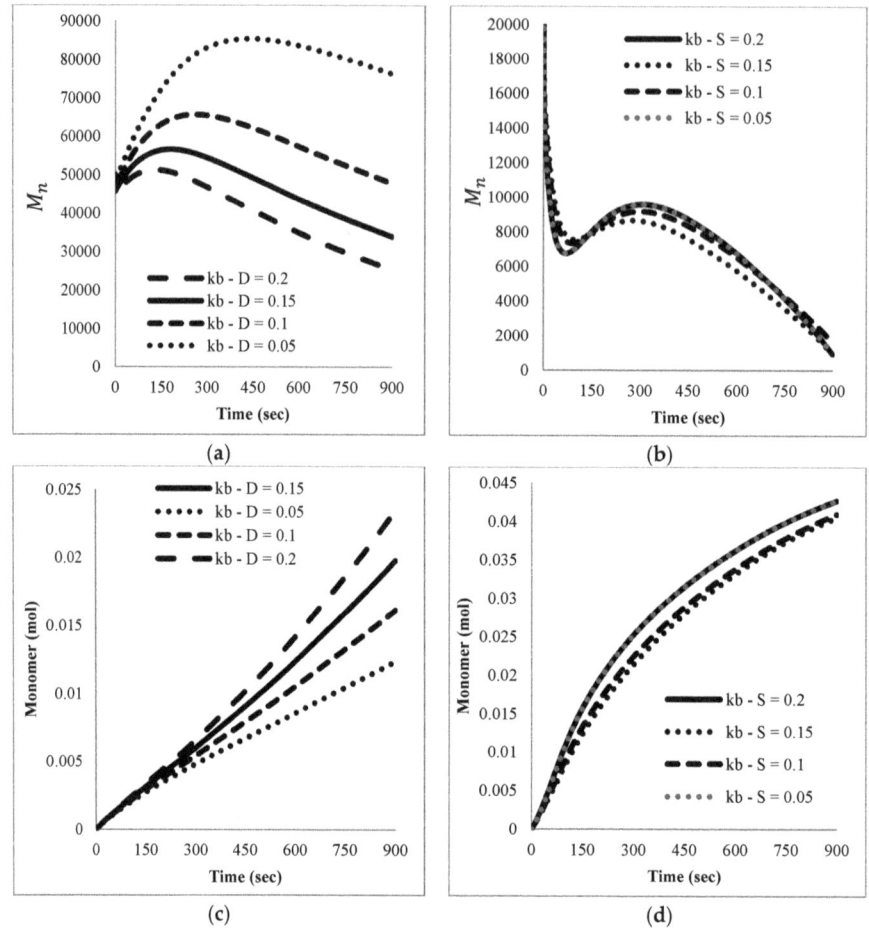

Figure 11. Effect of the variation of the mid-chain random scission coefficient, k_b, value on the pyrolysis M_n evolution of (**a**) dead polymer and (**b**) dormant polymer, and on the monomer generation of (**c**) dead polymer and (**d**) dormant polymer.

Notice that in all of the previous analyzed figures, the values of the constants tested can vary by several orders of magnitude, with no clear indication as to why the curves for the shown values were selected. The general criterion used to show curves for specific ranges of values of the kinetic constants was to maximize the sensitivity of the model to the parameters analyzed; that is, the curves were more sensitive for the ranges of values shown, and for others not shown the sensitivity was very small or almost null.

4.2. Base Case Simulation

As discussed before, given the difficulties to fit kinetic parameters for this system and the lack of detailed quantitative information from experiments, no attempt was made to compare in detail the simulations with the experimental data. The goal at this stage was just to obtain agreement of the simulations with gross measurements of reaction progress, specifically the monomer generation and the time of reaction. In this section simulation results for the base case are presented and discussed. In a more detailed study, it would be convenient to gather data of M_n evolution with time, but this is experimentally difficult as it implies sampling in rather short reaction times and separate the polymeric species from the reaction mixture. This will be attempted in a future study. Figures 12 and 13 show the outputs of the simulations with the base case. The base case was assumed to have a M_n of 50,000 with dispersity of 1.9 (same as the FRP case in Table 1). For the dormant polymer a livingness of 100% was assumed. As commented above, the goal was to find a set of kinetic parameters that at least semi-quantitatively could reproduce the observed experimental trends; in particular, the decrease of M_n and the generation of around 30–40 wt.% of monomer. Figure 12 shows the decrease of M_n (a) for both the dead and the dormant polymer cases, as well as the generation of monomer (b). The M_n produced from the degradation of dormant polymer goes down to very low values, while that produced from dead polymer remains relatively high (around 35,000), although, as discussed below, this is mainly live polymer in relatively low concentration. More monomer is generated in the case of dormant polymer, as experimentally observed; out of the initial polymer present, the model predicts the conversion to monomer of about 43% (vs. 38.5% experimental) of the total polymer for the pyrolysis of dormant polymer (0.092 initial moles expressed as moles of monomer in the polymeric chain), and of 23% for the pyrolysis of dead polymer (vs. 33% experimental). The difference only qualitatively agrees with the experimental observation; the quantitative deviation can be explained by the oversimplification of the model with respect to the final small molecules formed, such as styrene dimers and trimers, which are not explicitly taken into account in the model, in particular when they are formed by recombination of monomer molecules.

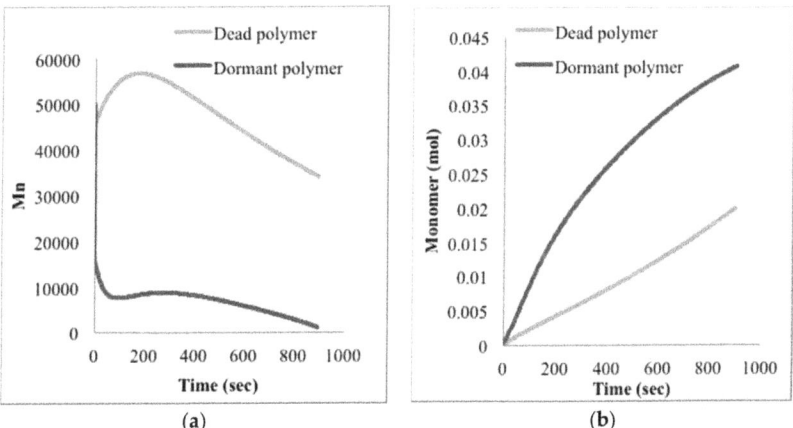

Figure 12. (a) M_n evolution, and (b) monomer generation for the pyrolysis of dead and dormant polymer predicted by the model with the base set of kinetic constants of Table 5.

Figure 13a,b show the predicted evolution of live and dead polymer concentration (zero-th moment) and of the first moments for both populations during the pyrolysis of dead polymer, indicating that the model predicts a very rapid activation of the dead polymer in the first instants of the reaction followed by reverse propagation and mid-chain breaking reactions. Initially, the first moment of dead polymer is around 0.092 mol, but this is not visible in plot (b) because in less than 1 s this polymer is activated and converted into live (activated) polymer which is gradually degraded. Similarly, plots (c) and (d) show the predicted evolution of the zero-th and first moments, respectively, of the four polymer populations existing during the pyrolysis of dormant polymer. Although more complex than in the former case, the overall dynamics of the process are equivalent. Looking at plot (d) all the dormant polymer is rapidly converted into live polymer or live polymer with a nitroxide-end, and these two populations undergo reverse propagation decaying gradually, but faster than in the case of dead polymer pyrolysis. In the process some dead polymer is formed, but in concentrations several orders of magnitude lower than the live polymer.

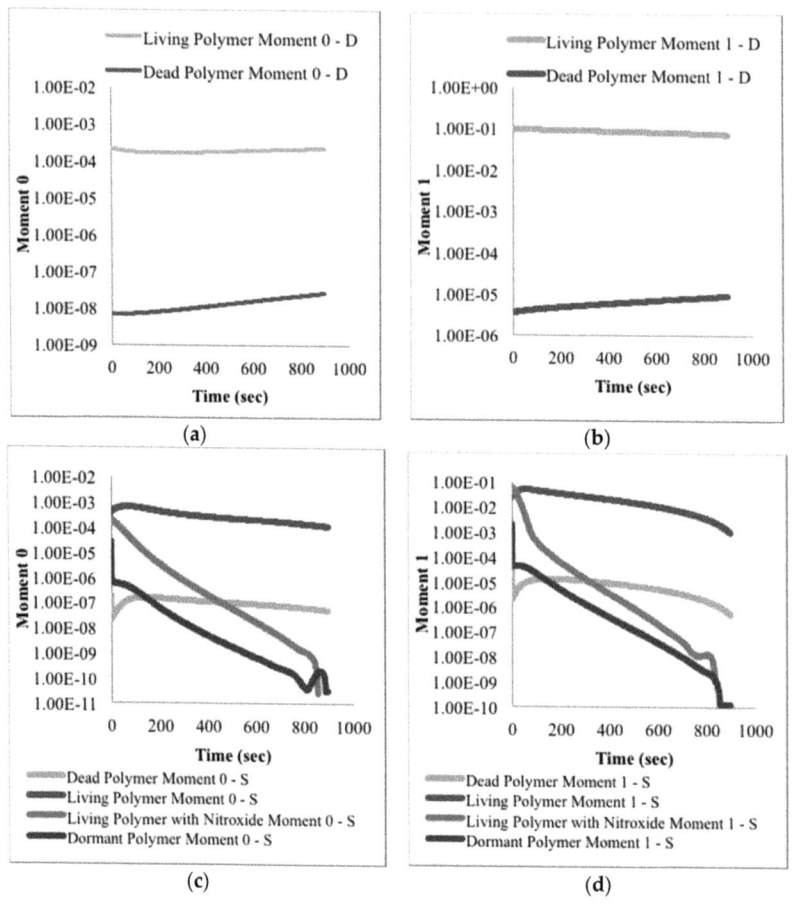

Figure 13. Model predictions for (**a**) zero-th moment evolution and (**b**) first moment evolution of live and dead polymer for the pyrolysis of dead PS. Also, model predictions for (**c**) zero-th moment evolution and (**d**) first moment evolution of live, dead, dormant, and live with a nitroxide-end polymer starting from dormant PS.

5. Conclusions

A simple process for the thermal pyrolysis of PS has been presented. PS-T (dormant PS) undergoes thermal pyrolysis and monomer formation more effectively than conventional (dead) PS, presumably due to the additional activation produced at the nitroxide end at temperatures above the T_c of PS, supporting the hypothesis of this work. With the nitroxide-functionalized PS the monomer production increases from ~33 to ~38.5 wt.% with respect to the initial polymer, an efficiency enhancement of roughly 15–16%. A kinetic and mathematical model was developed to explain the observed differences, and, after some parameter fitting, the model is capable of produce semi-quantitative agreement with the experimental observations regarding time of reaction and monomer production, starting either from dead or dormant polymer. It is a difficult task to develop a predictive model for polymer pyrolysis due the complex polymer structure, the multiple reactions that take place and the large amount of intermediate compounds with different size and compositions that are formed in this intricate process, but with the help of population balance equations and the method of moments it is in principle possible to track the formation of the pyrolysis products and, in particular, the evolution of monomer.

Author Contributions: Conceptualization and methodology, E.S.-G., A.O.-Q. and A.M.-A.; software, E.S.-G. and A.O.-Q.; validation, A.M.-A. and A.O.-Q.; formal analysis, E.S.-G.; resources, E.S.-G.; writing—original draft preparation, A.O.-Q. and E.S.-G.; writing—review and editing, E.S.-G. All authors have read and agreed to the published version of the manuscript.

Funding: This research was partially funded by Consejo Nacional de Ciencia y Tecnología México (CONACYT), grant numbers 256358 and 278829.

Conflicts of Interest: The authors declare no conflict of interest.

References

1. Saldívar-Guerra, E.; Vivaldo-Lima, E. Polymers and polymer types. In *Handbook of Polymer Synthesis, Characterization and Processing*; Chapter 1; Saldívar-Guerra, E., Vivaldo-Lima, E., Eds.; John Wiley and Sons: Hoboken, NJ, USA, 2013.
2. McCrum, N.G.; Buckley, C.P.; Bucknall, C.B. *Principles of Polymer Engineering*, 2nd ed.; Chapter 1; Oxford Science Publications: Oxford, UK, 1997.
3. Villalobos, M.A.; Debling, J. Bulk and solution processes. In *Handbook of Polymer Synthesis, Characterization and Processing*; Chapter 13; Saldívar-Guerra, E., Vivaldo-Lima, E., Eds.; John Wiley and Sons: Hoboken, NJ, USA, 2013.
4. Tullo, A.H. Fighting ocean plastics at the source. *Chem. Eng. News* **2018**, *96*, 28–34.
5. Scott, A. Not-so-fantastic plastic. *Chem. Eng. News* **2018**, *96*, 16–18.
6. Achilias, D.S.; Andriotis, L.; Koutsidis, I.A.; Louka, D.A.; Nianias, N.P.; Siafaka, P.; Tsagkalias, I.; Tsintzou, G. Recent advances in the chemical recycling of polymers (PP, PS, LDPE, HDPE, PVC, PC, Nylon, PMMA). In *Material Recycling Trends and Perspectives*; Achilias, D., Ed.; InTechOpen: London, UK, 2012.
7. Al-Salem, S.M.; Antelava, A.; Constantinou, A.; Manos, G.; Dutta, A. A review on thermal and catalytic pyrolysis of Plastic Solid Waste (PSW). *J. Environ. Manag.* **2017**, *197*, 177–198. [CrossRef] [PubMed]
8. David, C. Thermal degradation of polymers. In *Comprehensive Chemical Kinetics*; Bamford, C.H., Tipper, C.F.H., Eds.; Elsevier Scientific Publishing Company: New York, NY, USA, 1957; Volume 14, pp. 1–173.
9. Guyot, A. Recent developments in the thermal degradation of polystyrene—A review. *Polym. Degrad. Stab.* **1986**, *15*, 219–235. [CrossRef]
10. Grassie, N.; Kerr, W.W. The thermal depolymerization of polystyrene. Part 1—The reaction mechanism. *Trans. Faraday Soc.* **1957**, *53*, 234–239. [CrossRef]
11. Faravelli, T.; Bozzano, G.; Colombo, M.; Ranzi, E.; Dente, M. Kinetic modeling of the thermal degradation of polyethylene and polystyrene mixtures. *J. Anal. Appl. Pyrolysis* **2003**, *70*, 761–777. [CrossRef]
12. Malhotra, S.L.; Hesse, J.; Blanchard, L.P. Thermal decomposition of polystyrene. *Polymer* **1975**, *16*, 81–93. [CrossRef]
13. Lehrle, R.S.; Peakman, R.E.; Robb, J.C. Pyrolysis-gas-liquid-chromatography utilised for a kinetic study of the mechanisms of initiation and termination in the thermal degradation of polystyrene. *Eur. Polym. J.* **1982**, *18*, 517–529. [CrossRef]
14. Madorsky, S.L. Rates of thermal degradation of polystyrene and polyethylene in a vacuum. *J. Polym. Sci.* **1952**, *9*, 133–156. [CrossRef]
15. Guaita, M. Thermal degradation of polystyrene. *Br. Polym. J.* **1986**, *18*, 226–230. [CrossRef]
16. Odian, G. *Principles of Polymerization*; John Wiley & Sons, Inc.: Hoboken, NJ, USA, 2004.
17. Stevens, M.P. *Polymer Chemistry: An Introduction*, 3rd ed.; Oxford University Press: New York, NY, USA, 1999.
18. Audisio, G.; Bertini, F. Molecular weight and pyrolysis products distribution of polymers. I. Polystyrene. *J. Anal. Appl. Pyrolysis* **1992**, *24*, 61–74. [CrossRef]

19. Beyler, C.L.; Hirschler, M.M. Beyler_Hirschler_SFPE_Handbook_3. In *SFPE Handbook of Fire Protection Engineering*; National Fire Protection Association: Quincy, MA, USA, 2002; pp. 111–131.
20. Liu, Y.; Qian, J.; Wang, J. Pyrolysis of polystyrene waste in a fluidized-bed reactor to obtain styrene monomer and gasoline fraction. *Fuel Process. Technol.* **2000**, *63*, 45–55. [CrossRef]
21. Simard, Y.D.M.; Kamal, M.R.; Cooper, D.G. Thermolysis of Polystyrene. *J. Appl. Polym. Sci.* **1995**, *58*, 843–851. [CrossRef]
22. Jellinek, H.H.G. Thermal degradation of polystyrene and polyethylene. Part III. *J. Polym. Sci.* **2003**, *4*, 13–36. [CrossRef]
23. Ebert, K.H.; Ederer, H.J.; Schroder, U.K.O.; Hamielec, A.W. On the kinetics and mechanism of thermal degradation of polystyrene. *Makromol. Chem.* **1982**, *183*, 1207–1218. [CrossRef]
24. Richards, D.H.; Salter, D.A. Thermal degradation of vinyl polymers II—The synthesis and degradation of polystyrene containing thermally weak bonds. *Polymer* **1967**, *8*, 139–152. [CrossRef]
25. Anderson, D.A.; Freeman, E.S. The kinetics of the thermal degradation of polystyrene and polyethylene. *J. Polym. Sci.* **1961**, *54*, 253–260. [CrossRef]
26. Wegner, J.; Patat, F. Thermal degradation of polystyrene. *J. Polym. Sci. Part C* **1970**, *31*, 121–135. [CrossRef]
27. Paisley, M.A.; Litt, R.D. Monomer Recovery from Polymeric Materials. U.S. Patent 5,326,919, 5 July 1994.
28. Luo, X.; Yie, F.; Tan, W. Method for Retrieving Styrene Monomer from Discarded Polystyrene Scrap through Pyrolytic Reduction. U.S. Patent 5,072,068, 10 December 1991.
29. Lee, S.; Gencer, M.A.; Fullerton, K.L.; Azzam, F.O. Depolymerization Process. U.S. Patent 5,386,055, 31 January 1995.
30. Tippet, J.; Butler, J.; Assef, J.; Ashbaugh, J.; Clark, J.; Duc, M.; Cary, J.-B. Depolymerization of Plastic Materials. U.S. Patent 9,650,313, 15 May 2017.
31. Northemann, A. Recovery of Styrene from Waste Polystyrene. U.S. Patent 5,672,794, 30 September 1997.
32. Yoon, B.T.; Choi, M.J.; Kim, S.B.; Lee, S.B. Method for Preparing Fermentable Sugar from Wood-Based Biomass. U.S. Patent 8,449,725, 24 March 2013.
33. Nicolas, J.; Guillaneuf, Y.; Lefay, C.; Bertin, D.; Gigmes, D.; Charleux, B. Nitroxide-mediated polymerization. *Prog. Polym. Sci.* **2013**, *38*, 63–235. [CrossRef]
34. Roland, A.I.; Schmidt-Naake, G. Thermal degradation of polystyrene produced by nitroxide-controlled radical polymerization. *J. Anal. Appl. Pyrolysis* **2001**, *58–59*, 143–154. [CrossRef]
35. Saidel, G.M.; Katz, S. Dynamic analysis of branching in radical polymerization. *J. Polym. Sci. Part A-2* **1968**, *6*, 1149–1160. [CrossRef]
36. Marongiu, A.; Faravelli, T.; Ranzi, E. Detailed kinetic modeling of the thermal degradation of vinyl polymers. *J. Anal. Appl. Pyrolysis* **2007**, *78*, 343–362. [CrossRef]
37. Kruse, T.M.; Woo, O.S.; Broadbelt, L.J. Detailed mechanistic modeling of polymer degradation: Application to polystyrene. *Chem. Eng. Sci.* **2001**, *56*, 971–979. [CrossRef]
38. Kruse, T.M.; Woo, O.S.; Wong, H.W.; Khan, S.S.; Broadbelt, L.J. Mechanistic modeling of polymer degradation: A comprehensive study of polystyrene. *Macromolecules* **2002**, *35*, 7830–7844. [CrossRef]
39. Lemoine-Nava, R.; Flores-Tlacuahuac, A.; Saldívar-Guerra, E. Optimal operating policies for the nitroxide-mediated radical polymerization of styrene in a semibatch reactor. *Ind. Eng. Chem. Res.* **2006**, *45*, 4637–4652. [CrossRef]
40. Beuermann, S.; Buback, M. Rate coefficients of free-radical polymerization deduced from pulsed laser experiments. *Prog. Polym. Sci.* **2002**, *27*, 191–254. [CrossRef]
41. Sterling, W.J.; Walline, K.S.; McCoy, B.J. Experimental study of polystyrene thermolysis to moderate conversión. *Polym. Degrad. Stabil.* **2001**, *73*, 75–82. [CrossRef]
42. Simha, R.; Wall, L.A.; Blatz, P.J. Depolymerization as a chain reaction. *J. Polym. Sci.* **2003**, *5*, 615–632. [CrossRef]

© 2020 by the authors. Licensee MDPI, Basel, Switzerland. This article is an open access article distributed under the terms and conditions of the Creative Commons Attribution (CC BY) license (http://creativecommons.org/licenses/by/4.0/).

Article

New Aspects on the Modeling of Dithiolactone-Mediated Radical Polymerization of Vinyl Monomers

Anete Joceline Benitez-Carreón [1], Jesús Guillermo Soriano-Moro [2], Eduardo Vivaldo-Lima [1,3,*], Ramiro Guerrero-Santos [4] and Alexander Penlidis [3]

1. Departamento de Ingeniería Química, Facultad de Química, Universidad Nacional Autónoma de México, Ciudad de Mexico 04510, Mexico; anjobeca_23@hotmail.com
2. Centro de Química, Instituto de Ciencias, Benemérita Universidad Autónoma de Puebla (BUAP), Edif. IC8, Ciudad Universitaria, Puebla 72570, Mexico; jesus.soriano@correo.buap.mx
3. Institute for Polymer Research (IPR), Department of Chemical Engineering, University of Waterloo, Waterloo, ON N2L 3G1, Canada; penlidis@uwaterloo.ca
4. Centro de Investigación en Química Aplicada (CIQA), Blvd. Enrique Reyna 140, Saltillo, Coahuila 25140, Mexico; ramiro.guerrero@ciqa.edu.mx
* Correspondence: vivaldo@unam.mx; Tel.: +52-55-5622-5226

Received: 14 October 2019; Accepted: 7 November 2019; Published: 10 November 2019

Abstract: A kinetic model for the dithiolactone-mediated radical polymerization of vinyl monomers based on the persistent radical effect and reversible addition (negligible fragmentation) was used to calculate the polymerization rate and describe molar mass development in the polymerization of methyl methacrylate at 60 °C, using 2,2-azobisisobutyronitrile (AIBN) as an initiator, as well as dihydro-5-phenyl-2(3H)-thiophenethione (DTL1) and dihydro-2(3H)-thiophenethione (DTL2) as controllers. The model was implemented in the PREDICI commercial software. A good agreement between experimental data and model predictions was obtained.

Keywords: dithiolactones; RAFT polymerization; kinetic modeling; vinyl monomers; methyl methacrylate

1. Introduction

Reversible deactivation radical polymerization (RDRP) techniques have become important in the last three decades [1]. They provide versatile routes to synthesize polymers with tailored architectures [2]. The most studied methodologies which have emerged are nitroxide-mediated polymerization (NMP), atom transfer radical polymerization (ATRP), and reversible addition–fragmentation chain-transfer (RAFT) polymerization [3]. However, other RDRP techniques such as iodine-transfer polymerization (ITP) [4], telluride-mediated polymerization (TERP) [5], and organostibine-mediated polymerization [5] have also been proposed.

RAFT polymerization is considered one of the most successful RDRP techniques. In RAFT polymerization, propagating free radical molecule **1** adds to RAFT agent **2** (see Figures 1 and 2), thus producing intermediate radical **3** (one-arm adduct), which undergoes β-scission, yielding back the reactants or producing dormant polymer (macro RAFT agent) **4** and radical R˙ [6]. The main equilibrium reactions consist of the formation of intermediate radical **6** (two-arm product) and its fragmentation into free radical **1** and dormant polymer **4** (see Figure 3). In the "intermediate radical termination" (IRT) model [7], an additional reaction is considered, namely the cross-termination between free radical **1** and the intermediate radical **6**, thus producing a star-shaped polymer (three-arm dead polymer), as seen in Figure 4 [6]. It is worth mentioning that further investigation has led to the assumption of side reactions between the three-arm adduct **7** with a propagating radical **1** to yield intermediate radical **6** and a dead polymer [8]. Variations of the IRT model have also been considered: reversible IRT [9] and the occurrence of cross-termination of the intermediate radical with oligomeric radicals only (the IRTO model) [10]. A reversible reaction between intermediate radical **6** and the original RAFT agent **2**, producing a secondary intermediate radical that may undergo further RAFT process, has also been proposed [11].

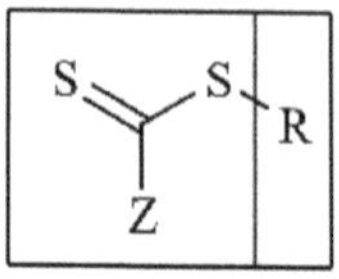

Figure 1. General chemical structure of a reversible addition–fragmentation chain-transfer (RAFT) agent (denoted as AB in the modeling equations below).

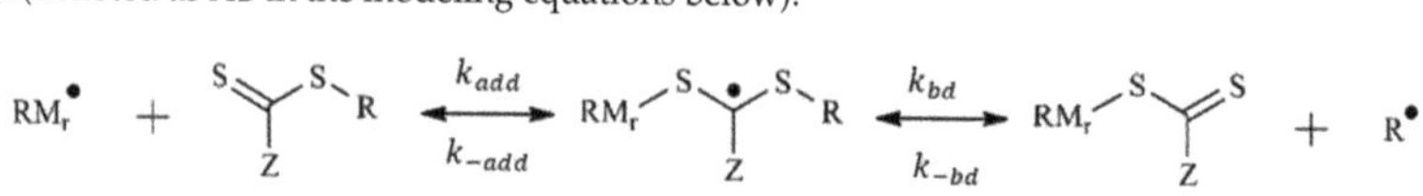

Figure 2. Reversible chain transfer to the RAFT agent (pre-equilibrium).

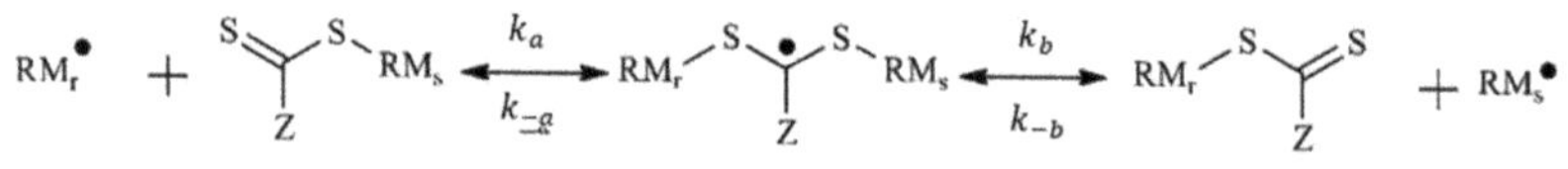

Figure 3. Chain equilibration (addition–fragmentation).

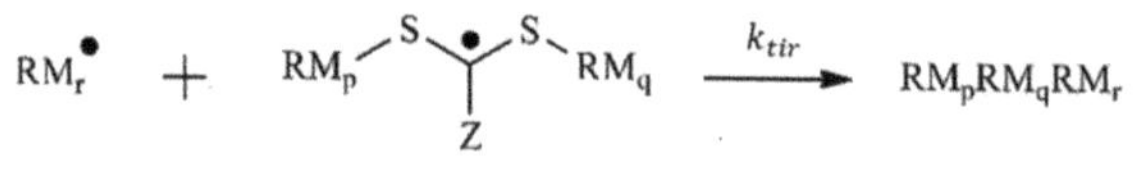

Figure 4. Intermediate radical termination (IRT).

Dithiolactones are the cyclic counterpart to dithioester and thiocarbonylthio compounds, which are used as RAFT agents. However, as seen in Figure 5, there is no fragmentation in dithiolactone-mediated free-radical polymerizations (DTLMP), and the addition reaction is reversible [12,13].

In a previous publication from our group, we modeled the DTLMP of styrene using 2,2′-azobisisobutyronitrile (AIBN) as the initiator and γ-phenyl-γ-butyrodithiolactone as the controller [14].

In this contribution, we address the DTLMP of methyl methacrylate (MMA) using dihydro-5-phenyl-2(3H)-thiophenethione—whose common name is γ-phenyl-γ-butyrodithiolactone (referred to as DTL1 in the remainder of our contribution)—and dihydro-2(3H)-thiophenethione—whose common name is γ-butyrodithiolactone (referred to as DTL2 in the remainder of our contribution)—as controllers, from both experimental and mathematical modeling perspectives.

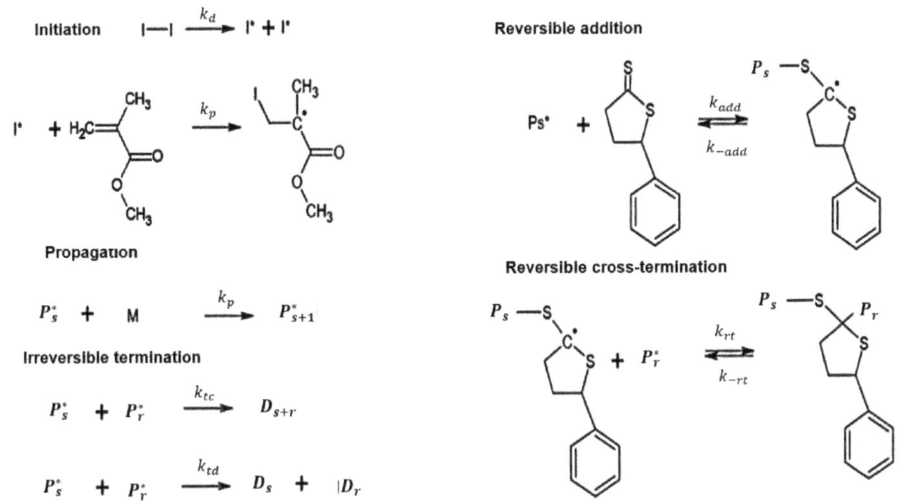

Figure 5. Polymerization scheme for dithiolactone-mediated (DTLM) radical polymerization of vinyl monomers.

2. Experimental

As mentioned above, the system studied in this contribution was the polymerization of MMA using AIBN as initiator and DTL1 as well as DTL2 as controllers. The polymerizations were carried out at 333.15 K (60 °C) using two molar ratios of DTL1 and DTL2 (monomer:controller:initiator ratios of 300:2:1 and 300:4:1) [13,15].

2.1. Polymerization Procedure

A stock solution of AIBN in MMA was prepared first, then split into several heavy-wall glass tubes, and mixed with calculated amounts of controller (DTL1 or DTL2). Mixtures were degassed by three freeze–thaw–pump cycles, sealed, and heated at 333.15 K in a thermostated oil bath. A reference experiment was carried out using the original stock solution without the chain transfer agent. Polymers were isolated by precipitation in methanol. Monomer conversion was gravimetrically determined by duplicate, in all cases [13].

2.2. Measurement of Molar Mass Characteristics

Molar mass distributions were determined by size exclusion chromatography (SEC) or gel permeation chromatography (GPC) with a Hewlett Packard (HP) modular system comprising an auto injector, a Polymer Laboratories 5.0 μm bead-size guard column, and by three linear PLgel columns (106, 105 and 103 Å), with differential refractive index detector (HP 147 A) and ultraviolet (UV) detector. The eluent was tetrahydrofuran (THF) at a flow rate of 1 cm^3 min^{-1} at 313.15 K. The system was calibrated using narrow poly(methyl methacrylate) standards (ranging from 620 to 1.52×10^6 g mol^{-1}). Data acquisition was performed using Polymer Laboratories GPC software. All measurements were obtained by duplicate [13].

3. Modeling

3.1. Polymerization Scheme

The polymerization scheme used for our simulations was based on the reaction mechanism proposed by Soriano-Moro et al. for DTL1 [13,15], shown in Figure 5. This polymerization scheme corresponded to a RAFT polymerization without fragmentation. Its implementation in the PREDICI software package of CiT [16,17], which was the numerical tool to carry out the simulations, is shown in Table 1. The initiation, propagation and bimolecular radical termination correspond to conventional free radical polymerization [14,18,19]. The first reversible addition [20], second reversible addition (which is the reaction between two molecules of intermediate radicals generating a new type of adduct (see Figure 6)) [13], and reversible cross-termination are reactions due to the presence of dithiolactone controller molecules, as explained above [13–15]. Also shown in Table 1 are parameter sources [14,18–23]. It can be observed in Table 1 that our values of $K = k_{add}/k_{-add}$ for DTL1 and DTL2, which were in the range of 10^5–10^6 m^3 kmol^{-1} for conventional RAFT agents, were slightly low.

Table 1. Polymerization scheme and kinetic rate constants used.

Reaction	Step Pattern	k (m^3 kmol^{-1} s^{-1}, Unless Otherwise Stated)	DTL1 [Source]	DTL2 [Source]
Initiator decomposition	$I \rightarrow I^\bullet + I^\bullet$	k_d (s^{-1})	4×10^{-6} [21]	4×10^{-6} [21]
First propagation	$I^\bullet + M \rightarrow P^\bullet_{s=1}$	k_i	833 [18]	833 [18]
Propagation	$P^\bullet_s + M \rightarrow P^\bullet_{s+1}$	k_p	685.9 [22,23]	685.9 [22,23]
First reversible addition	$P^\bullet_s + DTL \rightarrow P_sDTL^\bullet$	k_{add}	2.3×10^3 [this work]	8.1×10^3 [this work]
	$P_sDTL^\bullet \rightarrow P^\bullet_s + DTL$	k_{-add}	2.622×10^{-2} [this work]	8.6×10^{-2} [this work]
Reversible cross-termination	$P^\bullet_s + P_rDTL^\bullet \rightarrow P_sDTLP_r$	k_{rt}	5×10^2 [14]	5×10^3 [this work]
	$P_sDTLP_r \rightarrow P^\bullet_s + P_rDTL^\bullet$	k_{-rt}	2×10^{-4} [this work]	2×10^{-4} [this work]
Irreversible termination	$P^\bullet_s + P^\bullet_r \rightarrow D_s + D_r$	k_{td}	3.40×10^7 [22,23]	3.40×10^7 [22,23]
	$P^\bullet_s + P^\bullet_r \rightarrow D_{s+r}$	k_{tc}	9.78×10^7 [22,23]	9.78×10^7 [22,23]
Second reversible addition	$P_sDTL^\bullet + P_rDTL^\bullet \rightarrow P_sDTLDTLP_r$	k_{es}	2×10^2 [this work]	2×10^2 [this work]
	$P_sDTLDTLP_r \rightarrow P_sDTL^\bullet + P_rDTL^\bullet$	k_{-es}	4×10^{-4} [this work]	4×10^{-4} [this work]

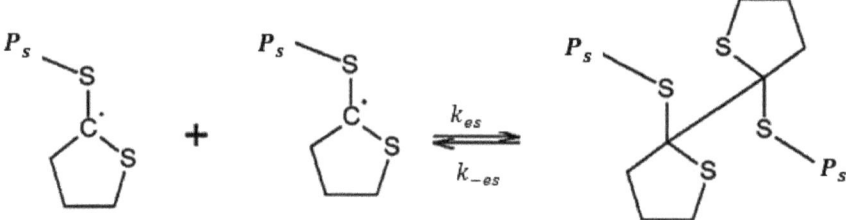

Figure 6. Second reversible addition reaction.

3.2. Diffusion-Controlled (DC) Effects

Free-volume theory was used to account for diffusion-controlled (DC) effects in (conversion-dependent) kinetic rate constants. The expressions used for DC-effects are summarized in Table 2 [14,18]. β_i in the expressions of Table 2 are the free-volume "overlap" parameters for the i-th reaction, with i accounting for termination, propagation, forward and reverse addition (first and second), and reversible cross-termination; T and Tg_i are the reaction temperature and glass transition temperature of component i, respectively; α_i is volumetric expansion coefficient for species i; V_i and V_t denote the volumes of species i and total volumes, respectively; v_{f0} and v_f denote fractional free volume at initial conditions and at calculation time, respectively. The free-volume parameters are summarized in Table 3. As mentioned earlier, the simulations were carried out using the PREDICI software of CiT, version 11.3.0.

Table 2. Mathematical expressions for diffusion-controlled (DC)-effects.

Reaction	Mathematical Expression	β Parameters for DC-Expressions [Source]
Propagation	$k_p = k_p^0 Exp[-\beta_p(\frac{1}{V_f} - \frac{1}{V_{f0}})]$	0.01 [14]
First reversible addition	$k_{add} = k_{add}^0 Exp[-\beta_{add}(\frac{1}{V_f} - \frac{1}{V_{f0}})]$	0.01 [14]
	$k_{-add} = k_{-add}^0 Exp[-\beta_{-add}(\frac{1}{V_f} - \frac{1}{V_{f0}})]$	0.01 [14]
Reversible cross-termination	$k_{rt} = k_{rt}^0 Exp[-\beta_{rt}(\frac{1}{V_f} - \frac{1}{V_{f0}})]$	0.45 [this work]
	$k_{-rt} = k_{-rt}^0 Exp[-\beta_{-rt}(\frac{1}{V_f} - \frac{1}{V_{f0}})]$	0.45 [this work]
Irreversible Termination	$k_{tc} = k_{tc}^0 Exp[-\beta_{tc}(\frac{1}{V_f} - \frac{1}{V_{f0}})]$	0.45 [14]
	$k_{td} = k_{td}^0 Exp[-\beta_{td}(\frac{1}{V_f} - \frac{1}{V_{f0}})]$	0.45 [14]
Second reversible addition	$k_{es} = k_{es}^0 Exp[-\beta_{es}(\frac{1}{V_f} - \frac{1}{V_{f0}})]$	0.55 [this work]
	$k_{-es} = k_{-es}^0 Exp[-\beta_{-es}(\frac{1}{V_f} - \frac{1}{V_{f0}})]$	0.55 [this work]
Fractional Free Volume	$V_f = \sum_{i=1}^{N}[0.025 + \alpha_i(T - T_{gi})]\frac{V_i}{V_t}$	

Table 3. Other free-volume parameters.

Parameter	Value	Reference
α_{MMA} (K^{-1}), T_{gMMA} (K)	0.001, 143	[24], [25]
α_{PMMA} (K^{-1}), T_{gPMMA} (K)	0.0048, 392	[24], [25]
α_{DTL1} (K^{-1}), T_{gDTL1} (K)	0.0001, 173.15	[14], [14]
α_{DTL2} (K^{-1}), T_{gDTL2} (K)	0.0001, 173.15	This work, this work

3.3. Parameter Estimation Strategy

Literature values were used for k_d, k_i, k_p, k_{tc} and k_{td} (see Table 1). Given the wide variation range on reported k_{tc} and k_{td} values for MMA and the large effect of the ratio k_{tc}/k_{td} on the evolution of M$_n$ versus conversion profiles, a careful evaluation (mimicking a sensitivity analysis) of the reported values was carried out. The best results were obtained with the parameters reported in the Watpoly software database [22]. Once these parameters were fixed, a more detailed parameter sensitivity analysis was conducted with the remaining parameters, the ones related to the RDRP behavior of dithiolactones (k_{add}, k_{-add}, k_{rt}, k_{-rt}, k_{es}, and k_{-es}). The parameters accounting for DC-effects were assumed to be similar to the ones estimated for the polymerization of styrene using DTL1 [14] and the same for both controllers (DTL1 and DTL2). They are summarized in Table 2 (last column).

The overall strategy used in our study is similar to the one used by Gomez-Reguera et al. [26].

4. Results and Discussion

4.1. Polymerization of MMA Using DTL1

The effect of DTL1 concentration on polymerization rate, expressed as conversion versus time, number average molar mass (M$_n$), and molar mass dispersity (Đ), is shown in Figure 7. As expected, it can be observed in Figure 7 that increasing DTL1 concentration resulted in a slower polymerization rate, a lower molar mass, and a slightly lower Đ.

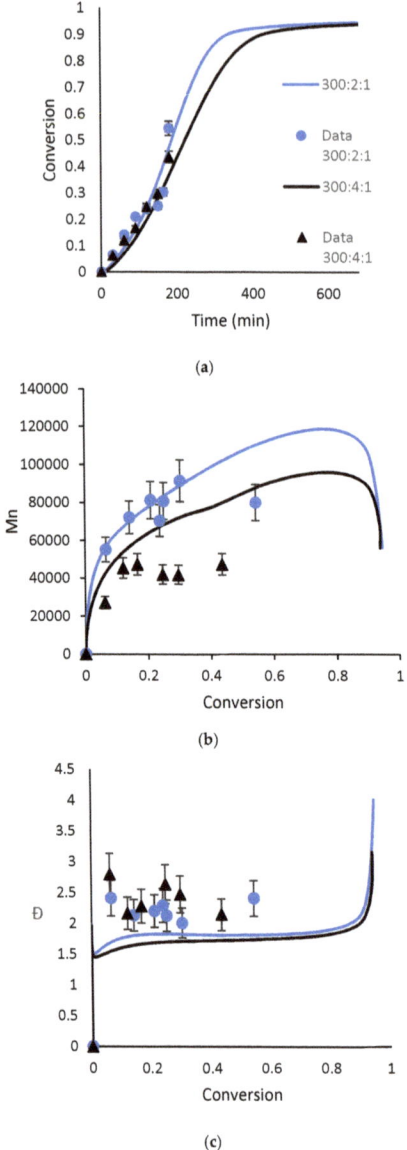

Figure 7. Comparison of model predictions and experimental data of (**a**) monomer conversion versus time, (**b**) number average molar mass versus conversion, and (**c**) Đ versus conversion. [methyl methacrylate (MMA)]:[γ-phenyl-γ-butyrodithiolactone (DTL1)]:[2,2-azobisisobutyronitrile (AIBN)]=300:2:1 for EXP 1 and 300:4:1 for EXP 2.

Overall, a fair agreement can be observed in Figure 7 between the calculated and experimental profiles, with some deviations. The evolution of polymerization rate and molecular weight development were captured well by the model in the case of 300:2:1, but significant deviations were observed for both profiles in the case of 300:4:1. Several attempts to improve the agreement between calculated and experimental profiles of M_n versus conversion for the case of 300:4:1 by fine-tuning some of the kinetic rate constants or DC effect parameters were made. If the agreement was improved, the changes resulted in the worsening of the agreement for the conversion versus time profiles, as well as the obtainment of a very poor performance for the case of 300:2:1. This is not uncommon in modelling attempts when multi-response data points are used with the same set of parameters. Therefore, the behavior of M_n after 20% monomer conversion observed in the case of 300:4:1 suggests either a high(er) experimental error occurring in this case, or, more likely, the presence of side reactions not considered in the polymerization scheme when the controller content was high. The deviations are also evident in Figure 7 for the case of Đ versus conversion. Again, this points towards the direction of side reactions affecting the data—reactions which are unaccounted for in the polymerization mechanism and, hence, in the subsequent mathematical model. It is also possible that the parameters associated to DC effects may have become conversion dependent [26]. A similar overall performance was observed in the case of polymerization of styrene using DTL1 [13,14].

A comparison of calculated and experimental full molar mass distributions (MMDs) for the cases using DTL1 is shown in Figure 8. The calculated MMDs of Figure 8 are narrower than the experimental ones. This result agrees with the Đ versus conversion profiles of Figure 7, where calculated Đ values are lower than the experimental ones. However, the time evolution of the MMDs and the effect of controller on both M_n and the spread of the MMD were well captured by the model. As stated earlier, the disagreement between calculated and experimental profiles of Đ versus conversion and time evolution of the MMDs may be attributed to chain transfer side reactions [21]. Several possible alternatives and changes in the polymerization scheme, including reversible fragmentation (full RAFT case) and other side reactions, were indeed simulated in this study, but no significant improvement was obtained.

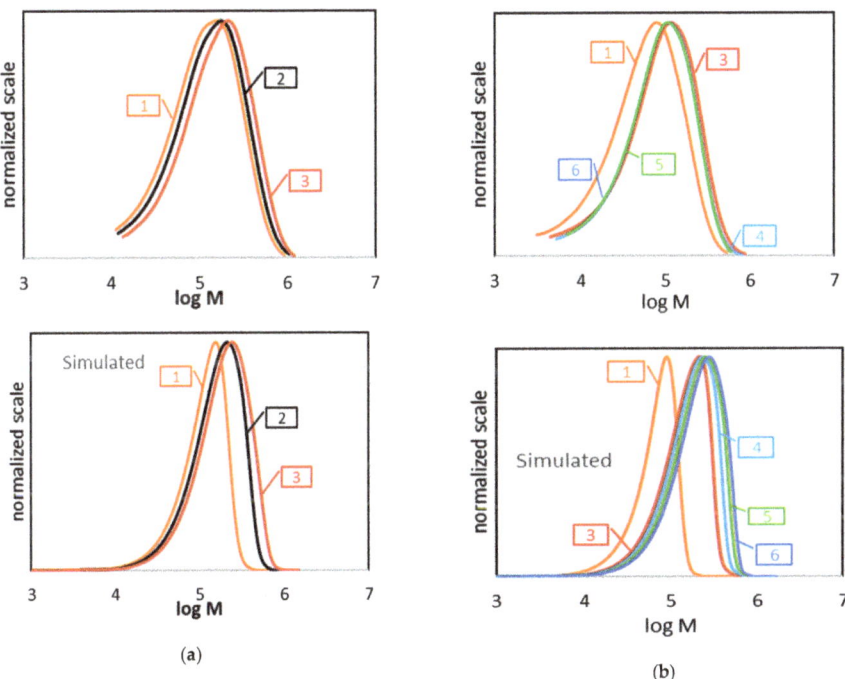

Figure 8. Evolution of molar-mass distributions for polymerizations using DTL1 (experimental and simulated): (a) 300:2:1 and (b) 300:4:1. Labels: "1" = 30, "2" = 60, "3" = 90, "4" = 120, "5" = 150, "6" = 180 min.

In our previous investigation on the polymerization of styrene using DTL1 or benzyl dithiopropianate (BDP), overall radical concentrations were measured using electro-spin resonance (ESR) spectroscopy. Overall radical concentrations in the order of 10^{-7}–10^{-6} kmol m^{-3} were obtained [13,15]. Though the model used included the calculation of concentrations of all the species involved in the polymerization scheme, no comparison between calculated and experimental profiles of overall radical concentration versus time was presented. Moad et al. [27] pointed out that the reversible addition mechanism for the DTLMP of vinyl monomers proposed herein (see Figure 5) seems unlikely since, for control, most of the propagating species would need to be present at the dormant state (polymer product in the forward direction of the reversible addition reaction of Figure 5, or species P$_s$DTL$^\bullet$ in Table 1). They added that ESR experiments (triangles in Figure 9) showed that a high concentration of radicals was not present during polymerization, thus implying that the mechanism would suggest that the final product should also have the P$_s$DTL$^\bullet$ structure. In order to gain more insight regarding this issue, we carried out simulations of overall radical concentration versus time for both styrene and MMA DTLMPs, using DTL1 for styrene and MMA, as well as DTL2 for MMA. The results using DTL1 are shown in Figures 9 and 10. (Results using DTL2 are shown in Section 4.2).

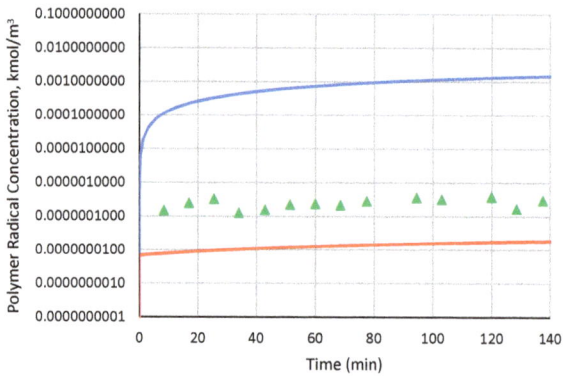

Figure 9. Comparison of experimental [13,15] and calculated profiles of polymer radical concentration versus time for the dithiolactone-mediated free-radical polymerizations (DTLMP) of styrene using DTL1: overall polymer radical (blue solid line); polymer radicals excluding P$_s$DTL$^\bullet$ (red solid line); experimental data (green triangles). Monomer:controller:initiator: 300:2:1.

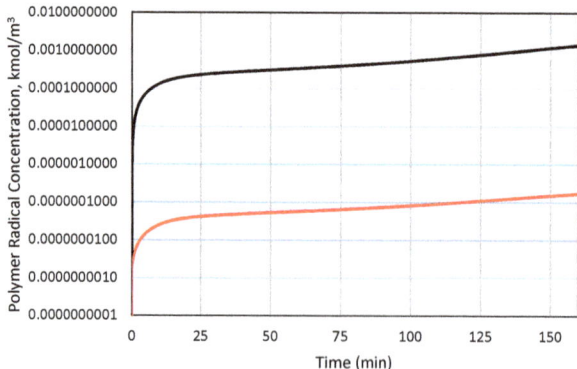

Figure 10. Calculated profiles of polymer radical concentration in the free radical polymerization of MMA using DTL1: overall polymer radicals (black solid line) and polymer radicals excluding dormant polymer P$_s$DTL$^\bullet$ (red solid line).

It can be observed in Figure 9 that the predicted overall concentration of polymer radicals (blue solid line, which includes "dormant" polymer $P_sDTL^\bullet$) exceeded by almost four orders of magnitude the experimental profile. It can also be observed in that figure that the calculated concentration of polymer radicals excluding $P_sDTL^\bullet$ was one order of magnitude lower than the experimental profile of the overall polymer radical concentration. This result indicates that the level of control experimentally obtained would require a much higher content of dormant polymer $P_sDTL^\bullet$, as pointed out by Moad et al. [27]; this suggests that our proposed polymerization scheme may be incomplete, pointing again towards other side reactions in the mechanism, as postulated above during the discussion of the experimental results of Figures 7 and 8.

As mentioned earlier, no measurements of polymer radical concentration versus time were obtained for the case of the DTLM polymerization of MMA. However, calculated profiles were generated, and the ranges are like those of Figure 9. Figure 10 shows calculated profiles of overall polymer radical concentration (black solid line) and polymer radicals excluding the "dormant" population ($P_sDTL^\bullet$, red solid line). It can again be observed that most of the polymer radical population is made out of dormant polymer $P_sDTL^\bullet$, which allows the polymerization to be controlled.

4.2. Polymerization of MMA Using DTL2

The effect of controller concentration on the polymerization rate and molar mass development for the polymerization of MMA using DTL2 was also addressed. It can be observed in Figure 11 (upper plot) that the agreement between the calculated and experimental profiles of conversion versus time is very good for the case of 300:2:1, but large deviations were observed for the case of 300:4:1. It is worth noting that the experimental profile corresponding to a higher RAFT content (grey triangles) resulted in higher conversions than the case of 300:2:1, which was contrary to the expected performance. The calculated profiles, however, showed the correct trend with polymerization rate decreasing as RAFT content increased. This result suggests that our experimental profile of conversion versus time for the case of 300:4:1 may have been in error. The agreement between the experimental and calculated profiles of M_n versus conversion, on the other hand, is very good for both RAFT concentration levels (see middle plot of Figure 11). It is interesting to notice that the predicted profile of Ð versus conversion for the case of 300:4:1 also agrees very well with the experimental one, as observed in the bottom plot of Figure 11. The agreement between the calculated and experimental profiles of Ð versus conversion for the case of 300:2:1, on the other hand, is again poor, as with DTL1.

The polymerization of MMA using AIBN and DTL1 or DTL2 controllers showed hybrid behavior between RDRP and conventional chain transfer radical polymerization. This hybrid behavior has also been observed in the RAFT polymerization of MMA using cumyl phenyldithioacetate [20]. This hybrid behavior strongly depends on temperature; at temperatures below 45 °C, the conventional character dominates, whereas controlled behavior dominates at higher temperatures [20].

The comparison of calculated and experimental MMDs shown in Figure 12 confirms the information provided by the middle and bottom plots of Figure 11. It can be observed that the calculated MMDs for the case of 300:2:1 were narrower than the experimental ones. In the case of system 300:4:1, the agreement between the calculated and experimental MMDs is very good, except for the experimental MMD at very low conversions, which was much broader, verifying a dominant conventional chain transfer behavior in that case, as alluded to above.

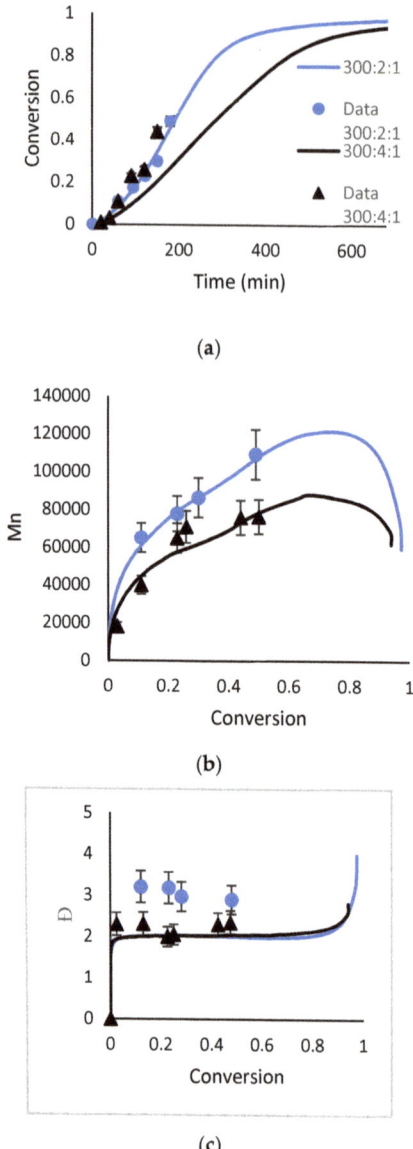

Figure 11. Comparison of model predictions and experimental data for (**a**) monomer conversion versus time, (**b**) number average molar mass versus conversion, and (**c**) Đ versus conversion. [MMA]:[dihydro-2(3)-thiophenethione (DTL2)]:[AIBN] = 300:2:1 for EXP 1 and 300:4:1 for EXP 2.

Finally, the calculated profiles of overall polymer radical concentration and concentration of dormant polymer ($P_sDTL^\bullet$) versus time for the case of DTLM polymerization of MMA using DTL2$^\bullet$ are shown in Figure 13. As in the case using DTL1 earlier, it can be observed that large amounts of dormant polymer were required to obtain adequate control.

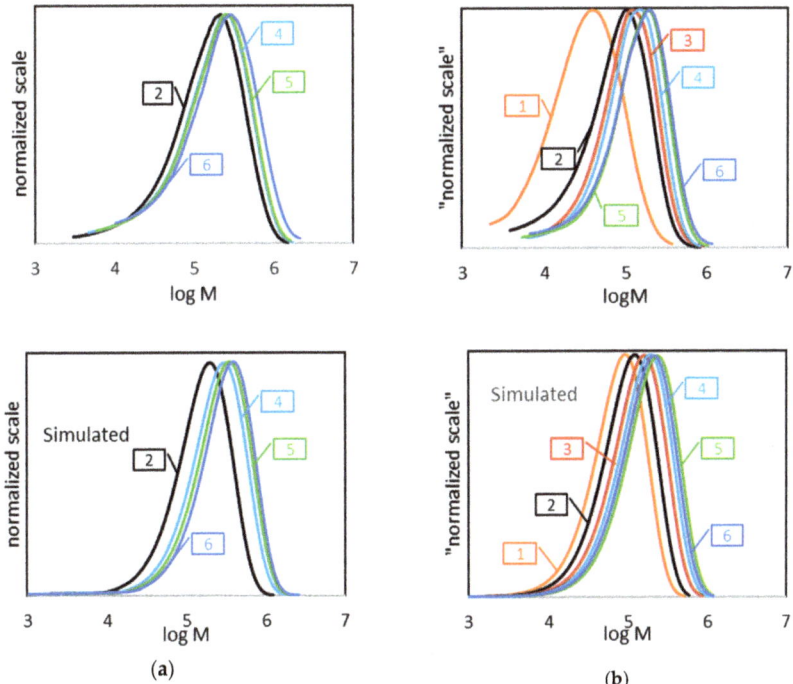

Figure 12. Evolution of the molar-mass distributions with DTL2 (experimental shown at the top and simulated at the bottom): (**a**) 300:2:1 and (**b**) 300:4:1. Labels: "1" = 40, "2" = 60, "3" = 90, "4" = 120, "5" = 150, "6" = 180 min.

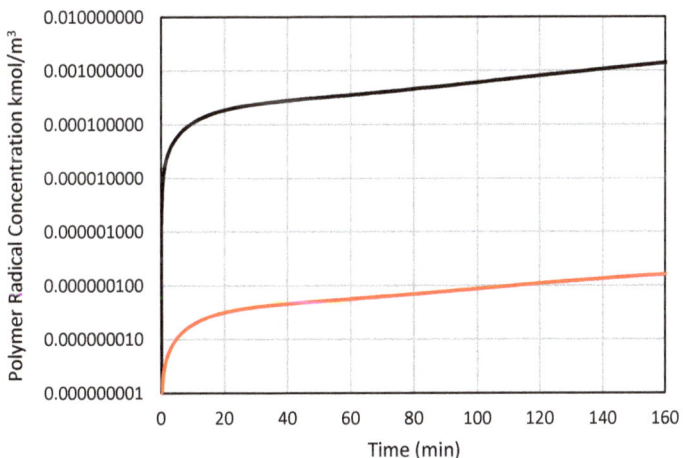

Figure 13. Calculated profiles of polymer radical concentration in the free radical polymerization of MMA using DTL2: overall polymer radicals (black solid line) and polymer radicals excluding dormant polymer $P_sDTL^\bullet$ (red solid line).

4.3. Some Thoughts on Future Research Steps

Masoumi et al. [6] used a sequential Bayesian Monte Carlo model discrimination (SBMCMD) method to discriminate between two rival RAFT models, the so-called slow-fragmentation (SF) method [28] and one type of IRT model [29]). They found that if one of the two competing models represents the 'real' mechanism behind the RAFT process, the SBMCMD framework can identify the correct model by analyzing the data from the most informative experiments designed by the model discrimination steps. However, for the system studied in that paper, if the 'real' model was not known beforehand, the method was not able to discriminate between the SF and IRT models, based on polymerization rate and molar mass development data (conversion versus time and molar mass averages versus conversion experimental profiles). The bottom-line conclusion is that additional independent experimental information (e.g., concentrations of intermediate products) is needed to get more conclusive results.

As stated in Section 4.1, we carried out additional simulations under different mechanistic variations, including trials using the full RAFT model (the inclusion of fragmentation), but no significantly better results were obtained. More comprehensive results in that direction will be presented by our group in future contributions.

5. Conclusions

The free radical polymerization of MMA in the presence of AIBN and DTL1 or DTL2 at 333.15 K was studied experimentally and modeled using a polymerization scheme where only reversible addition was considered (fragmentation neglected). The agreement between the experimental and modeling profiles of conversion versus time, average molar mass and full MMDs can range from fair to very-good. Our results complement our previous study on the polymerization of styrene using AIBN and DTL1.

Though the polymerization scheme based on reversible addition only seems to be adequate to describe the overall performance of dithiolactone-mediated free radical polymerizations, described typically by polymerization rate and average molar mass development, the fact that the global description of the system, considering a wide range of experimental conditions and independent responses was difficult to obtain by using a single set of model parameters, suggests that the proposed polymerization scheme may be incomplete.

Our modeling results suggest that high concentrations of a dormant polymer ($P_sDTL^\bullet$) are required to provide adequate control of the polymerization. This result further suggests that the polymerization scheme may be incomplete. This is an issue that requires further experimental and mechanistic considerations in the area of DTLMP.

Author Contributions: A.J.B.-C. performed all the simulations and parameter estimation procedure, guided by E.V.-L. and A.P., R.G.-S. and J.G.S.-M. conceived the experimental research. J.G.S.-M. conducted all experiments and characterization. E.V.-L. and R.G.-S. provided technical direction to the project. A.J.B.-C. wrote the paper with feedback from J.G.S.-M., while E.V.-L. and A.P. revised the different intermediate versions until the final document.

Funding: This research was funded by: (a) Consejo Nacional de Ciencia y Tecnología (CONACYT, México), MEng scholarship granted to A.J.B.-C.; (b) DGAPA-UNAM, Projects PAPIIT IG100718, IV100119, and PASPA sabbatical support to E.V.-L. at the University of Waterloo, in Ontario, Canada; (c) Facultad de Química-UNAM, research funds granted to E.V.-L. (PAIP 5000–9078); (d) VIEP-BUAP, research funds granted to J.G.S.-M. and (e) Natural Sciences and Engineering Research Council (NSERC) of Canada.

Conflicts of Interest: The authors declare no conflict of interest.

References

1. Jenkins, A.D.; Jones, R.G.; Moad, G. Terminology for reversible-deactivation radical polymerization previously called "controlled" radical or "living" radical polymerization (IUPAC Recommendations 2010). *Pure Appl. Chem.* **2009**, *82*, 483–491. [CrossRef]
2. Luo, Y.D.; Chiu, W.Y. Synthesis and kinetic analysis of DPE controlled radical polymerization of MMA. *J. Polym. Sci. Polym. Chem.* **2009**, *47*, 6789–6800. [CrossRef]

3. Vivaldo-Lima, E.; Jaramillo-Soto, G.; Penlidis, A. Nitroxide-Mediated Polymerization (NMP). In *Encyclopedia of Polymer Science and Technology*, 4th ed.; Mark, H.F., Ed.; John Wiley & Sons: New York, NY, USA, 2016; pp. 1–48.
4. Kim, K.; Kim, Y.; Ko, N.R.; Choe, S. Effect of molecular weight on the surface morphology of crosslinked polymer particles in the RITP-dispersion polymerization. *Polymer* **2011**, *52*, 5439–5444. [CrossRef]
5. Yamago, S. Development of organotellurium-mediated and organostibine-mediated living radical polymerization reactions. *J. Polym. Sci. Part A Polym. Chem.* **2016**, *44*, 1–12. [CrossRef]
6. Masoumi, S.; Duever, T.A.; Penlidis, A.; Azimi, R.; López-Domínguez, P.; Vivaldo-Lima, E. Model discrimination between RAFT polymerization models using sequential Bayesian methodology. *Macromol. Theory Simul.* **2018**, *27*, 1800016. [CrossRef]
7. Monteiro, M.J.; de Brouwer, H. Intermediate Radical Termination as the Mechanism for Retardation in Reversible Addition-Fragmentation Chain Transfer Polymerization. *Macromolecules* **2001**, *34*, 349–352. [CrossRef]
8. Buback, M.; Vana, P. Mechanism of Dithiobenzoate-Mediated RAFT polymerization: A missing reaction Step. *Macromol. Rapid Commun.* **2006**, *27*, 1299. [CrossRef]
9. Barner-Kowollik, C.; Buback, M.; Charleux, B.; Coote, M.L.; Drache, M.; Fukuda, T.; Goto, A.; Klumperman, B.; Lowe, A.B.; Mcleary, J.B.; et al. Mechanism and Kinetics of Dithiobenzoate-Mediated RAFT Polymerization. I. The Current Situation. *J. Polym. Sci. Part A Polym. Chem.* **2006**, *44*, 5809–5831. [CrossRef]
10. Konkolewicz, D.; Hawkett, B.S.; Gray-Weale, A.; Perrier, S. RAFT Polymerization Kinetics: Combination of Apparently Conflicting Models. *Macromolecules* **2008**, *41*, 6400–6412. [CrossRef]
11. Li, C.; He, J.; Liu, Y.; Zhou, Y.; Yang, Y. Probing the RAFT Process Using a Model Reaction between Alkoxyamine and Dithioester. *Aust. J. Chem.* **2012**, *65*, 1077–1089. [CrossRef]
12. Inuit, T.; Yamanishi, K.; Sato, E.; Matsumoto, A. Organotellurium-Mediated Living Radical Polymerization (TERP) of Acrylates Using Ditelluride Compounds and Binary Azo Initiators for the Synthesis of High-Performance Adhesive Block Copolymers for On-Demand Dismantlable Adhesion. *Macromolecules* **2013**, *46*, 8111–8120. [CrossRef]
13. Soriano-Moro, J.G.; Rico-Valverde, J.C.; Enriquez-Mendrano, F.J.; Maldonado-Textle, H.; Vivaldo-Lima, E.; Acosta-Ortiz, R.; Guerrero-Santos, R. Toward a Living Radical Polymerization of Styrene by Using Dithiolactone as a New Type of Mediating Agent. *Macromol. Rapid Commun.* **2008**, *29*, 80–85. [CrossRef]
14. Soriano-Moro, J.G.; Jaramillo-Soto, G.; Guerrero-Santos, R.; Vivaldo-Lima, E. Kinetics and Molecular Weight Development of Dithiolactone-Mediated Radical Polymerization of Styrene. *Macromol. React. Eng.* **2009**, *3*, 178–184. [CrossRef]
15. Soriano-Moro, J.G. Síntesis y Caracterización de Ditiolactonas y su Empleo Como Agentes de Transferencia en la Polimerización RAFT. Ph.D. Thesis, Programa de Doctorado en Tecnología de Polímeros, CIQA, Saltillo, México, February 2008.
16. Wulkow, M. Computer Aided Modeling of Polymer Reaction Engineering-The Status of Predici, 1-Simulation. *Macromol. React. Eng.* **2008**, *2*, 461–494. [CrossRef]
17. Hungenberg, K.D.; Wulkow, M. *Modeling and Simulation in Polymer Reaction Engineering: A Modular Approach*; Wiley-VCH: Weinheim, Germany, 2018; pp. 92, 256.
18. Beurmann, S.; Buback, M. Rate coefficients of free-radical polymerization deduced from pulsed laser experiments. *Prog. Polym. Sci.* **2002**, *27*, 191–254. [CrossRef]
19. Pallares, J.; Jaramillo-Soto, G.; Flores-Cataño, C.; Vivaldo Lima, E.; Lona, L.M.F.; Penlidis, A. A Comparison of Reaction Mechanisms for Reversible Addition-Fragmentation Chain Transfer Polymerization Using Modeling Tools. *J. Macromol. Sci. A Pure Appl. Chem.* **2006**, *43*, 1293. [CrossRef]
20. Barner-Kowollik, C.; Quinn, J.F.; Nguyen, T.L.U.; Heuts, J.P.A.; Davis, T.P. Kinetic investigations of reversible addition fragmentation chain transfer polymerizations: Cumyl phenyldithioacetate mediated homopolymerizations of styrene and methyl methacrylate. *Macromolecules* **2001**, *34*, 7849–7857. [CrossRef]
21. Jaramillo-Soto, G.; Castellanos-Cardenas, M.L.; García-Moran, P.; Vivaldo-Lima, E.; Luna-Barcenas, G.; Pendilis, A. Simulation of Reversible Addition-Fragmentation Chain Transfer (RAFT) Dispersion Polymerization in Supercritical Carbon Dioxide. *Macromol. Theory Simul.* **2008**, *17*, 280–289. [CrossRef]
22. Jung, W. Mathematical Modeling of Free-Radical Six-Component Bulk and Solution Polymerization. Master's Thesis, Department of Chemical Engineering, University of Waterloo, Waterloo, ON, Canada, 2008.

23. Gao, J.; Penlidis, A. A Comprehensive Simulator Database Package for Reviewing Free-Radical Copolymerizations. *J. Macromol. Sci. Part C Polym. Rev.* **1998**, *38*, 651–780. [CrossRef]
24. Vivaldo-Lima, E.; García-Pérez, R.; Celedón-Briones, O.J. Modeling of the Free-Radical Copolymerization Kinetics with Crosslinking of Methyl Methacrylate/Ethylene Glycol Dimethacrylate Up to High Conversions and Considering Thermal Effects. *Evista Sociedad Química México* **2003**, *47*, 22–33.
25. Faldi, A.; Tirrell, M.; Lodge, T.P.; von Meerwall, E. Monomer Diffusion and the Kinetics of Methyl Methacrylate Radical Polymerization at Intermediate to High Conversion. *Macromolecules* **1994**, *27*, 4184–4192. [CrossRef]
26. Gómez-Reguera, J.A.; Vivaldo-Lima, E.; Gabriel, V.A.; Dubé, M.A. Modeling of the Free Radical Copolymerization Kinetics of n-Butyl Acrylate, Methyl Methacrylate and 2-ethylhexyl Acrylate Using PREDICI. *Processes* **2019**, *7*, 395. [CrossRef]
27. Moad, G.; Rizzardo, E.; Thang, S.H. Living Radical Polymerization by the RAFT Process—A Second Update. *Aust. J. Chem.* **2009**, *62*, 1402–1472. [CrossRef]
28. Barner-Kowollik, C.; Quinn, J.F.; Morsley, D.R.; Davis, T.P. Modeling the reversible addition–fragmentation chain transfer process in cumyl dithiobenzoate-mediated styrene homopolymerizations: Assessing rate coefficients for the addition–fragmentation equilibrium. *J. Polym. Sci. Part A Polym. Chem.* **2001**, *39*, 1353–1365. [CrossRef]
29. Wang, A.R.; Zhu, S. Modeling the reversible addition–fragmentation transfer polymerization process. *J. Polym. Sci. Part A Polym. Chem.* **2003**, *41*, 1553–1566. [CrossRef]

© 2019 by the authors. Licensee MDPI, Basel, Switzerland. This article is an open access article distributed under the terms and conditions of the Creative Commons Attribution (CC BY) license (http://creativecommons.org/licenses/by/4.0/).

MDPI
St. Alban-Anlage 66
4052 Basel
Switzerland
Tel. +41 61 683 77 34
Fax +41 61 302 89 18
www.mdpi.com

Processes Editorial Office
E-mail: processes@mdpi.com
www.mdpi.com/journal/processes